WCDMA — Requirements and Practical Design

To the people who call a spade a spade.
RT

I would like to dedicate this book to Katherine and William.
JPW

WCDMA — Requirements and Practical Design

Edited by

Rudolf Tanner and Jason Woodard
UbiNetics Ltd, UK

John Wiley & Sons, Ltd

Copyright © 2004   John Wiley & Sons Ltd, The Atrium, Southern Gate, Chichester,
West Sussex PO19 8SQ, England

Telephone   (+44) 1243 779777

Email (for orders and customer service enquiries): cs-books@wiley.co.uk
Visit our Home Page on www.wileyeurope.com or www.wiley.com

Other Wiley Editorial Offices

John Wiley & Sons Inc., 111 River Street, Hoboken, NJ 07030, USA

Jossey-Bass, 989 Market Street, San Francisco, CA 94103-1741, USA

Wiley-VCH Verlag GmbH, Boschstr. 12, D-69469 Weinheim, Germany

John Wiley & Sons Australia Ltd, 33 Park Road, Milton, Queensland 4064, Australia

John Wiley & Sons (Asia) Pte Ltd, 2 Clementi Loop #02-01, Jin Xing Distripark, Singapore 129809

John Wiley & Sons Canada Ltd, 22 Worcester Road, Etobicoke, Ontario, Canada M9W 1L1

Wiley also publishes its books in a variety of electronic formats. Some of the content that appears
in print may not be available in electronic books.

British Library Cataloguing in Publication Data

A catalogue record for this book is available from the British Library

ISBN 0-470-86177-0

Typeset in 10/12pt Times by TechBooks, New Delhi, India
Printed and Bound in Great Britain by TJ International Ltd, Padstow, Cornwall.
This book is printed on acid-free paper responsibly manufactured from sustainable forestry
in which at least two trees are planted for each one used for paper production.

Contents

Preface

During 2001 we surveyed the literature for books describing developing a *Wideband Code Division Multiple Access* (WCDMA) modem. We knew the book by Young [1] which is a well written textbook with an emphasis on RF, from an engineer with many years industrial hands-on experience. Another good book, with an emphasis on digital signal processing, is by Frerking [2], who gives enough theoretical background to solve practical problems in modem design. Both books cover the technology and engineering aspects well for frequency and time division duplex systems, but not code division multiple access modems.

After the release of the first version of the 3GPP standard, Release 99, it was possible to make the content of the standards available in a more accessible form. The book by Holma and Toskala [3] has become a common reference throughout our industry and beyond. This book describes the standards very well with perhaps a slight emphasis on network operation.

In early 2002, the editors felt that it would also be useful to have a book which describes not only how WCDMA works, but also provides some information regarding the design of the WCDMA *Access Stratum* (AS) in a handset. We suggested writing such a book, and the contributing authors offered to help us. The result is a book on WCDMA for the *Frequency Division Duplex* (FDD) mode of the *Universal Mobile Telecommunications System* (UMTS), with an emphasis on the development of a handset. The reader is expected to be familiar with basic communications theory, *Direct Sequence-Code Division Multiple Access* (DS-CDMA), and some digital signal processing concepts. These could not be treated here in depth since the size of the book is limited and the engineering issues are complex. However, we hope that we have been able to include enough detail on the standards, and on the operation of the access stratum in a handset, to support the practical and engineering aspects of WCDMA modem design and implementation.

This book is not intended to fully define the operation and implementation of a WCDMA mobile terminal, but to provide the reader with sufficient information to understand the issues associated with this, and to make the normative 3GPP specifications more accessible.

Acknowledgements

The editors and the authors would like to thank the following people for reviewing the chapters and providing valuable suggestions, in alphabetical order: Andrew Alexander, Francoise Bannister, Chris Esplin-Jones, Juan Garcia, Stephen Goodwin, Nick Hallam-Baker, Moritz Harteneck, Jimmy James, Gopi Krishna, Tong Liew, Richard Littlewood, Anne McAleer, Mark Norris, Debbie Opuko, Machiel Posthumus, Sergei Senin, Jon Thackray, Battula Venkateswara and John Wilson.

The editors would like to thank the team at Wiley for their support and help, a special thank you goes to Sophie Evans for her assistance and Mark Hammond for his patience with us.

Last but not least, special thanks to Katherine Woodard and Christine Tanner, for their help and patience with us while we dedicated a significant amount of our free time to this project over the last eighteen months.

Abbreviations

Use as defined by 3GPP.

3GPP	3rd Generation Partnership Project
3GPP2	3rd Generation Partnership Project 2
AbS	Analysis by synthesis
AC	Alternating current
ACELP	Algebraic code-excited linear predictive
ACK	Acknowledgement
ACLR	Adjacent channel leakage ratio
ACP	Adjacent channel power
ACS	Add compare select or Add compare select
ADC	Analogue to digital conversion
AFC	Automatic frequency control
AGC	Automatic gain control
AI	Acquisition indicator
AICH	Acquisition indicator channel
AM	Acknowledged mode
AMC	Adaptive modulation and coding
AMD	Acknowledged mode data
AMPS	Advanced mobile phone service
AMR	Adaptive multi rate speech codec
AP	Access preamble
AP-AICH	Access preamble acquisition indicator channel
ARB	Arbitrary waveform generator

ARIB	Association of radio industries and business
ARPU	Average revenue per user
ARQ	Automatic repeat request
AS	Access stratum
ASC	Access service class
ASN.1	Abstract syntax notation no.1
AWGN	Additive white Gaussian noise
BB	Baseband
BCCH	Broadcast control channel
BCH	Broadcast channel
BER	Bit error ratio
BLER	Block error rate
BMC	Broadcast/multicast control
BO	Buffer occupancy
bps	bits per second
BPSK	Binary phase shift keying
BRP	Bit rate processing
BS	Base station
BTFD	Blind transport format detection
C/I	Carrrier/interference ratio
C-RNC	Controlling radio network controller
C-RNTI	Cell-RNTI
C-SAP	Control SAP
CBMC	Control BMC interface
CBS	Cell broadcast service
CC	Call control or certification criteria
CCC	CPCH control commands
CCCH	Common control channel
CCTrCH	Coded composite transport channel
CD	Collision detection
CD/CA-ICH	Collision detection/channel assignment indicator channel
CDMA	Code division multiple access
CDP	Collision detection preamble
CFN	Connection frame number
CID	Context identifier
CMAC	Control MAC interface
CN	Core network
CPCH	Common packet channel
CPDCP	Control PDCP interface
CPHY	Control physical layer interface
CPICH	Common pilot channel
CQI	Channel quality indicator
CR	Chip rate

CRC	Cyclic redundancy check
CRLC	Control RLC interface
CRP	Chip rate processing
CS	Circuit switched
CSICH	CPCH status indicator channel
CTCH	Common traffic channel
CW	Continuous wave
CWTS	Chinese wireless telecommunications standard
D-RNC	Dynamic radio network controller
DAC	Digital to analogue conversion
DC	Direct current
DCCH	Dedicated control channel
DCH	Dedicated channel
DCR	Direct conversion receiver
DCS	Digital cellular standard
DD	Delay diversity
DECO	Decorrelation detector
DL	Downlink
DLL	Delay locked loop
DPCCH	Dedicated physical control channel
DPCH	Dedicated physical channel
DPDCH	Dedicated physical data channel
DRAC	Dynamic resource allocation procedure
DRNC	Drift radio network controller
DRP	Digital reference point
DRX	Discontinuous reception
DSCH	Downlink shared channel
DSP	Digital signal processing
DSSS	Direct sequence spread spectrum
DTCH	Dedicated traffic channel
DTX	Discontinuous transmission
DUT	Device under test
EDGE	Enhanced datarate for the GSM evolution
EFR	Enhanced full rate speech codec
EFSM	Enhanced finite state machine
EIRP	Equivalent isotropic radiated power
ENOB	Effective number of bits
ETSI	European telecommunications standard institute
EVM	Error vector magnitude
FACH	Forward access channel
FB	Feedback
FCC	Federal communications commission
FDD	Frequency division duplex

FEC	Forward error correction
FFT	Fast Fourier transform
FHT	Fast Hadamard transforms
FIR	Finite impulse response
FM	Frequency modulation
GCF	General certification forum
GERAN	GSM/EDGE radio access network
GMSK	Gaussian minimum shift keying
GPRS	General packet radio system
GPS	Global positioning system
GSM	Global system for mobile communications
GSM-MAP	GSM-mobile applications part
H-RNTI	HS-DSCH radio network temporary identifier
HS-DSCH	High speed downlink shared channel
HS-PDSCH	High speed physical downlink shared channel
HS-SCCH	High speed shared control channel
HARQ	Hybrid automatic repeat request
HC	Header compression
HDF	Hard decision feedback detector
HHO	Hard handover
HLS	Higher level scheduling
HNN	Hopfield neural network receiver
HPSK	Hybrid phase shift keying
HSCSD	High speed circuit switched data
HSDPA	High speed downlink packet access
HS-DPCCH	High speed dedicated physical control channel
ICI	Interchip interference
ID	Indentifier
IE	Information element
IF	Intermediate frequency
IFFT	Inverse fast fourier transform
IMS	IP multimedia subsystem
IMSI	International mobile subscriber identity
IMT-2000	International mobile telecommunications 2000
IP	Internet protocol
IPR	Intellectual properly rights
IP2/3	Intercept point 2/3
IPDL	Idle periods in the downlink
IQ	In-phase quadrature-phase
IR	Incremental redundancy
IS-95	Interim standard 1995
ISDN	Integrated services digital network
ISPP	Interleaved single-pulse permutation

ITP	Initial transmit power
kbps	kilo bits per second
L1	Layer 1 (physical layer)
L2	Layer 2 (data link layer)
L3	Layer 3 (network layer)
LAN	Local area network
LC	Inductor–capacitor (filter)
LCS	Location-based services
LI	Length indicator
LLR	Log likelihood ratio
LMU	Location measurement unit
LNA	Low noise amplifier
LO	Local oscillator
LOS	Line of sight
LPC	Linear predition coding
LSB	Least significant bit
LSF	Line spectrum frequency
LSN	Last sequence number
LSR	Lossless SRNS relocation
M-LWDF	Modified largest weight delay first
MAC	Medium access control
MAI	Multiple access interference
MAP	Maximum a posteriori
Mbps	Mega bits per second
Mcps	Mega chips per second
MCS	Modulation and coding scheme
MF	Matched filter
MGW	Media gateway
MIB	Master information block
MIMO	Multiple input multiple output channel
Mips	Million instructions per second
ML	Maximum likelihood
MLP	Multilayer perceptron neural network
MLSE	Maximum likelihood sequence estimator
MMSE	Minimum mean square error detector
MOSFET	Metal oxide silicon field effect transistor
MR	Maximum rate
MRC	Maximum ratio combining
MSB	Most significant bit
MSC	Mobile switching centre
MSE	Mean square error
MUD	Multiuser detection
NACK	Negative acknowledgement

NAS	Non access stratum
NBAP	Node B application part
NF	Noise figure
NLMS	Normalized least mean square
NNB	Neural network based receivers
Node B	3GPP term for a base station
NT	Notification
OCNS	Orthogonal channel noise source/simulator
OFDM	Orthogonal frequency division multiplexing
OLPC	Outer loop power control
OOR	Out of range
OSI	Open system interconnection
OTD	Orthogonal transmit diversity
OTDOA	Observed time difference of arrival
OVSF	Orthogonal variable spreading factor
P-CCPCH	Primary common control physical channel
P-CPICH	Primary commom pilot channel
P-SCH	Primary synchronization channel
PA	Power amplifier
PAPR	Peak to average power ratio
PBP	Paging block periodicity
PBU	Piggybacking unit
PC	Power control
PCCH	Paging control channel
PCH	Paging channel
PCP	Power control preamble
PCPCH	Physical common packet channel
PCS	Personal communication system
PDC	Pacific digital cellular
PDCP	Packet data convergence protocol
PDP	Packet data protocol
PDSCH	Physical downlink shared channel
PDU	Protocol data unit
PED	Polynomial expansion detector
PER	Packet error rate
PF	Proportionally fair
PhCH	Physical channel
PHY	Physical layer interface
PIC	Parallel interference cancellation detector
PICH	Page indicator channel
PID	Packet identifier value
PL	Puncturing limit
PLMN	Public land mobile network

PN	Pseudo noise
PPP	Point-to-point protocol
PRACH	Physical random access channel
PS	Packet switched
PSTN	Public switched telephone network
QAM	Quadrature amplitude modulation
QoS	Quality of service
QPSK	Quadrature phase shift keying
RAB	Radio access bearer
RACH	Random access channel
RB	Radio bearer
RC	Resistor–capacitor (filter)
RF	Radio frequency
RL	Radio link
RLC	Radio link control
RNC	Radio network controller
RNN	Recurrent neural network receiver
RNTI	Radio network temporary identity
ROHC	Robust header compression
RPP	Recovery period power
RR	Round robin
RRC	Radio resource controller
RRM	Radio resource manager
RSC	Recursive systematic convolutional code
RSCP	Received signal code power
RSSI	Received signal strength indicator
RTD	Relative time difference
RTP	Real time protocol
RTT	Round trip time
R&TTE	Radio and telecommunications terminal equipment
Rx	Receive
S-CCPCH	Secondary common control physical channel
S-CPICH	Secondary common pilot channel
S-RNC	Serving radio network controller
S-SCH	Secondary synchronization channel
SAD	Simulated annealing detector
SAI	Service area identifier
SAP	Service access point
SAS	Stand alone service
SaW	Stop and wait (ARQ type)
SAW	Surface acoustic wave (filter)
SCDMA	Space code division multiple access
SCH	Synchronization channel

SD	Spectral distortion
SDF	Soft decision feedback detector
SDU	Service data unit
SEGSNR	Segmental SNR
SF	Spreading factor
SFN	System frame number
SFR	Spreading factor reduction
SGSN	Serving GPRS support node
SHO	Soft handover
SIB	System information block
SIC	Serial interference cancellation detector
SIM	Subscriber identity module
SIR	Signal to interference ratio
SMLC	Serving mobile location centre
SMS	Short message service
SN	Sequence number
SNR	Signal to noise ratio
SOC	System on a chip
SOHO	Softer handover
SOVA	Soft output viterbi algorithm
SR	Selective repeat or Symbol rate
SRNS	Serving radio network subsystem
SSDT	Site selection diversity TPC
STC	Space time coding
STTD	Space time transmit diversity
SVD	Singular value decomposition
SVQ	Split vector quantization
TA	Type approval
TACS	Total access communication system
TB	Transport block
TBS	Transport block size
TSF	Technical construction file
TCP	Transmission control protocol
TCP/IP	TCP/Internet protocol
TCTF	Target channel type field
TDD	Time division duplex
TDMA	Time division multiple access
TF	Transport format
TFC	Transport format combination
TFCI	Transport format combination indicator
TFCS	Transport format combination set
TFRI	Transport format resource indicator
TFS	Transport format set

TM	Transparent mode
TMD	Transparent mode data
TMP	Time multiplexed pilot
TPC	Transmit power control
TrCH	Transport channel
TSG	Technical specification group
TSN	Transmission sequence number
TSTD	Time switched transmit diversity
TTA	Telecommunications technology association
TTC	Telecommunications technology committee
TTI	Transmission time interval
Tx	Transmit
TxD	Transmit diversity
U-RNTI	UE radio network temporary identity
UARFCN	UTRA absolute radio frequency channel number
UDP	User datagram protocol
UE	User equipment
UL	Uplink
UM	Unacknowledged mode
UMD	Unacknowledged mode data
UMTS	Universal mobile telecommunications system
URA	User registration area
USIM	Universal subscriber interface module
UTRA	UMTS terrestrial radio access
UTRAN	UMTS terrestrial radio access network
UWB	Ultra wideband
UWCC	Universal wireless communications consortium
VA	Viterbi algorithm
VCO	Voltage controlled oscillator
VS	Volterra series detector
VSA	Vector signal analyser
WCDMA	Wideband code division multiple access
WG	Working group
WiFi	Wireless Fidelity
WLAN	Wireless local area network
ZF	Zero forcing

1

Introduction

Rudolf Tanner and Jason Woodard

1.1 EVOLUTION AND REVOLUTION OF MOBILE TELEPHONY

We start with two individual perspectives on the advances in mobile telephony. The first in Section 1.1.1 is from Mike Pinches, a former director at Vodafone UK. This section discusses advances in cellular communications from a network operator's point of view. The second, in Section 1.1.2, discusses the evolution of cellular communications and handsets from a more technical point of view.

1.1.1 The Cellular Revolution

Mike Pinches

The third generation of mobile phone technology (3G) is arguably the most complex multinational collaborative project that the world has ever seen. Throughout virtually all developed nations, cellular equipment manufacturers, mobile network operators, silicon suppliers and software developers have been working together to bring about the successful deployment and commercial introduction of 3G. Although there have been some delays from the initial target schedules, the historical precedents in the development of the wireless industry suggest that the current problems associated with 3G roll out will be overcome and its long term commercial success is assured.

The move from *Global System for Mobile* (GSM) to WCDMA is nothing short of a global technological revolution. However, this is not the first time the industry has gone through such major change; the introduction of 2G standards such as GSM also posed technical challenges and involved substantial multinational collaboration. The first wireless

WCDMA – Requirements and Practical Design. Edited by R. Tanner and J. Woodard.
© 2004 John Wiley & Sons, Ltd. ISBN: 0-470-86177-0.

generation analogue cellular telephones were introduced in most of the developed world during the 1980s. In the UK, the first two operators were Vodafone (a subsidiary of Racal Electronics) and Cellnet, which was controlled by BT. In 1983, they worked together with the UK Department of Trade and Industry and chose to introduce the *Total Access Communication System* (TACS) standard, which was closely based on US developed *Advanced Mobile Phone Service* (AMPS) technology, primarily to minimize risks of any delays to initial revenue generation.

AMPS networks were already undergoing live testing in the USA, but slightly different frequency bands and channel spacing were essential to meet existing European specifications. However, less than 20 % of the existing developments needed adaptation for this and the result was that both networks were able to open service ahead of schedule in January 1985.

Less than two years had elapsed from selecting the technology to launching the services. By the end of 1985, both networks had substantially exceeded their own market forecasts. They were working hard to expand capacity to cope with overloads at peak times and service quality was sometimes suffering. However, by then it was already very clear that the cellular market would be a major success story.

1.1.1.1 GSM, the Second Generation
By 1987 there were five incompatible first generation analogue systems operating across Europe. The British TACS standard was rapidly becoming the most successful and was being adopted in other countries outside Europe. However, it was becoming obvious that any vision of a harmonized Europe would need international roaming and therefore a common mobile standard throughout Europe. It was also becoming apparent that escalating market demand was going to be very difficult to meet with the finite spectrum available.

The prospect of any of the existing national standards becoming accepted for adoption throughout Europe was certainly remote, if not impossible. But for once, technology could both solve a political problem and meet a real need at the same time.

Cellular networks offer high capacity by reusing each available frequency many times. Analogue technology can only offer a limited degree of reuse before interference starts to cause dropped calls and 'crossed lines'. However, digital technology could allow much greater reuse and, at the same time, offer more capacity with higher security through digital speech coding. It could support many new data services and, even more importantly, it was politically acceptable to all.

So the GSM concept was born. European equipment manufacturers, network operators, research establishments and politicians set out to work together to tackle arguably the most complex multinational high technology collaborative project ever. The agreed objective was to launch the first GSM services in most European capitals in July 1991.

However, the development of GSM was not without its problems. The technical standards were at least ten times more complex than TACS and virtually everything had to be designed from scratch. Almost everyone involved underestimated the software complexity. Silicon chip technology at that time was only just capable of meeting the processing requirements and developments overran. The process for ensuring terminal interoperability through independent type approval was seriously delayed. In July 1991, a number of networks 'launched' GSM services but, since there were virtually no handsets available, these launches made little impact and were widely criticized.

For the next three years, progress was slow, but early bugs were ironed out and more handsets gradually appeared. But some still questioned whether the operators' substantial

GSM network investments would ever be recovered. Hand portable analogue phones had become common, but the first GSM portable phones were bigger, cost more and had shorter battery life. Only the dedicated international traveller would seriously consider having one, although as soon as they became available, many promptly bought them.

Despite this slow start, GSM phones gradually became smaller, cheaper and more widely available. Over 380 million Europeans now carry one, often regarding it as one of their most treasured personal possessions. GSM has also gone on to become by far the world's leading mobile phone standard, with over 70 % of the world market and over 850 million users. In the UK, the TACS service finally closed in 2001.

GSM's real development phase lasted some six years, or more than three times longer than TACS. Not a bad achievement for a challenge that was at least 50 times more complex. And the early critics have long been well and truly silenced.

1.1.1.2 *2G in the USA and Japan*

GSM started life as a pan European concept, with the clear aim of ensuring that existing European technology suppliers should play a major role in manufacturing for the European markets. Many initially argued that a protectionist approach should be adopted to keep GSM truly European. However, others believed that the active participation of major US or Japanese companies, such as Motorola, NEC and Panasonic, was essential if GSM was to have a future as a major global standard.

The issue of intellectual property became a major factor in this debate when Motorola took a particularly strong position in defence of its substantial *Intellectual Property Rights* (IPR) portfolio. The outcome was a gradual shift for GSM towards a truly open global standard in which all were able to participate on a common basis.

Whilst this debate was developing, technology evolution was also happening elsewhere. The 900 MHz frequency band used initially for GSM in Europe was not available in the US and, in any case, there were strong pressures from US manufacturers to respond to the debate by developing a competitive 2G standard for the US market.

The result was the development and introduction of a narrowband CDMA standard known as IS-95. Ironically, CDMA had been considered as a candidate for GSM in 1987, but had been ruled out at that time as being unproven and too ambitious to meet a target launch of mid 1991.

The launch history of IS-95 in the USA suggests this decision was correct, because there were many teething problems and the technology took longer than GSM to stabilize. Today, GSM has about six times as many users worldwide as IS-95. And incidentally, over 22 million of the world's GSM users are now in North America!

In Japan, there had been a strong focus by NTT on developing all mobile standards themselves. This 'closed standard' approach had led to a very insular situation where Japanese manufacturers had, just for once, failed almost completely in promoting Japan's own mobile technology outside Japan.

By the mid 1990s, it was clear that the existing *Pacific Digital Cellular* (PDC) standard could not meet escalating demands for more capacity and higher bandwidth. Although a number of Japanese manufacturers had developed GSM phones for world markets, it was clear that Japan needed to try a new approach.

1.1.1.3 *The Third Generation (3G)*

By the late 1990s, the very success of GSM was again raising questions about the future need for yet more spectrum. And technology fragmentation was again an issue, but this time at a global level between Europe, North America and Japan as the search for third generation solutions began.

The Japanese started work very early on 3G with the clear objective of securing a major global role after their comparative failures with previous generations.

Meanwhile, the GSM community was initially focused on developing GSM's circuit and packet switched data services, using HSCSD and GPRS. Both are limited to maximum data rates of less than 50 kbps and neither can support video telephony. Any large scale adoption of such services would also further increase pressures on the available spectrum since the same spectrum is shared between voice and data services.

The addition of EDGE technology to GSM can support data rates up to 128 kbps, particularly over the downlink to the mobile terminal, but again only by reducing available voice capacity. So it was clear that more spectrum would have to be made available.

In the USA, there was an obvious potential evolution towards a wider bandwidth CDMA system.

So multinational collaboration was again applied to identify and agree a suitable technology that could be used with new spectrum to provide more capacity, offer intercontinental roaming and support high bandwidth services. But this time, the potential seemed to exist for a truly global common solution.

The universally accepted result was the WCDMA technology standard. CDMA technology had been briefly considered as a candidate for GSM in 1987, but was ruled out at that time as being unproven and much too ambitious for a 1991 target launch. This decision was undoubtedly correct, because the full complexity of WCDMA is now very much clearer. The specifications that have only recently been completed are perhaps ten times longer again than for GSM, and are almost all entirely different.

However, the holy grail of a single global standard still proved to be unachievable. Although Europe and Japan agreed to converge on a common WCDMA standard, the USA both used different frequencies and wanted an evolution path from IS-95. Hence it adopted a variant of WCDMA known as CDMA-2000.

So the transition from GSM to WCDMA represents another development challenge that is much more complex even than GSM. It represents a huge technological undertaking, which few in the press community have fully recognized. The lessons to be drawn from TACS and GSM are that evolutionary development can run to plan, but the revolutionary equivalent usually doesn't.

3G is clearly revolutionary and its development has not run to plan. A number of 3G network launches have been scheduled, only to be quietly postponed whilst ongoing technical issues are ironed out. But history would suggest that perhaps there is nothing wrong with 3G development; it is just that the predictions were wrong.

There are many examples of our own industry having suffered from collective self delusion. Launching GSM in 1991 might be seen as one, although the target undoubtedly focused many minds! Indeed, there are examples in many other industries too, frequently involving both strong competition and advanced technology. It sometimes takes an outspoken small boy to publicly observe that the emperor is not fully dressed, because no important courtier would wish to be seen breaking the party line.

A further handicap for 3G has come from those governments, including the UK, who chose to auction 3G spectrum at what turned out to be the peak of a technology boom. These auctions raised almost embarrassingly large sums from operators who, with hindsight, not surprisingly found the prospect of facing no future to be even worse than paying much more than they expected.

These sums can only be recovered in the long run from mobile phone users, although a fair number of commentators persist in suggesting otherwise. At least today's historically low interest rates are assisting operators to spread out the impact. However, the sums raised may have increased the pressure to announce early launch dates.

1.1.1.4 *The Future for 3G*   3G provides a natural successor to GSM by building upon GSM's proven SIM card roaming model, yet offering much more spectrum for voice services whilst enabling a much wider variety of data and multimedia services.

Some commentators, no doubt with vested interests in the fixed line market, have argued that WiFi access technology will kill 3G before it is even born. While the wireless *Local Area Network* (LAN) market certainly has a future, it will never provide anything like the same geographical coverage as 3G and the network investment economics are far from proven. In other words, the two technologies are likely to be complementary.

From a commercial perspective, there is a clear cut economic case to be made for 3G. While the licence and infrastructure investments have been immense, 3G provides long term reductions in the cost of carrying both voice and data traffic, at least in urban and suburban areas. The ability to increase *Average Revenue Per User* (ARPU) by providing new value added data services will ensure these investments are recouped, probably sooner than many expect.

It is important not to underestimate the challenges that lie ahead for the 3G industry. The most pressing topics include resolving interoperability issues (both amongst operators and handset manufacturers), agreement on standards evolution and ensuring that the customer's service quality expectations are met.

1.1.2 The Growth of Cellular Technology

Mark Norris

Radio communications technology has been around for some considerable time, however in the 1960s and 1970s the use of two way radio was largely restricted to the professional services (military/police) using private mobile radio networks based on large 'town sized' cells. As improved signalling control and automation was introduced, simplifying the user control of the handset and connecting the radio network to the *Public Switched Telephone Network* (PSTN), it became possible to offer a fully mobile cellular system with automatic dialling and roaming across cell boundaries.

The first analogue cellular system (AMPS) started in 1983 in the United States and was quickly followed by similar narrowband FM systems (NMT-450 and 900) in Europe and Scandinavia, and a UK based system (TACS). Technically these systems were similar, being based on narrowband FM modulation schemes with 25–30 kHz channel spacing. The main differences lay in their control mechanisms, especially in the algorithms used to handle the handover mechanism as the caller moved between cells. These 'first generation' analogue systems proved popular in the commercial market, with large increases in subscriber numbers. Soon it became clear that national coverage, call quality and network capacity were key performance targets required for successful operation of cellular networks.

Both the market pull and the technology push for better performance led to the development and introduction of 'second generation' systems. In the USA, a digital system D-AMPS was introduced as an overlay on the AMPS band, however in Europe a more dramatic change was introduced with the all digital GSM system in 1992. Advantage was taken from new digital signal processing technology that could now be fitted to a handset without consuming too much power. The wider channel spacing of 200 kHz selected to support multiple users (8 or 16) in a *Time Division Multiple Access* (TDMA) scheme, also enabled equalizers to compensate for multipath reflections in the RF environment.

Take up of the GSM service was rapid in European countries in the 1990s and it soon became adopted throughout the world. Its secure high quality voice service, compatibility with *Integrated Services Digital Network* (ISDN) and particularly its international roaming capabilities were popular in the market, and the emergence of its SMS text *Short Messaging Service*, as a fashionable application was a surprise to many in the industry.

1.1.2.1 *Market Drivers*

The move towards third generation cellular mobile systems largely has been triggered by the huge commercial success of GSM and other second generation systems across the world. The market for mobile telephony expanded with ever increasing penetration figures, including children as potential customers and a new desire for digital services. The main commercial drive for 3G networks was to offer significantly higher performance services to these customers, and to offer a mixture of service types ranging from standard voice to streaming video content. Having seen improved speed performance from Internet fixed line dial up connections, it was natural to expect the same from a wireless connection, including broadband services.

The other key driver came from network operators who have to provide these services in a cost effective manner. The primary targets of network coverage and cell capacity became targets for new technical solutions, and public health concerns over handset radiation pushed down power levels. It was clear that third generation networks would need a great number of small sized cells to offer such high capacity services.

1.1.2.2 *Handset Trends*

The commercial success of GSM over the last decade has led to increased competition amongst handset manufacturers, and a drive towards cheaper and higher performance handsets. Today's handset models are now commodity items, given away in exchange for a contractual commitment to a service provider. The large number of handset models available address a segmented market, from the cheap handsets to fashion/games models and high end business phones.

Technology improvements have helped develop additional market features for multiband operation (up to four bands for GSM) and multimode operation (3G/GSM), which are now expected as standard in new handsets.

The example in Figure 1.1 of handset size over three generations of mobile telecoms, shows how both the market and technology have helped push the progress of handset development. This trend-line graph collected from handset data over the years, illustrates both the rapid development in the early years of a new system, and the longer term effects from general technology improvement. The smallest handset sizes have now shrunk to about 60 ccm and are restricted by the physical attributes of the human body, such as finger size and the distance between the ear and mouth.

1.1.2.3 *Technology Drivers*

The growing popularity of second generation cellular systems and their establishment as a worldwide system has provided investment into new

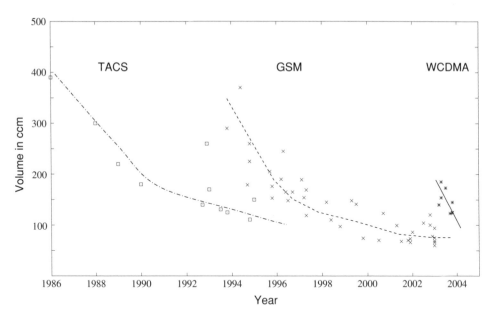

Figure 1.1. Evolution of handset sizes.

technologies. The availability of new manufacturing techniques has pushed component size down towards the practical limits of pick and place machines. Even in the difficult area of RF technology larger components such as the high Q-factor filters have been replaced by resonators derived from *Surface Acoustic Wave* (SAW) filter technology, both for RF and IF circuitry. No longer do we see large helical resonator structures in the front end of the handset radio.

Similar improvements have been seen in the semiconductor industry with the 0.6 micron geometry seen in early GSM handsets improving to 0.13 microns in today's baseband chips with a similar rise in performance. This is one of the factors considered when new standards are developed containing further sophistication in their digital processing requirements. It is difficult finding the right balance between performance improvements and the practical ability to pack enough processing power within a small/cheap handset, so the technology is always being stretched and GSM handset progress illustrates this well. Early GSM handset designs had to use digital hardware for the Viterbi equalizer component in order to achieve a practical solution without consuming too much power. However, it was not long before DSP processor power passed the key point of 13 Mips and the whole equalizer could be implemented in software along with the speech codec. Later extensions of DSP processing power were then used to advantage by improving the quality of the speech codec and adding handset differentiators such as speech recognition and memo storage. Today's GSM handsets contain DSPs operating at more than 100 Mips.

Even a simple measure such as chip count illustrates the technology progress achieved in GSM. Early handsets had about 8–11 chips in the baseband and a further six chips in the radio. Today the total count can be as low as three or four, with one device for each of the main functions including digital and analogue baseband processing chips (sometimes combined into one), a combination memory chip and a single chip RF transceiver.

Today these GSM performance numbers seem small, but similar trends can be seen taking place in 3G handset implementation. Hardware accelerators are often built into the processors and there is much discussion about battery power consumption in 3G handsets. There has also been increased emphasis on cheaper flash memory devices that are needed to support high capacity storage for multimedia applications.

1.1.2.4 Technical Choices in Cellular Standards

Decisions made during the standardization process arise from a mixture of sources from national bodies, telecommunications companies and academic research. Considering the underlying technology, there are some constraints that have to be satisfied for any cellular system. Performance issues for cellular radio systems are primarily dominated by the air interface and its unreliability. It is difficult tracking a signal in the low quality RF environment seen at cell boundaries or in congested areas, and compensation also must be applied for other effects such as fading, multiple reflections and Doppler shifts.

The fundamental choices of RF channel bandwidth, transmitter power and highest traffic speed affect many other performance parameters such as cell range and capacity.

The channel bandwidth increase from TACS to GSM was 25 kHz to 200 kHz and was used to introduce TDMA and support multiple users per RF carrier. Not only was this a manipulation of the time bandwidth product, but also the increased bandwidth allowed other compensation techniques to be used. With an over the air bit period of 3.69 μs it was possible to 'sound' the RF environment using a training sequence of 26 bits in every GSM burst. This enabled the equalizer to compensate for bad multipath environments where reflections can introduce delay spreads up to 10 μs and beyond. Interestingly, it was only a late change in the GSM standard that moved the training sequence from the start of the burst to its more optimal position in the burst centre.

The move from second to third generation systems has introduced a further increase in channel bandwidth and an over the air chip rate of 3.84 Mcps for 3GPP. A layered code spreading method is used to support both multiple users and variable traffic service rates in an efficient coding scheme. Compared to GSM the 3GPP standard achieves a higher cell capacity but has the added capability of supporting mixed traffic types ranging from circuit switched voice to broadband packet services. Also in the 3GPP standard, additional advantage is taken from the wideband channel to compensate for multipath signal components. In 3GPP the fundamental measurement resolution is significantly improved over GSM by using a wideband channel and a chip period of 260 ns. The separate multipath components seen in a highly reflective environment can be resolved and coherently combined in a RAKE receiver (see Chapter 3) to gain a diversity improvement in signal quality. The same receiver method is also used in 3GPP to support soft handover from one cell to the next by combining separate rays from each cell that can be heard.

Comparisons of network capacity improvements from GSM to WCDMA are often made both from theory and practical measurement. A generally accepted multiplier of 2–3 can be used as a rough guide, but there are many other benefits that should be considered as well. Not least of these is the ability to run a single frequency network in WCDMA, which gives considerable advantage to the network operator by simplifying network planning. Cell reuse patterns using seven frequencies in TACS were replaced by four frequency reuse when frequency hopping was activated in GSM networks, and these are now followed by a single frequency 3G network. The cell size in a 3G network is also smaller than GSM, which increases the network capacity, especially for high speed services. This change, combined with the ability to run base stations asynchronously, helps the operator add fill ins and capacity improvements in a relatively straightforward manner. Capacity extension by adding a second frequency is also simple to implement in 3GPP, allowing an overlay mapping of macro and micro cells without replanning the local network.

1.1.2.5 Summary   There has been significant change in the cellular standards over past years, both in establishing the wireless telecommunications market and in the improvements made in digital communications technology. This has been backed up by ongoing development of digital baseband technology, allowing a significant amount of signal processing power to be fitted in a small portable device.

Changes in the standards from GSM to 3GPP, combined with progress in device technology and newly developed ideas for signal processing, have all helped improve the efficiency and capacity of cellular networks and have introduced a wide variety of mobile services to the end user. We can only expect these advances to improve with time to match the increasing demand for higher bandwidth services.

1.2   THE THIRD GENERATION PARTNERSHIP PROJECT

Harri Holma and Antti Toskala, Nokia Networks, Finland

This section presents an overview of the *Third Generation Partnership Project* (3GPP), the global standardization organization for third generation networks. The background, organization, standard releases and future evolution of 3GPP are covered in this section.

1.2.1   3GPP Background

The standardization of first generation analogue systems was done by national standard bodies or by a group of a few countries. The second generation digital systems, such as GSM in Europe, were standardized using a regional approach. Later, the GSM standard was adopted by several other regions including the Americas.

The third generation standardization work started in regional standardization organizations in Europe (ETSI), in Korea (TTA), in Japan (ARIB/TTC) and in the USA (T1P1). In order to obtain a harmonized global standard, a single standardization body was required. The new organization, called the third generation partnership project, was created in 1998 by the regional standardization bodies. The partners agreed on joint efforts for the standardization of UTRA, now standing for *Universal Terrestrial Radio Access*. The standardization of different system generations is shown in Figure 1.2.

Companies such as manufacturers and operators are members of 3GPP through the respective standardization organization to which they belong. Later, during 1999, CWTS (the *China Wireless Telecommunication Standard Group*) also joined the 3GPP.

3GPP also includes market representation partners, such as: the GSM Association; the UMTS Forum; the Global Mobile Suppliers Association; the IPv6 Forum and the *Universal Wireless Communications Consortium* (UWCC).

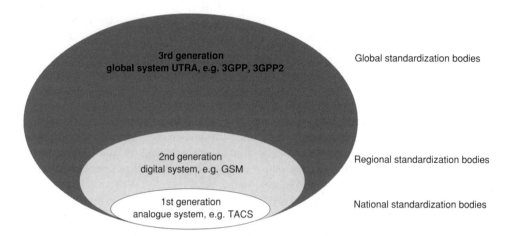

Figure 1.2. Global third generation standard created by 3GPP.

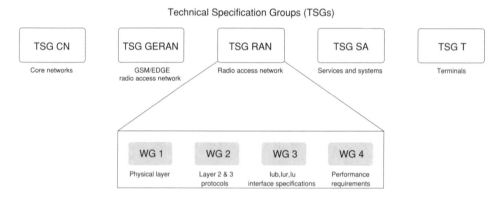

Figure 1.3. 3GPP organization.

1.2.2 3GPP Standardization Organization

3GPP is organized into five *Technical Specification Groups* (TSGs). The most relevant groups for UMTS system design are TSG CN for *Core Network* standardization and TSG RAN for *Radio Access Network* standardization. TSG RAN is divided into four *Working Groups* (WGs). The detailed technical work is carried out by the working groups and at the TSG level the specifications and changes to the specifications are then approved. The TSG level has an important role of ensuring coordinated work in the working groups both for the work already ongoing and also for the work to be initiated on new items.

After the first phase, the GSM/EDGE standardization work was also moved to 3GPP in order to ensure harmonized GSM/EDGE and WCDMA standards. The current 3GPP organization is illustrated in Figure 1.3.

Following the creation of 3GPP, IS-95 evolution work was also moved to a similar organization named 3GPP2. The number of regional standardization bodies is fewer than in 3GPP as there are less regions involved in IS-95 evolution work. More details on 3GPP and 3GPP2 organizations can be found from [4] and [5] respectively.

The work in 3GPP is organized around work items which, depending on the topic, may cover one or more working groups. Whenever a new topic is to be started, a work item proposal is first presented and approved at the TSG level before working groups are tasked to tackle the issue. In the case of a larger issue, typically a feasibility study is done first to identify both the potential of the new technology being proposed and the impact on terminals and network elements. A good example is *High Speed Downlink Packet Access* (HSDPA), where a feasibility study was conducted during 2001 and following the encouraging results in terms of performance and resulting complexity, a work item was established for Release 5. As HSDPA was a large topic there was not only one work item, but a separate one was established corresponding to each working group in TSG RAN. The work items for WG1, WG2 and WG3 were finalized in March 2002 and the last issues in the Release 5 performance work were concluded in June 2003. The delay in the performance requirements was inevitable as detailed simulations cannot be started before the physical layer details are complete. A big topic like this requires activities in several working groups and one needs to convince a sufficient number of the participating companies, equipment vendors and operators of the benefits of the technology one is proposing.

The normal way of making decisions in 3GPP is by consensus, thus before finalization of the work and approval of the specification changes, all the objections and concerns raised during the work should be covered by the proponents of a particular technology. In some rare cases voting has been used, where the 3GPP rules require a 71 % majority for a technical issue to be passed. In 3GPP the IPR policy requires that essential patents are declared to respective regional standards organizations and that licensing needs to be done on a reasonable basis. The actual licensing negotiations are not part of the 3GPP but are covered by bilateral or multilateral negotiations amongst participating companies.

1.2.3 3GPP Standard Releases

The 3G standardization was a rapid process, with 3GPP producing the first full version of the WCDMA standard at the end of 1999. This release, called Release 99, contains all the necessary elements to meet the requirements for IMT-2000 technologies including 2 Mbps data rate with variable bit rate capability, support of multiservice, a flexible physical layer for easy introduction of new services, quality of service differentiation and efficient packet data. Also, the necessary functionality for the GSM/WCDMA intersystem handovers is part of the Release 99 standards, both for WCDMA and GSM.

The next version of the standard release, Release 4, was completed in March 2001. The main new feature is in the circuit switched core network where Release 4 allows the split of the user plane and the control plane of the *Mobile Switching Centre* (MSC) to *Media Gateway* (MGW) and to MSC server respectively. Release 4 also includes IP header compression which is important for the radio efficiency of IP based applications. On the radio access side, Release 4 contains mainly small improvements and clarifications over Release 99. In the area of *Time Division Duplex* (TDD) technology, the additional chip rate option of 1.28 Mcps was added for the TDD specifications in addition to the 3.84 Mcps TDD in Release 99.

3GPP Release 5 was completed in March 2002. The main new feature is HSDPA which pushes the peak data rates beyond 10 Mbps within the WCDMA bandwidth of 5 MHz. The HSDPA concept utilizes a distributed architecture where the packet data processing is brought to the base station, which allows faster packet scheduling and faster retransmissions.

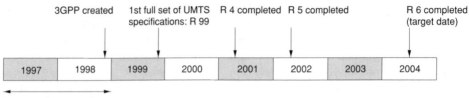

Figure 1.4. 3GPP standard releases.

HSDPA uses techniques such as adaptive modulation and coding, higher order modulation and retransmission combining in the receiver to improve the radio performance, as covered later in Chapter 12. HSDPA pushes practical user data rates beyond 1 Mbps and improves the spectral efficiency of WCDMA. More details on the HSDPA performance are covered in [6] and the references therein.

Other issues in Release 5 were support for the *Internet Protocol* (IP) based transport on the *UMTS Terrestrial Radio Access Network* (UTRAN) side, and in the core network side the introduction of the *IP Multimedia Subsystem* (IMS) for standards based IP service creation. The WCDMA frequency coverage was also extended to cover the 1900 MHz band in the Americas.

In general, the work related to a new frequency band is release independent, so that it is, for example, possible to make a Release 99 terminal for the 1900 MHz band even though the actual RF requirements and signalling extensions are only in the Release 5 specifications.

The timing of the 3GPP standards releases is illustrated in Figure 1.4, with the completion date indicating the time when the release content was frozen and all the required details were specified. Typically the performance and possible new RF requirements are completed slightly later. As the actual implementation and testing work by the different vendors becomes more advanced, typically some errors are detected. Necessary corrections and clarifications are then covered with change requests to the released specifications.

In the 3GPP documentation, Release 99 related standards have version number 3.x.x, Release 4 version 4.x.x and Release 5 version 5.x.x. The second digit increases when a TSG approves change requests to the specification after the first version has been created. All 3GPP standards are public, with a few exceptions in the security area, and can be obtained from [4].

1.2.4 3GPP Standards Evolution

3GPP work continues beyond Release 5. The following features are currently being studied in 3GPP for further releases including Release 6 and beyond:

- enhanced uplink *Dedicated Channel* (DCH), which basically studies similar methods to those used for HSDPA for the uplink operation. The focus is on the improved coverage and capacity of the uplink services, especially with packet based traffic. Also the delay related to initiating uplink packet transmission is an area that is being studied for improvements;

- new frequency variants of WCDMA, such as the utilization of both the 2.5 GHz and 1.7/2.1 GHz spectrum. The latter is relevant in the USA;

- advanced antenna technologies, including enhancements for the beamforming capabilities in the network side as well as transmit diversity technologies with one or two receiver antennas in the terminal. The latter work is being carried out under the *Multiple Input Multiple Output* (MIMO) term;

- the support of multicast services to provide efficient means for sharing common content between several users;

- in the area of core network, the focus is to add new features to the IMS;

- on the GERAN side, key areas are to improve GERAN radio flexibility for the support of the new services by developing the physical layer flexibility away from the traditional bit by bit specification of each data rate to be supported. Work is also being done on performance improvement features.

1.3 3GPP TERMINOLOGY

The signal path of a communications system is typically segmented into different layers. The ISO/ITU[1] network model, called the *Open System Interconnection* (OSI) model, provides a structure which divides the end to end communications link into seven different layers [7]. Figure 1.5 shows the different OSI layers labelled with the 3GPP terminology for the base station, which is referred to as the Node B, and the mobile device, which is referred to as the *User Equipment* (UE). Note that this book addresses Layer 1 to Layer 3 in the UE.

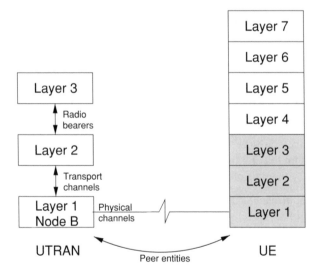

Figure 1.5. The open system interconnection system OSI mapped to a WCDMA system.

In the OSI model the purpose of each layer is as follows.

1. Layer 1 (L1), the physical layer, specifies air interface issues such as modulation and coding.

---

[1]International Standardisation Organisation, International Telecommunication Union

2. Layer 2 (L2), the data link layer, formats data into blocks and enables reliable transmission between adjacent nodes or peer entities.

3. Layer 3 (L3), the network layer, specifies addressing and routing.

4. Layer 4 (L4), the transport layer, is responsible for a reliable user to user data transfer.

5. Layer 5 (L5), the session layer, is, amongst others, concerned with security features.

6. Layer 6 (L6), the presentation layer, specifies the presentation of the data to the end user.

7. Layer 7 (L7), the application layer, defines the actual tool which interfaces with the user.

A commonly used term in this book is 'channel'. As this term can have different meanings, this can lead to confusion if the context is not clear. The different meanings of this term are summarized below, some of which are included in Figure 1.5.

- The propagation channel refers to the channel between the Node B antenna and the UE antenna;

- the frequency spectrum is divided into bands and channels. For example, band I is 60 MHz wide and contains several 5 MHz channels. A channel is used to refer to the frequency range used by a network operator who offers a WCDMA service;

- a *Physical Channel* (PhCH) is a data carrying channel over the air interface. Each PhCH has a unique spreading code to provide discrimination from the other PhCHs. See Chapter 3 for details. An active user in a cell can use dedicated PhCHs, common PhCHs, or both. A dedicated channel is a PhCH to a particular UE, while a common channel is shared amongst the UEs in a cell;

- the *Transport Channels* (TrCHs) are the channels offered by the physical layer to Layer 2 for the transfer of data. These TrCHs are mapped to PhCHs by *Bit Rate Processing* (BRP), see Chapter 4;

- logical channels are the channels offered by the *Medium Access Control* (MAC) sublayer within Layer 2 to higher layers. See Chapter 6 for details. Logical channels are defined by the type of information they carry;

- radio bearers are channels offered by Layer 2 to higher layers for the transfer of either user or control data.

Having defined some terminology necessary for us to continue, we now give a very brief overview of the operations carried out in the UE when receiving information from the network.

1.4 THE JOURNEY OF A BIT

This section is a preface to the technical chapters that follow. Without going into the technical details too much, we describe what happens to a bit during the course of the signal processing chain in a handset. This provides the reader with an overview of the whole signal chain. We assume that the bit is part of a voice (speech) service and may resemble a noise such as a 'peep'. Other applications are of course also possible, such as video streaming, picture messaging and Internet browsing.

We join the bit while it travels to the handset from the Node B antenna as an analogue radio wave, having been converted into this form in the Node B. After a journey through the air lasting perhaps a few microseconds, the radio wave hits the UE antenna and is turned into a current and a tiny voltage.

After the antenna, the bit passes the duplexer, a circuit which enables the receiver and the transmitter to operate simultaneously. This is typical for an FDD CDMA system, where the downlink (Node B to UE) and the uplink (UE to Node B) operate simultaneously, but use a different frequency. The duplexer directs the received signal to the radio's receive section. The analogue signal now enters a low noise amplifier which increases the signal strength so that it is suitable for further analogue processing.

The bit buried in the RF signal is now ready for extraction by means of demodulation. The carrier is removed by the mixer and the baseband signal extracted. The demodulator is split into two branches, in-phase and quadrature-phase, because WCDMA employs a complex signal. The analogue baseband signal now enters a low pass filter in order to remove the images introduced by the frequency down conversion process. Unfortunately a DC component is generated but this is removed by a high pass filter.

The bit now enters an adjustable amplifier stage, whose gain is controlled by the *Automatic Gain Control* (AGC) loop. The purpose of this stage is to ensure that the subsequent *Analogue to Digital Conversion* (ADC) device can resolve the analogue signal optimally. After the variable amplifier, the analogue signal is converted into a digital signal representation by the ADC. For the first time since we joined the bit on its journey, it is part of a digital word which comprises bits. Thus the bit is pleased with this transformation back into its natural domain. However, the digital signal is full of chatter from other cells and other users. A pulse shaping filter is employed to reduce this gossip. Our bit is now hidden in a digital word at a defined digital reference point.

So far, the bit carrying radio wave has been processed and conditioned, and converted into a referenced digital representation. However, the bit is hidden in the continuous data stream pouring out of the digital reference point into the subsequent baseband processing stage. Usually, the term baseband encompasses all functions that deal with the baseband signal from where it has been digitized to where the digital signal is passed on to Layer 2. However in this book we have chosen to limit the baseband to the functions between the ADC and the digital reference point. The next stop on the bit's route is the land of *Chip Rate Processing* (CRP).

At the digital reference point our bit is in the form of many digitized chips. This is the input to CRP, where these chips are fed into several blocks simultaneously. There are numerous functions processing the chips, one of which is the channel estimator. The output of the channel estimator is used to synchronize the receiver with the transmitter, and to allow the receiver to combine any multipath channel components. Other functions performed in the CRP block are Node B detection, known as searching, signal power measurements, frequency correction and signal power control. These functions are used to control the reception of our bit, but it does not pass directly through any of them. The CRP entity that it does pass through is the RAKE receiver, which extracts symbols representing the transmitted

Chip rate processing

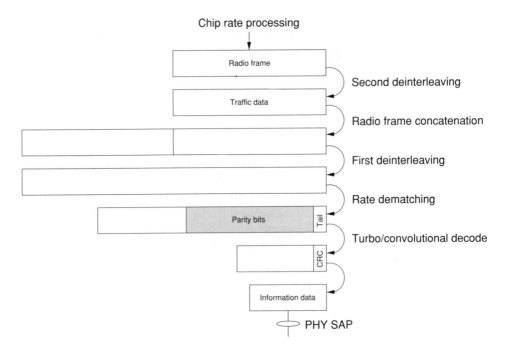

Figure 1.6. The bit rate processing data block conversion path.

bits out of the chip stream. The output of the RAKE receiver is a stream of symbols. Each symbol is separated into its in-phase and quadrature-phase component which yields two real numbers known as soft decisions.

However, at this stage we still do not know if our bit is a one or a zero since the information is still hidden in an encoded block of soft decisions. Hence, the data stream of soft decisions is forwarded into the BRP entity for further treatment. Figure 1.6 shows a simplified form of how the data is processed in the BRP entity, starting at the entry point where the soft decisions are obtained from the CRP, to the point where the decoded information bits enter Layer 2, which is the *Physical Service Access Point* (PHY SAP).

A CRP block of soft decisions is processed by several BRP stages, such as deinterleavers, transport channel demultiplexers, frame desegmentation, channel decoders and rate dematching. These are all described in detail in Chapter 4. While our bit travels through the BRP signal avenue, it enters the channel decoder. There are two different types of decoder possible, a Viterbi decoder or a Turbo decoder. The decoder output is a block of information data, known as a *Transport Block* (TB), accompanied by a much shorter *Cyclic Redundancy Check* (CRC) sequence. The CRC sequence is used later to determine if the data block has been received in error, which is used to determine the block error ratio and perhaps trigger a retransmission process of the same data block.

After BRP the TB enters the protocol avenue through the PHY SAP. The protocol avenue is split into a control plane (C-plane) and a user plane (U-plane), which is a separation to distinguish the control data messaging from the user data transfer. At the transmitter side, the input to each protocol entity is called a *Service Data Unit* (SDU), and the output from

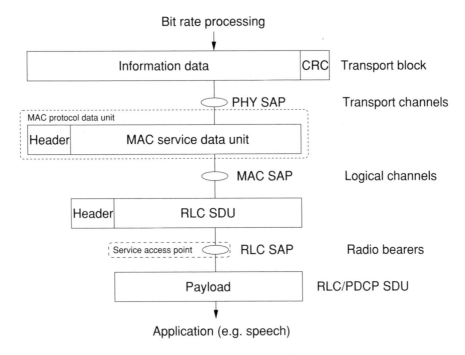

Figure 1.7. The Layer 2 and Layer 3 protocol signal path.

each entity is called a *Protocol Data Unit* (PDU). Figure 1.7 shows what happens to the
TB, and its position as part of SDUs and PDUs, while it passes through the protocol layers.

The first block on the road leading to the application is the *Medium Access Control*
(MAC) entity. The TBs from the BRP are MAC PDUs, which may contain a MAC header
as well as the logical channel SDU input information to the MAC. The MAC reads the header
and processes its PDU accordingly. It maps the transport channels onto logical channels
and vice versa, and forwards its SDUs to the next block in the receiver chain, the *Radio
Link Control* (RLC) entity. The MAC SDU again may contain a header and an RLC SDU.
The RLC entity reads any RLC header and processes its PDU. RLC is, amongst other tasks,
responsible for retransmissions, which is another method for reducing the packet error rate.
Depending on the type of data contained in the SDU, RLC will forward it to other protocol
entities or directly to the application.

Circuit switched user data can be passed from the RLC directly to the application. Other
user plane data packets remain in Layer 2. Data packets are passed to the *Packet Data
Convergence Protocol* (PDCP) entity. The RLC PDU again contains a header and a payload
SDU. The PDCP entity manages the transfer of IP packets over the air. It contains header
compression features to increase the spectral efficiency of the air interface. The final Layer 2
entity is the *Broadcast/Multicast Control* (BMC) entity, which manages services like the
popular *Short Messaging Service* (SMS). None of this concerns our bit which is passed
directly to the speech decoder from RLC.

All the previous entities need to be managed, otherwise our bit ends up at the wrong
application. The brain controlling the data flow is the *Radio Resource Control* (RRC) entity,

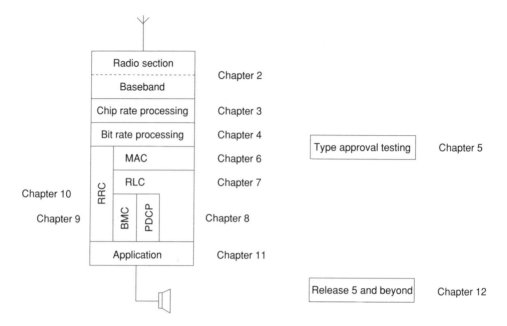

Figure 1.8. The structure of the book.

which belongs to the C-plane and Layer 3. The RRC entity has numerous jobs but the key task is to control the Layer 1 and 2 protocol entities.

After a ride on the protocol avenue, our bit has finally reached the speech decoder. It has lost a lot of friends on its journey, including various header and parity bits. In the speech decoder, the bit and many others are finally converted back to a sound after their long journey. The user hears the short *peep* sound. Fortunately he or she is probably unaware of the work our bit has had to do to produce this sound.

1.5 STRUCTURE OF THE BOOK

The previous section has already provided some insight into the content of the book. Figure 1.8 shows all the following chapters at a glance. The radio section and baseband signal processing is described in Chapter 2. The chip and bit rate processing are discussed in Chapters 3 and 4. Chapters 6, 7, 8 and 9 discuss the Layer 2 entities of MAC, RLC, PDCP and BMC. Chapter 10 discusses RRC. In addition, Chapter 5 addresses testing the Layer 1 block error rate performance against the 3GPP type approval requirements, and Chapter 12 discusses future technologies and HSDPA in particular. Finally, Chapter 11 discusses speech coding, which is one possible application of the WCDMA access stratum technology discussed in the rest of the book.

2

RF and Baseband Processing

Nick Parker, Mel Long, John Reeve, Marcello Caramma

2.1 INTRODUCTION

This chapter concentrates on the radio aspects of 3GPP's *Frequency Division Duplex* (FDD) version in the European frequency bands. The chapter examines the requirements imposed by the 3GPP specifications, describes a typical UMTS transceiver and looks at key performance requirements.

The *Radio Frequency* (RF) section of a handset is crucial to the *User Equipment* (UE) performance and hence to the success of a product. Poor quality reception can encourage customers to change their handset for a better one. Network operators, on the other hand, have an interest in maximizing cell radius in order to reduce the number of base stations to provide coverage. Both sides, customers and operators, also want attractive UE costs, while battery lifetime is generally of interest to the users. However, both objectives are difficult to achieve if additional signal processing or circuitry is required to recover the losses incurred by the RF front end.

The RF front end comprises, amongst other components, amplifiers, filters, oscillators and mixers (modulators) for both transmitter and receiver. The term *Baseband* (BB), as used in this chapter, refers to the (digital) signal processing blocks which are closely linked to the RF transceiver circuits. These include the *Analogue to Digital Conversion* (ADC) and *Digital to Analogue Conversion* (DAC) stages, the pulse shape filtering stage and the *Automatic Gain Control* (AGC) loop. The RF stage also is closely connected to the *Chip Rate Processing* (CRP) stage described in Chapter 3, in which the received signal is extracted for further processing and the transmitted signal is prepared for modulation. CRP may also control parts of the RF stage, such as the AGC loop and *Automatic Frequency Control* (AFC).

WCDMA – Requirements and Practical Design. Edited by R. Tanner and J. Woodard.
© 2004 John Wiley & Sons, Ltd. ISBN: 0-470-86177-0.

Good introductory textbooks on practical design and implementation issues are [1] for RF and [2] for baseband processing. Reference [8] addresses WCDMA and provides a more up-to-date survey on published RF designs, including a section on the relevant RF fundamentals.

This chapter is organized as follows. The first section introduces the 3GPP requirements for the receiver and transmitter, followed by an RF front end architectural overview. Details of the receiver RF and baseband design are covered in Sections 2.3 and 2.4. This is followed by the transmitter design which again is split into baseband and RF aspects covered in Section 2.5 and Section 2.6 respectively.

2.2   UMTS RADIO REQUIREMENTS

The radio transceiver architecture is of course neither described nor defined by 3GPP, but some of the mandatory properties have been defined in the main radio performance standards.

- TS25.101 [9] defines the UE radio and performance requirements;

- TS34.121 [10] specifies the type approval tests which also cover the radio performance.

This section summarizes a selected number of key specification points which are now defined in 3GPP TS25.101, some of which were discussed during the early days of 3GPP, for example in [11, 12]. But before the requirements are presented, a short briefing on a few important 3GPP signal power terms is presented in order to ease reading and understanding of the notation used in the subsequent sections.

Figure 2.1 depicts the different signal and noise power terms [9]. I_{or} denotes the total transmitted power spectral density of the downlink signal measured at the Node B (base station) antenna connector, while $\hat{I}_{or}$ is the total received power spectral density at the UE antenna connector (integrated over $(1 + \alpha)$ times the chip rate, where α denotes the roll off factor of the pulse shape filter). I_{oc} is the total interference and noise power spectral density of a band limited white noise source measured at the UE's antenna connector. Finally, I_o is the total received signal power spectral density at the UE antenna connector.

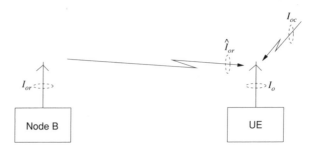

Figure 2.1. The 3GPP notation for the different signal powers.

Table 2.1   Key receiver RF requirements as specified in [9].

Parameter	Limit or test condition
Operating frequency	Band I: 2110–2170 MHz Band II: 1930–1990 MHz Band III: 1805–1880 MHz
Nominal duplex separation	Band I: 190 MHz Band II: 80 MHz Band III: 95 MHz
Operating level range (I_{or})	−25 dBm to −106.7 dBm
Sensitivity	Band I: −117 dBm Band II: −115 dBm Band III: −114 dBm
Adjacent channel selectivity	Equivalent to 33 dB for the defined test
Spurious emissions	< −60 dBm in RX frequency band at antenna
In-band blocking signals	±10 MHz, − 56 dBm modulated interference ±15 MHz, − 44 dBm modulated interference
Out-band blocking signals	Frequency range 1, − 44 dBm CW interferer Frequency range 2, − 30 dBm CW interferer Frequency range 3, − 15 dBm CW interferer For declared spurious frequency: − 44 dBm CW
Intermodulation	±10 MHz signal, − 46 dBm CW ±20 MHz signal, −46 dBm modulated

2.2.1 Receiver Performance Requirements

Table 2.1 lists a selected number of RF related requirements. For most tests specified in TS25.101, a *Bit Error Rate* (BER) of 0.001 for the *Dedicated Physical Channel* (DPCH) must be achieved with the appropriate test channel settings.

2.2.2 Transmitter Performance Requirements

Table 2.2 lists the key transmitter RF specification points. Note that the term channel refers here to the first adjacent 5 MHz UMTS frequency band, the first adjacent channel.

2.2.3 Frequency Bands and Channel Arrangements

The UMTS specification defines frequency division full duplex operation as the standard mode for voice and data communication. The simultaneous and continuous operation of the transmit and receive circuits is one of the most significant changes from the way GSM works for example.

The frequency bands for FDD operation are shown in Figure 2.2, with Band I (B1) as the principal allocation within Europe. The two 60 MHz allocations give a nominal duplex spacing of 190 MHz for B1, however, the specification also allows for variable duplex, i.e. where the spacing between transmit and receive is between 130 MHz and 250 MHz. A

Table 2.2 Key transmitter RF specifications [9].

Parameter	Limit or test condition
Operating frequency	Band I: 1920–1980 MHz Band II: 1850–1910 MHz Band III: 1710–1785 MHz
Nominal duplex separation	Band I: 190 MHz Band II: 80 MHz Band III: 95 MHz
Maximum output power and accuracy	Class 1: +33 dBm +1/−3 dB Class 2: +27 dBm +1/−3 dB Class 3: +24 dBm +1/−3 dB Class 4: +21 dBm ±2 dB
Minimum output power	Class 3: −50 dBm to +24 dBm Class 4: < −50 dBm to +21 dBm
Open loop power control accuracy	Normal: ±9 dB Extreme: ±12 dB
Transmit power switching period	±25 μs either side of relevant slot
Error vector magnitude	≤17.5 % RMS
Adjacent channel leakage power ratio	< −33 dBc for first adjacent channel < −43 dBc for second adjacent channel

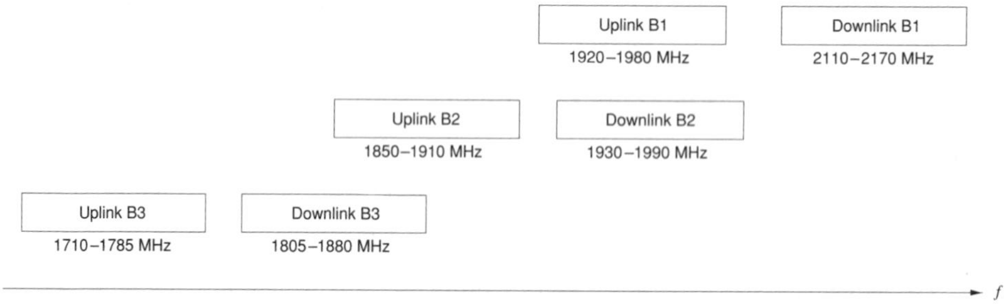

```
                              ┌──────────────┐        ┌────────────────┐
                              │  Uplink B1   │        │  Downlink B1   │
                              └──────────────┘        └────────────────┘
                              1920–1980 MHz            2110–2170 MHz

             ┌──────────────┐          ┌────────────────┐
             │  Uplink B2   │          │  Downlink B2   │
             └──────────────┘          └────────────────┘
             1850–1910 MHz              1930–1990 MHz

┌──────────────┐        ┌────────────────┐
│  Uplink B3   │        │  Downlink B3   │
└──────────────┘        └────────────────┘
1710–1785 MHz            1805–1880 MHz

────────────────────────────────────────────────────────────► f
```

Figure 2.2. FDD frequency band allocation.

variable duplex system imposes extra demands on the transceiver design as the transmit and receive synthesizers have to tune independently of one another. Band II (B2) reuses the existing *Personal Communication System* (PCS) band and is intended for use in the USA with UMTS and GSM systems coexisting within the spectrum. The duplex spacing is only 80 MHz for Band II and places more difficult requirements on the transceiver hardware. Additional spectrum has been highlighted for future use within the 2560–2670 MHz band [13, 14].

The channel numbering system is based on a raster of 200 kHz, with a nominal channel spacing of 5 MHz. The *UTRA Absolute Radio Frequency Channel Number* (UARFCN), the channel number N_{dl} with centre frequency f_{dl} for the downlink and N_{ul} with f_{ul} for the uplink respectively, are allocated according to Table 2.3 and Table 2.4 [9].

Table 2.3   UARFCN definition.

	UARFCN	Carrier frequency in MHz
Downlink	$N_{dl} = 5 f_{dl}$	$0 \le f_{dl} \le 3276.6$ MHz
Uplink	$N_{ul} = 5 f_{ul}$	$0 \le f_{ul} \le 3276.6$ MHz

Table 2.4   UARFCN definition (Band II additional channels).

	UARFCN	Carrier frequency in MHz
Downlink	$N_{dl} = 5(f_{dl} - 1850.1 \text{ MHz})$	$f_{dl} = 1932.5, 1937.5, \ldots, 1982.5, 1987.5$
Uplink	$N_{ul} = 5(f_{ul} - 1850.1 \text{ MHz})$	$f_{ul} = 1852.5, 1857.5, \ldots, 1902.5, 1907.5$

2.2.4 Radio Architecture Overview

The UE's radio access technology, which processes the signal from and to the air interface, is broken down into three chief components:

Radio transceiver comprises the RF section and the associated baseband signal processing stages, responsible for demodulation and modulation.

Chip rate processing entity comprises functions which are relevant to the proper functioning of the UE and the preprocessing of received data and processing of data suitable for modulation. A number of processes work without being controlled or managed by a layer above the physical layer, e.g. synchronization, while other operations require information and control from layers above, e.g. executing a compressed mode pattern. Its primary task is synchronization, spreading and despreading the signal.

Bit rate processing entity comprises the functionality which processes the actual information, from and to higher layers, suitable for transmission over the air interface via the chip rate processing block. Its primary task is one of signal protection and correction of errors.

This distinction appears arbitrary but is borne out of the structure of 3GPP and its specifications.

The radio transceiver hardware provides an interface between the antenna and the follow on digital signal processing entity. The block diagram of a typical WCDMA FDD (only) transceiver is shown in Figure 2.3, where the radio hardware has been split into four areas:

Receive RF filters, amplifies and down converts the RF signal received at the antenna.

Receive baseband filters and converts the analogue signal into the digital signal which can be processed by the chip rate processing entity.

Transmit baseband consists of blocks which process and convert the digitized signal to an analogue signal suitable for modulation.

Transmit RF modulates, up converts and amplifies the signal onto a high power RF carrier.

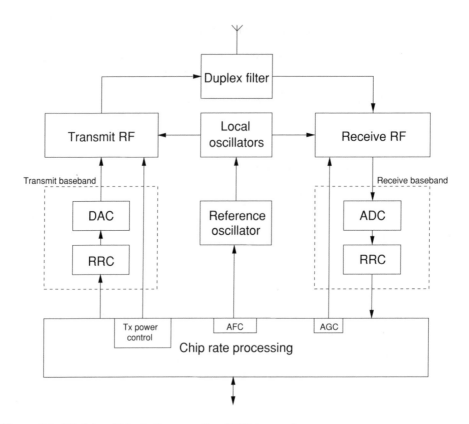

Figure 2.3. High level block diagram of an FDD transceiver.

Frequency control is achieved within the *Local Oscillator* (LO) [15] block. The wide bandwidth of the UMTS signal makes the phase noise performance requirements slightly easier than for GSM. A single LO synthesizer may be used for fixed duplex operation, however, for direct conversion RF architectures, it is more convenient to have two independent LO synthesizers, for the receiver and the transmitter. The implementation of compressed mode, see also Section 3.11, requires the receiver to tune to another channel to make a signal measurement. Fast synthesizer tuning is required to ensure that the measurement can be carried out within the defined time span. The frequency accuracy of the terminal is defined by the frequency reference and this is controlled using an AFC loop to maintain frequency synchronization with the basestation [16]. When operating using noncompressed mode, the GSM and UMTS circuits will be used simultaneously and the reference frequency must be chosen to meet the requirements of both systems.

The DAC and the ADC provide the interface to and from the chip rate block to the RF stages respectively. The transmit DAC must have sufficient dynamic range to meet the adjacent channel leakage requirements. The receive ADC must have sufficient dynamic range to cater for the signal peak to average ratio and residual blocking signals, and will also depend on the AGC algorithm employed in the receiver. Both converters operate at a multiple of the chip rate, typically between four to eight times, to achieve the system's filtering requirements and provide sufficient timing resolution for the RAKE receiver. The

Table 2.5   Typical duplex filter requirements.

Parameter	Requirement in dB
Transmit RF to antenna attenuation in the Tx band	< 1.5
Receive RF to antenna attenuation in the Rx band	< 2.5
Tx to Rx isolation	> 50

UMTS channel filters, the pulse shape filters, are implemented using digital *Root Raised Cosine* (RRC) filters in the baseband.

The transmit RF stage provides up conversion from a baseband signal to RF and provides power level control for the transmitter. The modulation employed for WCDMA is different from GSM in that it does not have a constant amplitude. Amplitude distortion in the signal processing components will result in unacceptable degradation of the modulation, consequently linear RF circuits must be used. Another problem with WCDMA systems occurs between mobiles that are at different distances from the base station. As many terminals share the same frequency band, a near mobile may block a far mobile if it transmits at a high power level. This problem is overcome using a power control loop in which the transmit power is adjusted 1500 times per second. Implementation of fast and accurate transmit power control circuits is vital to the performance of the UMTS system, see Chapter 3 for further details.

2.2.4.1 *The Duplexer: Connecting Transmitter and Receiver*   The receiver and transmitter must be isolated as the transmit signal will appear as a blocking signal at the receiver. This is typically achieved using a filter with two bandpass sections [11], called the duplex filter in Figure 2.3. The duplex filter should have a low insertion loss in the transmitter frequency band (Tx band), high transmitter to receiver isolation in the receiver frequency band, and a low insertion loss in the Rx band, and a high Tx to Rx isolation in the Tx band.

At the transmitter, the insertion loss can be reduced if the duplex filter attenuates noise stemming from the *Power Amplifier* (PA), while noise generated before the PA should be filtered in the transmitter's signal path. Another design aspect which requires attention is the transmit RF leakage which can have an impact in the Rx band because of receiver nonlinearities. Table 2.5 summarizes a set of duplex requirements.

The receive RF stage must be able to process multiple signals received on the same frequency. To allow separation of the wanted signal from the other cochannel signals, the receiver must maintain the level of orthogonality between the different code channels. The high peak to average power of the composite received signal requires linear RF signal processing, and attention must be paid to differential group delay within the receiver.

2.3 RECEIVER RF DESIGN

The heterodyne receiver is well established and used for a wide range of communication systems. They achieve good performance with respect to channel selectivity and sensitivity, at the expense of complexity and cost. However, recent developments in receiver technology and the requirement for multistandard (GSM/WCDMA) receivers operating in different frequency bands has favoured a *Direct Conversion Receiver* (DCR) architecture.

A direct conversion, or homodyne/zero *Intermediate Frequency* (IF), receiver converts an RF signal to a signal centred at zero frequency. The conversion employs a quadrature down conversion, using an in-phase and quadrature-phase local oscillator to allow the signal to be recovered, although spectral overlap occurs in the conversion. Using a single down conversion stage results in a simpler receiver and this translates directly into a cost advantage. The attendant difficulties that are often experienced with direct conversion receivers will be discussed in this chapter. A more general treatment of heterodyne, homodyne and other receivers can be found in reference [8].

2.3.1   Direct Conversion Receiver

Direct conversion receivers are becoming more popular for handset applications due to their lower component count, cheaper cost and simpler frequency plan, e.g. [17]. The elimination of an IF removes the need for a *Surface Acoustic Wave* (SAW) filter and an IF synthesizer (LO oscillator) and mixer. In principle, the RF signal processing chain for GSM and UMTS is the same for a direct conversion receiver allowing a single receiver to provide multimode operation [18]. In practice, the characteristics of the two systems are sufficiently different that such a solution is difficult. However, direct conversion receivers are well suited to UMTS in that the wideband signal reduces the impact of flicker noise and reduces the gain that must be provided in the baseband amplifiers [19]. Unlike a GSM signal, a UMTS signal can be correctly demodulated if the baseband processing includes a high pass filter. Any problems with DC offsets may then be eliminated using simple AC coupling in the receiver gain stages. Given the merits of lower cost and simpler implementation, it is likely that almost all commercial WCDMA receivers will use a direct conversion architecture and so the rest of this chapter will focus on direct conversion implementation for UMTS.

Figure 2.4 shows the block diagram of a simple direct conversion receiver for UMTS. In this example the antenna is routed to the GSM or UMTS circuits using a switch. This will allow compressed mode to be implemented. However, an alternative arrangement may be required for noncompressed mode. A duplex filter is used to sum the UMTS transmit signal and to provide isolation. Both these elements introduce loss in the receiver front end and consequently degrade the sensitivity. A *Low Noise Amplifier* (LNA) is positioned immediately after the duplex filter to recover the signal level. The gain of the LNA component is

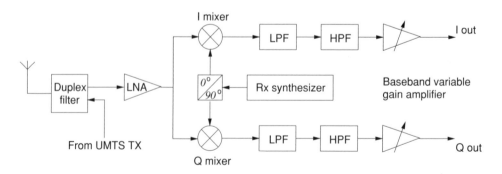

Figure 2.4. Typical direct conversion receiver.

chosen to be high enough that noise introduced in subsequent circuits has a minimal impact on the overall sensitivity.

The amplified RF signal from the antenna is converted directly to baseband using a mixer circuit. If a simple down conversion was used then all spectral components of the wanted signal would be mixed to a band of 0–1.92 MHz and the signal could not be correctly demodulated.

The receiver uses a pair of mixer circuits provided with in-phase (I) and quadrature-phase (Q) local oscillators so that positive and negative frequency components can be resolved by the follow on baseband processing. The output of the mixer comprises the baseband difference signal and the sum signal with a frequency of $2f_{LO}$, where f_{LO} is the local oscillator frequency. The output also includes a number of signals resulting from distortion.

The sum signal is not required and so a filter is used after each mixer so that it is attenuated to a low level. In practice, the frequency separation of the sum is so large that a simple low pass filter, e.g. a simple RC filter, will provide adequate rejection. As there is no narrowband filtering within the RF portion of the receiver, the selectivity is defined entirely by the steepness of the baseband filters. The filter may be implemented partially in the analogue processing and partially in subsequent digital signal processing stages, however, some analogue filtering is required to attenuate the level of blocking signals before further baseband amplification.

As there is no narrowband filtering within the RF portion of the receiver, the selectivity is defined entirely by the baseband filters. The signal at the output of the filter stage requires amplification to increase the level so that it matches the operating level of the ADC. As the input signal level will vary depending on the propagation conditions, the amplifier gain must be variable and is typically controlled using an AGC loop. The presence of DC offsets at the output of the mixer must be considered when designing the receiver. A high pass filter is included before the amplifiers to remove DC offsets and low frequency signals to prevent the amplifier output from becoming saturated when set to high gain. The presence of DC offsets at the ADC appears as a blocking signal and can seriously degrade the sensitivity of the receiver. To overcome this problem, the output of the baseband amplifier will typically be AC coupled to the ADC so that any offset voltage introduced in the amplifiers does not impact on performance.

2.3.2 Direct Conversion and Even-Order Distortion

Even-order distortion is the nemesis that awaits all direct conversion receiver designers. It is the problem of even-order distortion, and in particular second-order distortion, that marks the difference between superheterodyne receivers using a real IF and direct conversion receivers using a zero frequency IF. However, with careful design, the problems associated with even-order distortion can be managed [11].

Consider an input signal $V_{in}(t)$ comprised of two sinusoidal carriers with equal magnitude A and frequencies ω_1 and ω_2

$$V_{in}(t) = A \cdot \sin(\omega_1 t) + A \cdot \sin(\omega_2 t) \tag{2.1}$$

This input signal is applied to a signal processing device that exhibits second-order distortion. The device is modelled by a simple equation where the constant k defines the magnitude of the second-order component $V_{in}^2(t)$.

$$V_{out}(t) = V_{in}(t) + k(V_{in}(t))^2 \tag{2.2}$$

The second-order component $k(V_{in}(t))^2$ may be expanded to

$$V_2(t) = kA^2(\sin^2(\omega_1 t) + \sin^2(\omega_2 t) + 2\sin(\omega_1 t)\sin(\omega_2 t)). \tag{2.3}$$

Using the trigonometric identities

$$\sin^2(x) = 0.5 + 0.5\cos(2x)$$

and

$$\sin(x)\sin(y) = 0.5\cos(x - y) - 0.5\cos(x + y),$$

we arrive at

$$\begin{aligned} V_2(t) = kA^2 \{ &1 + 0.5\cos(2\omega_1 t) + \\ &0.5\cos(2\omega_2 t) + \\ &\cos(\omega_1 t - \omega_2 t) - \cos(\omega_1 t + \omega_2 t) \} \end{aligned} \tag{2.4}$$

Hence, the output comprises a DC component (first line in (2.4)), components at twice the frequency of the inputs, and components at the sum and difference frequency of the two tones. For a case where the input frequencies are 2140 MHz and 2141 MHz respectively, the output will contain components at 4282 MHz, 4281 MHz, 4280 MHz, 1 MHz and at DC. The process is equivalent to AM demodulation of the input signal, where the result is an interfering signal that appears in the channel band in the zero IF receiver. The downlink WCDMA modulated signals used by the UMTS system have a significant AM component and so second-order distortion will result in desensitization of the receiver if the signal processing components have an inadequate *second-order Intercept Point* (IP2). The problem is neatly avoided in a superheterodyne solution as these spectral components will be removed by the IF filters. However, direct conversion designers have a weapon to fight the problem in the use of balance (differential amplifier principle), and all zero IF receivers are implemented with balanced circuitry applied rigorously throughout [20]. The use of balanced circuitry reduces second-order distortion and allows the sensitivity specifications to be met without using components with excessive signal handling or bias current requirements.

2.3.3 Transmit Leakage and IP2

The requirements for the RF signal processing components in a direct conversion receiver are generally defined in terms of IP2 [8, 12]. From Equation (2.4), it can be seen that the level of the second-order components is proportional to the input amplitude squared. Hence, every 10 dB change in input signal level results in a 20 dB change in the level of the second-order components. At some theoretical point, the level of IP2 products must then be equal to the input signal level and this point is called the second-order intercept. Figure 2.5 shows how the value of the 2-tone IP2 may be calculated for the inband distortion component. In practice, the receiver IP2 may be determined by measuring the inband tone level at the output of the mixer, and then subtracting the mixer gain to refer this level back to an equivalent level at the mixer input to find the IP2.

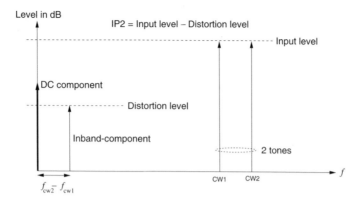

Figure 2.5. Determining IP2.

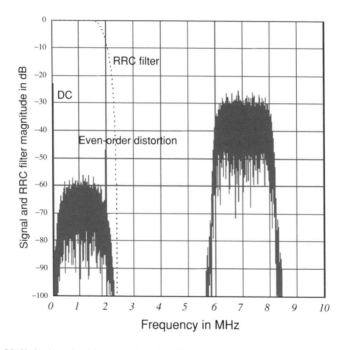

Figure 2.6. Uplink signal with second-order distortion.

The IP2 requirement for the receiver is dependent on the largest blocking signal present at the input. In practice, this is the UE's own transmitter. The largest out of band blocking signal defined in TS25.101 is a -15 dBm CW carrier, however, the duplex filter has significant rejection at the frequencies for which this test is specified. The transmit signal may have an amplitude in excess of $+25$ dBm and duplex filter Tx–Rx isolation figures are typically 45–55 dB, giving transmit leakage in the range of -20 dBm to -30 dBm at the LNA input. Figure 2.6 shows the typical output spectrum for a UMTS uplink signal with one DPCH subject to second-order distortion.

Table 2.6   3GPP receiver sensitivity power level requirements.

Operating band	Unit	DPCH_E_c	$\hat{I}_{or}$
I	dBm/3.84 MHz	−117	−106.7
II	dBm/3.84 MHz	−115	−104.7
III	dBm/3.84 MHz	−114	−103.7

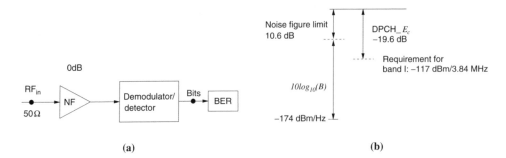

(a)   (b)

Figure 2.7. Noise figure analysis: (a) idealized receiver; (b) power level diagram.

It should be noted that most of the distortion power appears as DC and the use of a highpass filter after the mixer removes this signal before the baseband amplifiers [21]. The level of distortion products for the modulated WCDMA signal may be related to the level of second-order distortion for the 2-tone CW case to provide a simple method of measuring and specifying the system components. From simulations, the level of the distortion signal after filtering with an RRC filter is about −5.6 dB relative to the level of the inband signal shown in Figure 2.5 for equivalent input power. For example, if a CW carrier at 2140 MHz with a power of $P_{CW1} = 0$ dBm and one with $P_{CW2} = 0$ dBm at 2141 MHz results in a $P_2 = −40$ dBm at 1 MHz (with the P_2 level referred back to the point of application of the CW tones). The input 2-tone IP2 is +40 dBm for this example. A WCDMA signal with a level of +3 dBm at 2140 MHz applied to the same input would result in a $P_2 = −45.6$ dBm measured in a 1.92 MHz bandwidth and with a 25 kHz highpass filter (also referred to the input).

2.3.4 Receiver Sensitivity

The receiver sensitivity is defined in TS25.101 [9]. The sensitivity limits for the three current frequency bands are shown in Table 2.6, and, in each case, the test is conducted with a 12.2 kbps test reference channel and a BER of less than 0.1% must be measured.

An important parameter to consider during the design of a communication system is the receiver *Noise Figure* (NF). The power values for the DPCH_E_c contained in Table 2.6 may be used to determine a limiting worst case receiver noise figure to achieve a 0.1% BER. Figure 2.7(a) shows the block diagram of a theoretical receiver comprised of a noisy input amplifier with unity gain and a perfect noiseless demodulator section whose ouput is a sequence of bits. In this instance, the noise figure of the entire receiver is determined by the noise figure of the input amplifier and may be calculated as $NF = 10\log(SNR_{Amplifier\_in}/SNR_{Amplifier\_out})$ [1]. With reference to Section 2.4.1, a typical figure for the *Signal to Noise Ratio* (SNR) requirement DPCH_E_c/I_{oc} at the input of the demodulator to achieve a 0.1 % BER at the output is -19.6 dB. This figure is representative for the 3GPP test case defined for receiver sensitivity.

In the case of a band I system, the DPCH_E_c level at the input of the demodulator is -117 dBm and so, for a -19.6 dB SNR, the effective noise level at this point must be -97.4 dBm measured in a 3.84 MHz bandwidth. The thermal noise level of a 50 Ω system is -108 dBm measured in a 3.84 MHz bandwidth. The equivalent noise figure limit of the receiver is then 10.6 dB for a maximum limiting case where a 0.1% BER is just achieved. This limit is slightly higher than the 6–9 dB commonly reported, e.g. [6, 8], and does not include any implementation margin.

For a practical direct conversion receiver, the sum of all noise sources intrinsic in the RF circuits must be calculated. In addition to the thermal noise and the amplifier noise figure, the following additional noise contributions must be considered during the design:

1. Cochannel noise resulting from even-order distortion.

2. Phase noise of the receive local oscillator, e.g. [22]. The transmit leakage signal at the input of the receiver will mix with the receive local oscillator, and the phase noise at the duplex frequency spacing will appear as cochannel noise.

3. Transmit broadband noise, the phase noise and thermal noise of the transmit leakage signal appear in the receive channel for noise at the duplex spacing from the transmit carrier.

Once all the noise sources are added, the task of designing a UMTS compliant direct conversion receiver becomes significantly more challenging, and high quality filters with low loss and high isolation are required to achieve sufficient isolation between the transmitter and receiver. Table 2.7 shows an example calculation of the receiver noise. This calculation has been simplified and does not include allowances for other receiver impairments such as interchip interference and amplitude and phase imbalance [23]. The table calculates the total equivalent noise at the input of the receiver, and a value of -97.4 dBm would be expected to give a 0.1 % BER according to the limiting case example above. Practical aspects of measuring or verifying the noise figure are addressed in [24]. Note that for volume production, a manufacturing margin must also be added to the figures in Table 2.7.

2.3.5 Adjacent Channel Selectivity

The adjacent channel selectivity test in [9] defines the selectivity and rejection requirements of the channel filter, i.e. RRC pulse shape filter. An interfering signal is applied to the antenna port at an offset frequency of ± 5 MHz from a wanted signal and with a total power of 40.7 dB above the wanted signal I_{or}. The test is conducted well above the sensitivity level of the terminal with the transmitter operating, but with reduced power, while a BER of less than 0.1 % is required. The calculation of the filter selectivity may be obtained from a calculation similar to that presented in Table 2.7, but with the addition of three new items.

1. The interference in the baseband obtained from the interference level, less the filter rejection in the adjacent channel band.

2. Cochannel interference resulting from the receive LO phase noise mixing with the interferer.

3. Cochannel interference resulting from third-order distortion of the interferer.

Table 2.7   Noise calculation for a typical RF receiver stage.

Receiver block parameter	Value
Loss from antenna to LNA input	2.5 dB
Noise figure at LNA input	3 dB
Transmit power	24 dBm
Isolation from Tx to Rx LNA input	50 dB
2-tone IP2 at LNA input	+42 dBm
Rx phase noise at duplex offset	-85 dBc in a 3.84 MHz bandwidth
Noise from Tx at duplex offset	-80 dBc in a 3.84 MHz bandwidth
Thermal noise in 3.84 MHz bandwidth	-108 dBm (at 50 Ω)
Thermal noise (including NF) at antenna	-102.5 dBm
Noise from IP2 distortion	Tx leak at LNA $= -26$ dBm 2-tone P_2 at LNA $= -100$ dBm WCDMA P_2 at LNA $= -105.63$ dBm WCDMA P_2 at antenna $= -103.13$ dBm
Wideband noise from Tx	Tx leak at LNA $= -26$ dBm Noise at LNA $= -106$ dBm Equivalent noise at antenna $= -103.5$ dBm
Phase noise RX LO	Tx leak at LNA $= -26$ dBm Equivalent noise LNA $= -111$ dBm Equivalent noise at antenna $= -108.5$ dBm
Total equivalent noise at antenna connector	-97.8 dBm

The 3GPP specification provides an indicative figure of 33 dB for the selectivity required from the receiver. In practice, a slightly higher figure of channel filter rejection is typically required to achieve a BER of less than 0.1 % for the specified test parameters. For a receive architecture employing an IF, the selectivity is usually achieved using an IF SAW filter in addition to baseband channel filters implemented using active analogue circuits. However, using a direct conversion architecture requires that all the selectivity is achieved in the baseband filters. Component tolerance variation in practical analogue filter implementations means that a rejection of greater than 33 dB cannot be achieved without adjustment or calibration. The selectivity can be achieved by using a digital FIR filter in the baseband signal processing chain with the advantage that an accurate RRC response may be obtained.

The presence of the interfering signal means that if the entire channel filter is implemented using a digital filter, then the ADC would need sufficient dynamic range to handle the interferer and wanted signal. This adds an additional 40.7 dB ($I_{or} = -92.7$ dBm and interferer $= -52$ dBm [9]) or almost seven bits to the range of the converter and is very inefficient for cases where a large blocking signal is not present. In addition, the baseband amplifiers must have sufficient dynamic range to cope with the interferer and a solution without any filtering prior to the amplifiers will have a high current consumption. The channel filter is then split between two stages with some rejection achieved in analogue filters positioned at the output of the mixer and the balance in the baseband using a digital filter. The optimum split depends on the resolution of the ADC and the implementation of the baseband signal processing. A minimum of 15 dB to 20 dB rejection in the analogue filter provides a workable solution for a terminal using a seven or eight bit converter with a

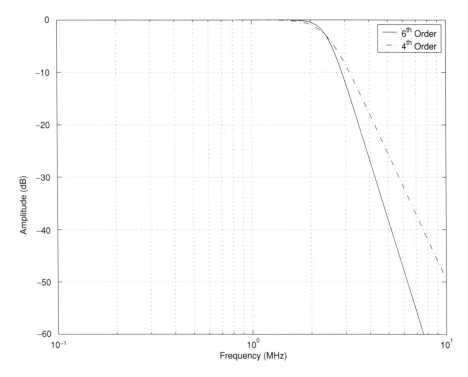

Figure 2.8. Rejection characteristics of two Butterworth filters.

sampling rate of 15.36 Msps. The required rejection can be achieved with a 5th or 6th-order Butterworth filter depending on the component tolerances used for the design. The response of such a filter is shown in Figure 2.8.

In addition to considering the adjacent channel interferer, the requirements of the blocking tests and the need to attenuate the transmit leakage signal must also be considered. However, the adjacent channel rejection requirement is the most difficult to satisfy. The presence of a large interfering signal as the result of transmitter leakage requires that the analogue filters are implemented early in the baseband signal processing chain to prevent the (variable gain) amplifiers from saturating.

The requirements of the anti alias filter, which must attenuate frequencies from the sampling frequency minus 3.84 MHz ($f_{\text{sampling}} - f_{\text{chip}}$), upwards, by approximately 50 dB is also achieved by the adjacent channel filter, for sample rates of four times chip rate and above, so no additional filtering is required in these cases.

2.3.6 Blocking and IP3

The blocking requirements defined in [9] are divided into three categories:

- inband;
- narrowband;
- out of band.

Table 2.8   Frequency Band I out of band blocking tests.

I_{blocking}(CW)		-44 dBm	-30 dBm	-15 dBm
Band I: F_{uw}	MHz	$2050 < f < 2095$	$2025 < f < 2050$	$1 < f < 2025$
Band I: F_{uw}	MHz	$2185 < f < 2230$	$2230 < f < 2255$	$2255 < f < 12750$

Inband blocking is tested with a WCDMA modulated interferer applied at an offset frequency of ± 10 MHz and ± 15 MHz from the wanted channel, i.e. carrier frequency f_c. The inband tests are very similar to that of the adjacent channel selectivity test, and the channel filter requirements may be found in the same way. However, a specific case to consider is where the receive ADC is sampling at four times the chip rate, or 15.36 MHz. This is a popular sampling scheme as it provides a minimum adequate timing resolution for the RAKE receiver. Signals at the input to the converter within the band 15.36 ± 1.92 MHz will be aliased and appear as cochannel interference. The ± 15 MHz blocking signal then falls broadly on the converter alias and a channel filter rejection of typically 60 dB is required to pass the test. The inband blocker must be attenuated prior to the ADC and so all the attenuation must be achieved in the analogue filter after the mixer. However, the analogue filter requirements for the adjacent channel selectivity test are more difficult than for the inband blocking test specification.

A different test for narrowband blocking applies only to UEs operating in the region II and III frequency bands. This test is designed to ensure adequate performance for the case where GSM and UMTS systems are sharing the same band. The interfering signal is a GMSK modulated carrier with an offset frequency of 2.7 MHz or 2.8 MHz from the wanted channel. The blocking signal consequently is partially 'in-channel' and filtering is difficult.

The out of band blocking tests in [9] are the most important when considering the design of a direct conversion receiver. The third-order linearity requirements of the receiver are defined by these tests which require a higher input *third-order Intercept Point* (IP3) than the intermodulation test defined. The discussion in the remainder of this section focuses on the frequency Band I requirements as an example, but similar cases are applicable to Bands II and III. The relevant requirements are summarized in Table 2.8.

The transmit circuits are active during the blocking tests and the output power is set to 3 dB or 4 dB below the full specified output power depending on the UE power class. The wanted signal is set 3 dB above the reference sensitivity as determined by the sensitivity measurement. The presence of transmit leakage at the receiver LNA input creates the potential for intermodulation with the CW blocking signal. Two particular cases need to be considered. The diagram of Figure 2.9 shows the situation where the blocking signal is applied at half the duplex offset above the transmit signal leakage. Due to the presence of a third-order nonlinearity in the receiver LNA, the CW signal intermodulates with the transmit leakage signal resulting in cochannel interference. For Band I, the frequencies which produce cochannel interference are in the range 2015 MHz to 2075 MHz. From Table 2.8, the level of the CW blocking signal is -15 dBm in the 2015–2025 MHz band and -30 dBm in the 2025–2050 MHz band, and these represent the most difficult requirements. The performance of the input RF filter is critical in these bands and a minimum of 15 dB rejection in the 2025–2050 MHz band and 30 dB in the 2015–2025 MHz band is necessary to reduce the level of the CW interferer to the same level as that of the inband blocking tests.

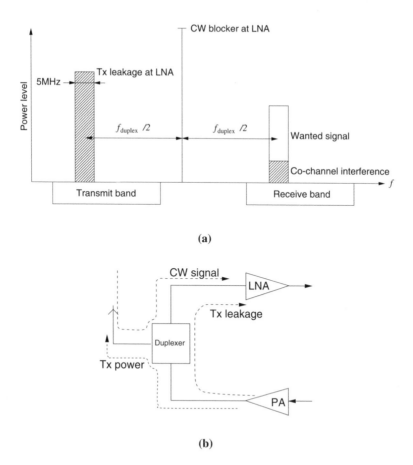

Figure 2.9. Near side blocking from P3. (a) spectrum view, as seen with a spectrum analyser; (b) circuit level view.

However, the transmit leakage signal may be as high as -26 dBm and the requirements on the third-order intercept are higher than for other tests. For a typical UMTS receiver, the required 2-tone IP3 will be in the range -10 dBm to -15 dBm at the LNA input, but this will obviously depend on the rejection achieved by the input RF filters.

The second case that must be considered is when the CW blocking signal is positioned below the transmit leakage signal by an offset frequency equal to the duplex spacing. Third-order distortion may result in additional interference in the 1730–1790 MHz band for fixed 190 MHz duplex or in the 1670–1850 MHz band for variable duplex operation. As the frequencies are at a greater offset from the receive band, a greater level of rejection may readily be achieved in the input RF filters. However, the level of interference resulting from intermodulation is higher in this case as the transmitted signal is now the near tone (nearby CW carrier)[1]. An RF filter rejection of 50 dB prior to the LNA is typically required to ensure the LNA input IP3 requirements are not greater than that for the near side blocking case.

---

[1] For the case where tones of unequal power are used in a two-tone third-order intermodulation measurement, the intermodulation product is greatest on the side with the higher tone level.

The blocking specifications are significantly easier to achieve for frequencies above the receive band, as third-order intermodulation products cannot be produced at the frequency of the wanted channel. However, the case of mixing products with the third and fifth harmonic of the receive local oscillator should be considered. A high level of odd-order harmonics is generated in the receive mixer due to the switching action of the transistors, and as a result, down conversion from the third and fifth harmonic may only be 10–20 dB below that from the fundamental frequency. For a Band I receiver, the third harmonic has a frequency in the band 6330–6510 MHz and the fifth harmonic has a frequency between 10 550–10 850 MHz. Both fall within the blocking specification frequency range. The rejection of typical handset RF filters falls significantly and becomes unreliable above 5 GHz. This is the result of unwanted resonances within the BPF filter and may mean that as little as 15–20 dB of isolation is achieved at the harmonic frequencies. These frequencies may be declared as spurious response frequencies [9], however, with some consideration of the design of the matching to the LNA, it is possible to ensure that there is enough rejection to pass the test with a −15 dBm blocking signal.

2.3.7 Spurious Emissions and LO Leakage

The requirements for receiver spurious emissions are defined in [9]. The use of a direct conversion receiver eases the problems of meeting this specification as the presence of only two synthesizers (LOs), operating at the transmit and receive frequencies, means that there is much less scope for spurious signal generation. However, the specification does place tight limits on emissions in the receive band which must be less than −60 dBm at the antenna port. As the receive local oscillator is operating within this band and at a relatively high level, good isolation is required to ensure that this requirement is met. Since filtering is not possible, this requires careful balancing of the mixers to maximize the LO to RF port isolation and design of the LNA for high reverse isolation. The duplex filter output and low noise amplifier will typically use balanced circuits to minimize signal pickup in addition to achieving a good IP2.

It is an advantage to minimize the amount of circuitry operating at the receive frequency to help reduce cross talk or feedback to the receive input. For this reason, the *Voltage Controlled Oscillator* (VCO) and the LOs (synthesizers) are often operated at twice the receive frequency and the resulting signal is divided by two to generate signals for the mixers. The process of dividing by two also provides a convenient way to generate accurate quadrature signals. A balanced oscillator may be used to generate signals at twice the receive frequency with a 180 degrees phase shift. After division by two, this results in signals with a 90 degree phase shift. The quadrature accuracy depends on the balance and level of harmonics of the original signal from the VCO. The use of oscillators operating at 4 GHz is convenient for integrated circuit implementation as inductors with adequate Q (resonance quality indicator [25–27]) may be fabricated on chip without requiring excessive chip area.

2.4 RECEIVER BASEBAND DESIGN

For the purpose of this section, a receive *Digital Reference Point* (DRP) has been defined at the output of the combined RF and baseband stage, see Figure 2.10. It is the task of the entire

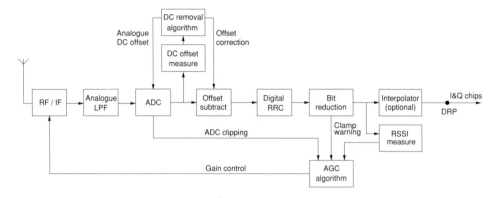

Figure 2.10. A typical baseband receiver signal processing chain.

receive chain, comprising RF, analogue and digital baseband, to maintain a digital signal at this digital reference point with enough fidelity (dynamic range) to allow the remaining digital signal processing stages to extract the wanted signal without undermining the performance. It must do so over component and manufacturing tolerances, UE characteristics and under various propagation conditions. A typical baseband receive signal processing chain is shown in Figure 2.10. The functions of each of the blocks are explained subsequently. It should be noted that this is a simplified diagram, with some details removed for clarity, such as amplifiers and control blocks.

The receive baseband is critical to the receiver's performance, because information is lost or corrupted within this block, and the receive RF can neither be recovered nor easily corrected. Also, the receiver BB functions yield some key parameters which are required to design the receiver's RF stage. Hence we need to focus first on areas such as:

- the receiver baseband demodulation performance;
- sampling rate requirements;
- analogue to digital conversion bit width requirements;
- interference suppression and filter requirements;
- chip rate processing bit width requirements;
- maximum allowed receiver noise contribution.

It is common practice in signal processing design to have a floating point and (maybe) a fixed point baseband simulation environment available in order to derive key parameters, optimize algorithms and measure the performance. Such a simulator is termed a link level simulator, because it may not include a multicell network environment. Simulations allow us then to benchmark a receiver design.

2.4.1 Baseband Demodulation Performance

The RF receiver tests defined in TS25.101 [9] require that either a BER or BLER limit is not exceeded for the specified test conditions. The receiver characteristics, including sensitivity,

Table 2.9   Relative channel powers for the 12.2 kbps
reference test channel.

Physical channel	Power ratio
P-CPICH	P-CPICH$\_E_c$ / DPCH$\_E_c$ = 7 dB
P-CCPCH	P-CCPCH$\_E_c$ / DPCH$\_E_c$ = 5 dB
SCH	SCH$\_E_c$ / DPCH$\_E_c$ = 5 dB
PICH	PICH$\_E_c$ / DPCH$\_E_c$ = 2 dB
DPCH	Test dependent power

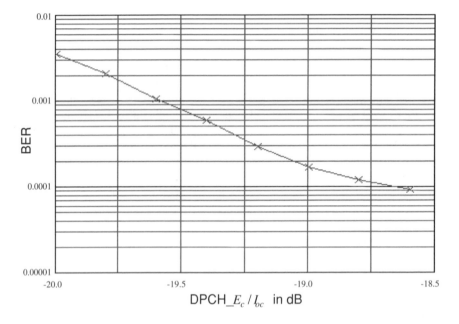

Figure 2.11. Plot of BER versus SNR showing the receiver's sensitivity at a BER = 0.001.

adjacent channel rejection and blocking must be met with a BER of less than 0.1 %, whilst the type approval performance tests with a static (i.e. AWGN) propagation channel and with the fading propagation channels are defined with a BLER limit.

All the receiver characteristics tests use a common 12.2 kbps reference test channel signal which is specified in Annex A in [9]. The signal power levels of the dedicated channels within the test channel set are defined in Annex C.3.1 and are reproduced in Table 2.9.

The specification is not easily applied to RF calculations based on the noise figure or SNR, however, the performance of the baseband receiver may be determined by simulation in a form that makes these calculations easier. The plot in Figure 2.11 shows a typical plot of BER against DPCH$\_E_c/I_{oc}$ for a convolutionally encoded data channel, the DPCH, in a static case, i.e. AWGN propagation channel. The plot is generated from simulation results and is indicative only, as the performance for a given UE will depend on the specific implementation of the receiver's digital signal processing architecture. The simulation uses the relative channel powers specified in [9] (Annex C.3.1) and includes the effects of scrambling and spreading. On the x-axis, DPCH_E_c is the power of the dedicated channel within the specified 12.2 kbps test channel and I_{oc} is the total noise power measured within the 3.84 MHz bandwidth channel filter. The value of DPCH_E_c/I_{oc} is then equivalent to the SNR and can be easily used in conventional RF system calculations. According to the example shown in Figure 2.11, an SNR level better than -19.6 dB for the DPCH_E_c/I_{oc} is required to achieve a BER below 0.1 %. Receiver performance results for fading propagation channels can be found in Chapter 5.

2.4.2 Pre ADC Signal Conditioning

This section covers the anti alias filtering and DC offset removal stages. Depending on the sampling frequency and the receiver architecture (direct or dual conversion to baseband or low IF) and the amount of RF/IF filtering already provided, the signal conditioning before the analogue to digital conversion can range from a complex, higher-order active filter, to virtually nothing other than ADC drivers. This is mainly due to filtering that may already be provided in the RF receive chain. For example, the requirement to reject adjacent channel signals may mean that no additional filtering is required for anti alias purposes. In general though, this is not the case, and some additional filtering at baseband is required.

A discussion of the requirements for blocking to eliminate DC offsets from the receiver can also be found in Section 2.3.2. However, when choosing the optimum corner frequency for the filter, a number of additional factors must be considered. All parts of the WCDMA signal spectrum are equal when considering descrambling and despreading of the wanted channel. Consequently, the removal of a small portion of the spectrum has a limited impact on the power of the received channel. The group delay associated with the filter will introduce *Inter Chip Interference* (ICI) and raise the level of nonorthogonal signal power, hence reducing the SNR of the wanted signal.

In addition to removing DC offsets, an additional highpass filter may be designed to remove other sources of interference. The baseband amplifiers, for example, will produce flicker noise. The level of the noise is inversely proportional to frequency and has a corner frequency, where the flicker noise is equal to the thermal (white) noise floor, of typically 1 kHz for bipolar devices, and up to 1 MHz for MOSFET amplifiers. The interfering signal resulting from second-order distortion of the transmit leakage signal occupies a bandwidth from 0–1.92 MHz. However, the radio link uses fast power control in both directions, which will result in AM spectral components at 1500 Hz and harmonics. The highpass filter may be designed to reject this interference. Finally, the receive and transmit phase noise tends to be highest in the first 20–30 kHz of the spectrum and this noise can also be filtered out prior to demodulation. While the highpass filter will remove the low frequency and DC components introduced before the ADC converters, the ADC converters themselves can have significant DC offsets which require removal. This can be accomplished either by a digital highpass filter, or a digital DC estimator and analogue feedback of a correction voltage.

2.4.3 Analogue to Digital Conversion

The conversion of the baseband analogue signal to a digital representation is a difficult balancing task [28], especially for a UE. The higher the sample rate, the higher the power consumption, but the simpler the baseband anti alias filtering, as the ratio of the sampling frequency to the wanted band edge, i.e. maximum baseband frequency, is larger. Higher sampling frequency means that more of the adjacent channel power is digitized and, assuming that the minimum amount of anti alias filtering is applied, this increases the number of ADC bits required, further increasing the power consumed.

2.4.3.1 Crest Factor   In CDMA, the transmitted signal is a superposition of many component signals, for example pilot channel, common control channels and dedicated channels [29]. The individual chips are added up and hence result in a vector signal. Obviously, the composite signal, i.e. each chip, can have a different amplitude and phase. Consequently, the signal amplitude is large if every chip has the same phase, since the vectors all add up. In reality, the varying nature of the information ensures that this is not the case. The occurrence of large amplitudes is more likely at the Node B because the downlink is synchronous and the pilot bits in the dedicated channel are all at the same location. To mitigate large peaks, the Node B can retard or advance the different dedicated channels. Still, the large numbers of users in a cell result in large peaks, hence this issue is very important to Node B manufacturers who are interested in minimizing the peak envelope and its rate occurrence to relax the requirements for their power amplifiers [30].

The signal characteristic, with respect to its dynamic range or *Peak to Average Power Ratio* (PAPR), is expressed through the Crest factor. Both terms are in common use and can be used interchangeably. Obviously, the Crest factor is a statistical figure because the composite signal can have all sorts of possible amplitude and phase combinations. Thus it is common practice to analyse the signal's ccdf (complementary cumulative distribution function). A Crest factor plot example is shown in Figure 2.12; this was generated with the commercially available product WinIQSIM™ [31].

Figure 2.12 shows the PAPR factor in dB against the probability of occurrence. For example, we can see that the PAPR is 10.94 dB at 0.01 %. Note that a CDMA system can tolerate some amount of signal clipping in the ADC due to the spread nature of the system. This means that signals exceeding the dynamic range of the ADC are saturated; however, clippings cause a change of the spectrum. Therefore, the amount of clipping can lie between 1 % and maybe up to 3–5 %, but should be assessed through simulations.

It has also been suggested that the Crest factor should not be used as a lone figure of merit, and that the signal power histogram should also be considered [32].

2.4.3.2 Sampling Rate Aspects   Selecting the sampling rate is usually governed by looking first at power consumption constraints since there are generally, from the technical point of view, some benefits of having a high sampling rate.

A low sampling rate is desirable from the power consumption point of view, but two and three times chip rate sampling typically require higher-order, low tolerance analogue filtering, due to both the amount of adjacent channel power which must be rejected and the low ratio of the sampling rate to the wanted maximum baseband frequency. A sampling rate of four times the chip rate (15.36 MHz) provides a good tradeoff between filter complexity,

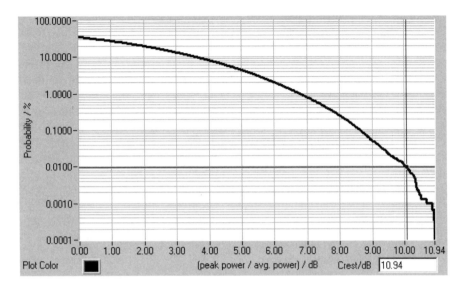

Figure 2.12. A typical Crest factor plot [31].

high sampling rate and power consumption. Five times chip rate sampling uses the standard 19.2 MHz reference frequency, eliminating the need for a low noise synthesizer (LO) to generate a four or eight times sampling rate clock. Compared to four times sampling rate, a five times scheme requires a slightly higher resolution and a complex interpolation to get to a binary multiple of chip rate, which is desirable for subsequent signal processing. An eight times chip rate sampling scheme allows all of the adjacent channel through the ADC, so making the preconditioning very simple, at the expense of more extensive digital filtering to remove the adjacent channel, and higher resolution ADC to represent the increased power in the wider bandwidth.

The clock frequency for the conversion must be generated with low phase noise [33] to ensure the jitter does not compromise the noise floor of the ADC [34]. This can be derived directly from a reference oscillator, which is more typical in multimode systems. It is also possible to digitize at a non chip-related frequency and interpolate the required chip sample stream.

2.4.3.3 *ADC Resolution Aspects*   The most demanding blocking case defined within the 3GPP specifications is the adjacent channel selectivity test defined within [9].

The specification states that a 0.001 BER must be achieved with a wanted received signal level $\hat{I}_{or}$ of −93 dBm and in the presence of an adjacent channel signal with a total power of −52 dBm. The blocking signal is a modulated carrier using an *Orthogonal Channel Noise Simulator* (OCNS) test channel. The receive analogue filter is assumed to achieve a minimum of 20 dB rejection of the blocking signal, see Section 2.3.5. The filter rejection is limited by component tolerance and the requirement to achieve a low level of interchip interference within the receiver. Therefore at the ADC, the level of the blocking signal is about 20 dB greater than the wanted signal. A further 7.3 dB of headroom is suggested to allow for the PAPR of the interfering signal. This ensures that the probability that the ADC will saturate

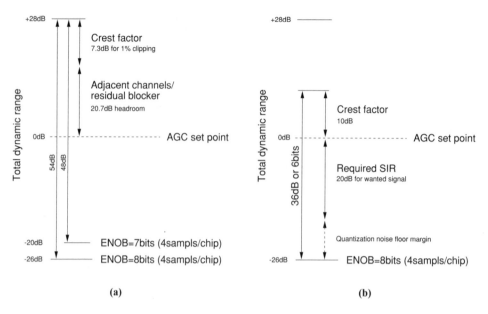

Figure 2.13. Estimation of the required ADC bit resolution. (a) Pre RRC; (b) Post RRC.

is less than 1 %. Two diagrams indicating the levels within the baseband processing, before and after the RRC filter, are shown in Figure 2.13(a) and 2.13(b) respectively.

The levels correspond to a converter of seven to eight *Effective Number of Bits* (ENOB) with four times oversampling, giving a dynamic range of 48 dB and 54 dB respectively. The use of oversampling gains one additional effective bit after filtering, as the quantization noise N_Q bandwidth is reduced. Figure 2.13(b) shows the dynamic range after truncation to six bits at the output of the digital RRC filter. The bit width K can also be derived from the total dynamic range according to $SNR_{\text{total}} = 6 \, \text{dB/bit} \cdot K + 1.76 \, \text{dB}$.

A larger headroom allowance is required at this point as the peak to average power of the remaining wanted signal can be higher than that of the OCNS test signal which only contains 16 active channels. The AGC set level point may be calculated once the ADC full scale peak to peak input voltage is known. For the reasons above, we assume here a digital reference point with six to eight bits per sample and an ADC sample rate of four times the chip rate. To improve the temporal resolution, especially of the RAKE receiver introduced in Section 3.4.2, we also assume here an eight times chip rate sample resolution, requiring the digital *Root Raised Cosine* (RRC) filter to interpolate from four to eight samples.

2.4.4 Receive Pulse Shape Filtering

The main function of this filter block is the RRC filtering of the received signal to achieve an aggregate raised cosine response when paired with the transmit filter in the Node B. Pulse shaping is a common practice to mitigate ISI [2]. 3GPP has defined a Nyquist type pulse shaping filter in TS25.101 [9] but not the filter taps; in contrast, other CDMA standards specify a non Nyquist filter and its filter taps and introduce ISI [35].

Raised cosine filters have zero crossings at chip intervals [2], so there is no *Inter Chip Interference* (ICI). The typical implementation of an RRC filter is a *Finite Impulse Response*

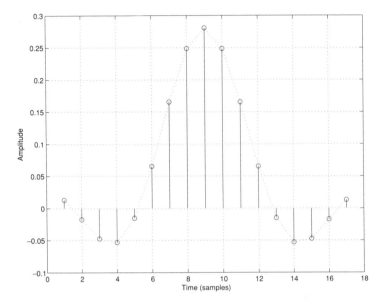

Figure 2.14. RRC filter coefficients.

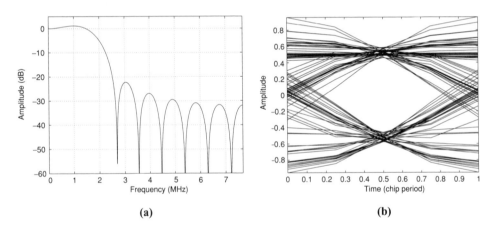

(a)

(b)

Figure 2.15. Characteristics of a nonwindowed RRC filter. (a) Frequency spectrum; (b) Eye diagram.

(FIR) digital filter, with a root raised cosine impulse response. A tradeoff between performance and complexity must be made between the fidelity (dynamic range) of the impulse response and the number of taps in the FIR filter. To achieve a true RRC filter with very low ICI would require a large number of taps. This would require a lot of resources (DSP or silicon area) for implementation. In any case, other factors, such as analogue filtering, will limit the achievable ICI. Figure 2.14 shows the coefficients of a 17-tap FIR filter which approximates the RRC, and the corresponding eye diagram 2.15(b), when cascaded with an ideal (long) transmit RRC filter, indicating that there is very little ICI. The corresponding frequency response is shown in Figure 2.15(a) and shows that the filter exhibits a visible inband ripple of about 0.5 dB and has poor stop band attenuation of about 22 dB from

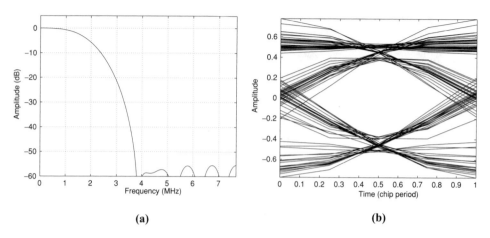

Figure 2.16. Possible characteristics of a windowed RRC filter. (a) Frequency spectrum; (b) Eye diagram.

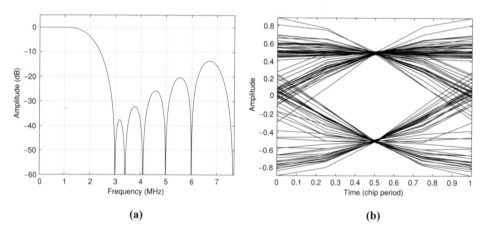

Figure 2.17. Characteristics of a typical RRC filter. (a) Frequency spectrum; (b) Eye diagram.

3.5 MHz. This can be improved by windowing the impulse response with a hand optimized function, as shown by Figure 2.16(a), which shows no significant inband ripple any more and an improved stop band rejection of at least 40 dB at 3.5 MHz. Unfortunately, the eye diagram in Figure 2.16(b) shows a significant degradation in the ICI. If we consider that the analogue adjacent channel filter depicted in Figure 2.4 has a cut off or corner frequency of approximately 2 MHz, and has a total attenuation of the adjacent channel of about 20 dB which starts at 2.5 MHz, it would be more effective for the digital filter to have more attenuation in the 2.5–3.5 MHz region, where the analogue filter is starting to roll off. The RRC candidates above have little attenuation in this region. By optimising the filter coefficients by hand, however, a compromise filter can be produced, which has a better eye pattern than the windowed FIR, and more attenuation in the critical 2.5–3.5 MHz region. The frequency response and the eye pattern of such a hand optimized filter are shown in Figures 2.17(a)

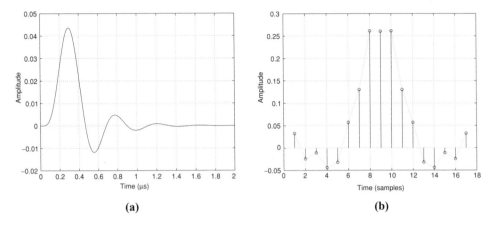

Figure 2.18. Impulse responses of two different pulse shape filters. (a) Butterworth filter impulse response; (b) RRC filter coefficients of filter 2.17 (a).

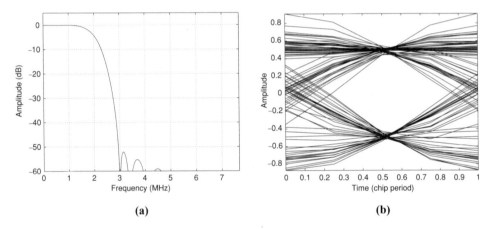

Figure 2.19. Characteristics of a combined Butterworth and RRC filter. (a) Frequency spectrum; (b) Eye diagram.

and 2.17(b). The tradeoff is obviously less stop band attenuation in the higher frequencies, but the analogue filter has significant attenuation at these frequencies to cover this area. The use of an analogue filter to provide part of the receiver's selectivity, i.e. filter's capability to extract the wanted signal, has the disadvantage that ICI will be introduced. The impulse response of a sixth-order Butterworth filter is shown in Figure 2.18(a).

The impulse response is asymmetric indicating that this is not a linear phase filter. The impulse response of the raised cosine filter and the Butterworth filter may be convolved to determine the overall channel impulse response for the case where an analogue filter is included in addition to a digital RRC filter in the receiver. The combined frequency response and eye pattern diagram are shown in Figure 2.19. The total level of ICI may be found by summing the power of the impulse response at all chip time offsets from the impulse response maxima. The ICI may then be expressed in decibels relative to the power of the impulse response maxima. Using a Butterworth filter it is possible to achieve a level of ICI lower than $-20\,$dB, or SNR = 20 dB, whilst also achieving greater than 20 dB of rejection of the first adjacent channel. The presence of ICI may be added to the system calculations as an additional source of nonorthogonal noise. For example, consider a received signal comprising a number of downlink physical channels such that the total power of the signal is 0 dBm. The signals are all from a single Node B and so there is a high level of orthogonality between the code channels. If the receiver has ICI of -20 dB then this can be modelled as a source of nonorthogonal noise with a total power of -20 dBm at the receiver input. As the amount of adjacent channel filtering and analogue filter compensation can change depending on the RF chipset, a baseband chip may have programmable filter coefficients, to permit a number of different RF architectures, filtering responses and chipsets to be used. The coefficients are then generated to match the components used in a specific UE.

The final step is upsampling and interpolation to maybe eight times chip rate, which is a straightforward step from the four times chip sampling rate initially assumed here. This block may not be needed, because oversampling may be carried out by the RRC filter through an efficient polyphase implementation, but can result in less complexity since samples having fewer bits have to be interpolated. Other sample rates, such as five times or three times chip rate, require a more complex interpolation function.

As noted earlier, the ADC requires more resolution if a significant portion of the adjacent channel is digitized, see also Section 2.3.5. After digital RRC filtering, this extra dynamic range is not required, so the resolution can be reduced to the six to eight bit range assumed at the digital reference point, which is more than has been suggested in [36]. Clamping, i.e. limiting the digital bit word to its extreme values, is used to prevent a bit wrap around and subsequently compromising the bit error rate, with a warning being provided to the software to indicate the overload condition upon which the AGC reacts.

2.4.5 Automatic Gain Control and Reference Point

It is not practical to make a receive signal chain that would be able to cope with the entire dynamic range of the received signal, which is in the order of at least 80 dB. Instead the finite dynamic range of the receive chain, which is typically limited by the ADC, is positioned in the best place for the current signal conditions by an automatic gain control loop, which changes the gain of the analogue stages in response to measurements of the digitized signal.

The aim is to maximize the SNR at the ADC output. This point must take into account the response time of the entire AGC loop, the statistical variation of the signal level over the response time due to propagation effects, and the characteristics of the basic measurement of the signal level. A block diagram of an AGC control loop on its own is shown in Figure 2.20.

It is assumed that the analogue filtering provides 20 dB of attenuation of the adjacent channel and a further 15 dB is required to meet the adjacent channel selectivity test. The combined Butterworth and RRC filter provides this attenuation. The power is measured at the output of the filter and used to control the variable amplifiers shown in Figure 2.4. Detection of the power after the digital RRC filter, prevents the AGC control loop from being influenced by out of band (blocking) signals. Hence, the AGC operates on the wanted inband signal only. A further advantage of implementing the AGC power measurement after the filter is that the level at this point is then maintained within a narrow range, allowing

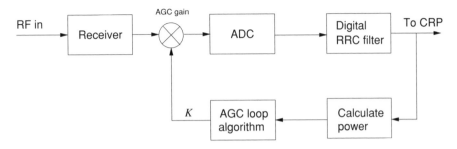

Figure 2.20. A typical automatic gain control loop (AGC).

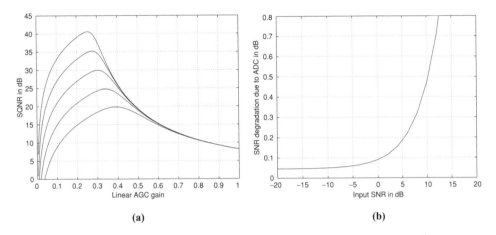

(a)   **(b)**

Figure 2.21. AGC/ADC performance curves. (a) Optimum AGC level for different ADC
resolutions, four bits (bottom curve) to eight bits (top curve); (b) SNR degradation for a four bit ADC.

truncation of the digital words to six bits at the digital reference point. The AGC loop can
operate slowly to maintain the long term average power constant at the output of the filter.
The inner loop fast power control will ensure that the power is maintained close to the
optimum operating level over a short duration.

Term K in Figure 2.20 is the AGC gain factor which must be optimized [36, 37]. The
goal of this optimization process is to minimize the overall SNR degradation at the ADC
output. Assume that the analogue signal coming from the receiver is normal distributed due
to the central limit theorem, hence $\mathcal{N}(0, \sigma^2)$. Then the signal is scaled after the AGC gain
and has variance $K^2\sigma^2$. The function to be optimized is:

$$10\log\left(\frac{SNR_{\text{in}}}{SNR_{\text{out}}}\right) = 10\log\left(1 + \frac{\sigma_Q^2}{K^2\sigma^2}(1 + SNR_{\text{in}})\right) \tag{2.5}$$

where σ_Q^2 denotes the quantization noise power. The optimum factor of K for a selected
number of ADC resolutions, namely four to eight bits, can be obtained from Figure 2.21(a).
The SNR loss at the optimum AGC gain K for a four bit ADC is depicted in Figure 2.21(b).

Sufficient headroom must be provided at the ADC input so that the largest anticipated adjacent channel signal will not cause saturation. Since the adjacent channel power is not noticed by the AGC loop, as it is removed by the RRC filter, the AGC cannot compensate by reducing the signal power at the ADC input.

2.4.6 Additional Receiver Signal Processing Functions

If a frequency reference is used which is not capable of being adjusted to achieve frequency lock with the Node B, an additional processing block is required, termed a frequency shifter or phase rotator, which is used to rotate continuously the phase of the received vector to compensate for the frequency error experienced in the mobile terminal.

2.5 TRANSMITTER BASEBAND DESIGN

The transmit chain consists of a similar set of blocks as the receive chain, covering interpolation, pulse shaping, timing adjust, DC offset compensation, modulator matching, analogue filtering and distortion filtering. Again, in common with the receive chain, the architecture is chosen as a result of finding a good set of compromises between a number of conflicting requirements.

The baseband signal processing [2] section in a UMTS mobile terminal comprises signal conversion, conditioning and control functions which, in the receive direction, take the analogue signals from the RF receive chain and provide an adequate digital representation of the band of interest. In the transmit direction, it must provide the transmit RF chain with an analogue representation of the digital transmit chain, without compromising the transmit spectrum mask. The operational and performance requirements of both the transmit and receive baseband chains are driven by a number of conflicting demands from the 3GPP specifications, the requirements of the transmit and receive RF architectures, the requirements of the digital signal processing, and the UE requirements. The baseband processing stages must provide an acceptable compromise between all these conflicting requirements.

2.5.1 Baseband Modulation

After the uplink data has been encoded and spread, as described respectively in Chapters 4 and 3, the signal must be modulated onto an RF carrier. This is achieved by forming a complex filtered signal from the control and data channels and modulating this onto the carrier using a vector modulator. The UMTS system uses a linear modulation scheme in which both the amplitude and phase of the signal vary. This achieves greater spectral efficiency but requires the use of linear RF components. In particular, the output power amplifier must be linear and the requirement to operate with output powers below the compression point reduces the efficiency and increases the power consumption. The 3GPP uplink modulation scheme employs *Hybrid Phase Shift Keying* (HPSK) [38], in order to relax the requirements posed on the power amplifier. HPSK exploits two signal properties, namely:

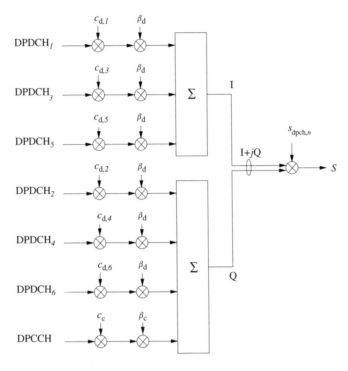

Figure 2.22. DPCH uplink channel configuration [29].

- by using a complex scrambling code, the power between the in-phase and quadrature component is equally distributed;

- using even-numbered Walsh codes, which consist of pairs of bits, the number of zero crossings is reduced by 50 % and hence a reduction in peak to average power ratio of about 1 to 1.5 dB is achieved.

Therefore, uplink modulation is different from the downlink modulation. The 3GPP uplink modulation scheme allows up to six *Dedicated Physical Data Channels* (DPDCHs) and one *Dedicated Physical Control Channel* (DPCCH) to be modulated onto the uplink carrier [29].

Figure 2.22 shows a full uplink configuration. For uplink data rates below 450 kbps, a single DPDCH is always used, in which case the in-phase component of the modulation is obtained from the DPDCH and the quadrature-phase component is obtained from the DPCCH. The data rate of the DPDCH can vary, whilst the data rate of the DPCCH is fixed and uses spreading code $c_{(256,0)}$, where 256 denotes the spreading factor and zero the code number. To maintain a constant E_b/N_0 relation between the DPDCH and DPCCH, the relative power can be adjusted using β_d and β_c weighting factors. The constellation diagram of a scrambled and modulated single DPDCH and DPCCH signal is shown in Figure 2.23.

For β_d set to one, β_c can be set with a weight of between 1 and 1/15 in steps of 1/15 giving a power relative to the DPDCH of between 0 dB and -23.52 dB. Note that subscript

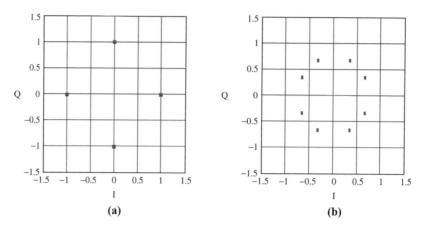

(a)   (b)

Figure 2.23. DPCH constellation diagrams for different β settings. (a) $\beta_d = \beta_c = 1$; (b) $\beta_d = 1$ and $\beta_c = 5/15$.

d denotes the data channel DPDCH and c the control channel DPCCH. For higher data rates additional DPDCH channels are added up to the limit of six, giving the maximum data rate of 2 Mbps. The additional (dedicated) data channels always have a spreading factor of four and use spreading codes 1, 2 and 3 with code 0 occupied by the DPCCH. As only three codes are available at this spreading factor, the codes are reused on the in-phase and quadrature-phase branches. The use of additional spreading codes significantly increases the peak to average power ratio of the modulated signal. To minimize this problem, new codes are only used after the existing code has been used on in-phase and quadrature-phase branches, hence, code $c_{(d,1)}$ is used on I and Q followed by code $c_{(d,2)}$ on I and Q and finally by code $c_{(d,3)}$.

The combined data and control channels are multiplied by a complex scrambling code. Phase transitions of 180 degrees in the modulated signal result in a higher peak to average power ratio and so the scrambling code is contrived such that the phase shift from one chip to the next is limited to ± 90 degrees. After multiplication with the spread channels, phase transitions from chip to chip are mostly ± 90 degrees, with a relatively small number of 180 degree shifts (zero crossings), resulting in a peak to average ratio for a single DPDCH which is lower than for standard QPSK [38].

The bandwidth of the modulated signal is defined by filtering the baseband I and Q signal with an RRC filter with a bandwidth of 1.92 MHz and $\alpha = 0.22$. A vector diagram of the filtered signal comprising a single data and control channel with $\beta_d = \beta_c = 1$ is shown in Figure 2.24(a). The vector diagram for other β values is similar, although the constellation points are no longer easily distinguishable, see Figure 2.24(b).

2.5.2 Pre Digital to Analogue Conversion Signal Processing

The main functions of this block are interpolation and RRC pulse shaping, typically integrated into a single digital filter block. Other functions are phase rotation in cases where the frequency reference is not controlled, timing adjustment to slew the timing of the transmit signals to match the Node B timing requirements, DC offset cancelling to compensate for DC offsets in the DAC and subsequent analogue blocks, and possibly compensation

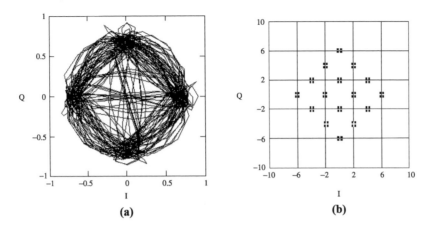

Figure 2.24. (a) Vector diagram obtained from an RRC filtered DPDCH channel; (b) Six channel DPCH constellation diagram, $\beta_d = 1$ and $\beta_c = 3/15$.

for RF chain imperfections. Digital modulation may also be required for RF architectures which do not use baseband IQ modulation. This block feeds the transmit digital to analogue converters with the samples at the correct rate and resolution.

The pulse shaping function produces an oversampled root raised cosine response, from the chip rate samples presented to it from the uplink processing functions. Since the RRC function has even symmetry, it is advantageous for the filter to have programmable coefficients, and allow asymmetric responses, to allow it to compensate for imperfections in the transmit analogue signal path. The impulse response and eye pattern of a typical RRC filter (not optimized) are shown in Figure 2.15(a) and 2.15(b) respectively.

The timing adjustment is typically implemented by sample drops and replication of the oversampled samples from the filter output. This is adequate if the DAC rate is four times the chip rate or higher, but below this rate, the Node B may lose lock on the signal due to the increased timing offset.

A controlled amount of offset may be added to the data to correct for DC offsets in the analogue chain up to the transmit modulator, and in the modulator itself, although this can be of limited use in some architectures as the offsets can be temperature and transmitter channel dependent.

2.5.3 Digital to Analogue Conversion

The DAC needs to be fast enough so that practical reconstruction filtering can reduce the sampling image to a level below the 3GPP specifications for adjacent channel leakage. The same applies to the resolution and spurious responses of the DAC, the quantization noise floor or spurious noise must not cause a frequency mask (profile) violation. A typical compromise is four times chip rate sampling, allowing simple analogue reconstruction filtering and straightforward timing adjustment. Higher sampling rates require more power, and while the transmit power amplifier is the dominant power drain in the uplink chain, a good compromise is still needed. Phase noise and jitter should also be controlled at the transmitter [34].

2.5.4 Post Conversion Processing

The DAC outputs must be filtered so as to reduce the image responses, quantization noise floor and spurious responses outside the wanted band without compromising the eye pattern of the wanted signal. This is made more difficult if a direct conversion RF architecture is used, as there is no further channel filtering in the RF chain. After conversion and reconstruction filtering, the power from 2.5 MHz upwards must be 40–50 dB below the peak of channel ($<$ 2.5 MHz) to achieve the adjacent channel leakage ratio. The RRC filter and DAC noise floor must be designed to meet this up to 1/2 sampling rate, after which the analogue filter must start to attenuate the rising image response. The filter should not compromise the EVM specification of the transmit chain, so its cut off point should be as high as possible while meeting the attenuation requirements.

If the DAC samples at four times chip rate, and the filter starts to roll off at two times chip rate, the filter must achieve about 36 dB of attenuation (includes approximately a 4 dB loss due to the DAC's frequency attenuation characteristics and the $\sin(x)/x$ attenuation) at four times chip rate (one octave) to attenuate the image response, corresponding to a sixth-order Butterworth filter. The complexity can be reduced by lowering the cut off frequency, or using a different design of filter, at the expense of the inband performance. If the latter route is chosen, then the inband response of the filter can possibly be corrected for in the digital RRC filter, as discussed earlier. Finally, the filter may be partitioned on either side of the baseband gain control block to improve the overall performance of the chain.

2.6 TRANSMITTER RF DESIGN

2.6.1 RF Up Conversion

This section looks at the conversion from baseband I and Q signals to a modulated RF signal. Two different schemes for achieving this are shown in Figures 2.25 and 2.26. These are different in that the scheme in Figure 2.25 modulates directly to the final RF carrier,

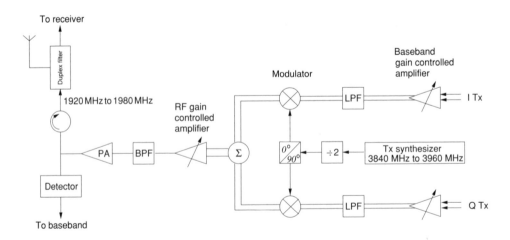

Figure 2.25. Direct conversion transmitter.

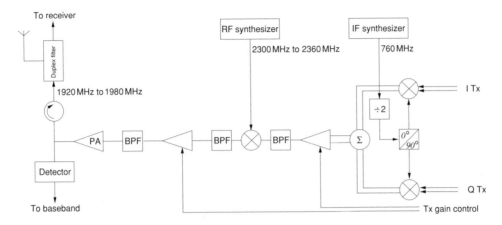

Figure 2.26. IF based transmitter.

whilst Figure 2.26 uses an intermediate frequency stage. Both schemes incorporate RF gain control which must have sufficient range to allow the output power to be controlled from the maximum power level down to -50 dBm. For a class 3 handset, a minimum of 74 dB control range is required. However, for a practical transmitter, more than 90 dB is required, allowing for component and manufacturing tolerance. The control can be split so that half of the range is implemented at the IF frequency and half at the RF frequency. However, for the direct conversion solution, almost all of the control range must be implemented at the RF output frequency. This requires a high level of isolation within the transmitter at 2 GHz and careful attention to the layout and distribution of gain control between two or more integrated circuit packages. A limited amount of gain control may also be applied by adjusting the levels of the baseband signal, however, due to the presence of DC offsets and carrier leakage within the modulator, this is typically less than 30 dB. The offsets and carrier leakage may be improved by applying correction to the baseband processing, however this is more difficult in the case of the direct modulator as the carrier leakage is frequency dependent. A further disadvantage of direct modulation is that most of the gain control range will also have a frequency response, whereas IF gain control is independent of the output frequency. If the frequency response varies with the gain control setting, then the direct modulation scheme requires more calibration data points to achieve the required output power accuracy.

The IF frequency may be chosen so that a single high frequency LO can be used for both receive and transmit sections. A fixed duplex spacing of 190 MHz is achieved with a receive IF of 190 MHz and a transmit IF of 380 MHz. Whilst some terminals use this simple scheme the specification allows for variable duplex offset, which requires two independent oscillators. A disadvantage of the fixed duplex spacing is that the transmitter must be disabled during periods where the receiver is retuned to search for nearby cells. This is supported using the uplink compressed mode where a gap is created during the transmission and the link is maintained by transmitting data at a higher rate and higher power for a period either side of the gap. The disadvantage of this operating mode is a reduction in the network capacity and so a terminal capable of variable duplex is advantageous from the point of view of the network operators. For a terminal using a direct conversion receiver and transmitter, each local oscillator will match the respective receive or transmit frequencies. Therefore, two separate oscillators are required and variable duplex spacing will be supported automatically.

2.6.2 Transmitter Direct up Conversion

The main advantage of direct modulation is its low cost and small size. These factors are extremely important as handset manufacturers are constantly looking to reduce the size and cost of handsets. A typical direct conversion transmitter is shown in Figure 2.25.

The local oscillator VCO (Tx synthesizer) is run at twice the transmit frequency. This has the benefit of reducing any coupling effects of the local oscillator directly to the transmit output and allows an accurate quadrature signal to be created using division by two. Further explanation of the local oscillator coupling issue is given in Section 2.6.6.

The baseband variable gain control amplifiers have differential I and Q inputs to reduce common mode noise and interference from other baseband circuits. The baseband gain control range is typically in the range of 20–30 dB and has to be set with a minimum resolution of 1 dB. One of the limiting factors in baseband gain control is the amount of local oscillator leakage that is present at the output of the modulator. As the signal is reduced through the baseband amplifier, the local oscillator leakage becomes more significant and the transmit signal *Error Vector Magnitude* (EVM) is degraded [39]. It is important that the local oscillator leakage is at least 10–15 dB below that of the wanted signal. If the carrier leakage level is too high it will not be possible to reduce the transmit power below the required −50 dBm minimum power level, or maintain the 1 dB step accuracy at the lower levels. To achieve a baseband control range of 20 dB, the modulator must have a carrier leakage less than −30 dB.

A low pass filter is employed at the output of the baseband amplifier. The filter must be designed to ensure that alias products from the DAC and noise from the DAC and baseband are adequately attenuated. It is important that the first and second adjacent channel noise and spurious emissions are more than 40 dB and 50 dB below the wanted signal respectively, as there is no other adjacent channel filtering in the transmitter after this baseband filter.

The baseband signal is converted directly to the RF frequency by the modulator. Low local oscillator leakage and linearity performance are important parameters for the modulator. Third-order distortion in the modulator will cause degradation in the transmitter's *Adjacent Channel Leakage power Ratio* (ACLR) performance. Improving modulator linearity invariably means increasing supply current, but as low current is always a key requirement for a handset, the distortion performance of the transmitter elements are carefully optimized to achieve the ACLR performance at minimum power consumption.

The RF gain control amplifier has a gain control in the range of 60 dB to 70 dB. Maximum output power level will be typically 0 dBm to 5 dBm. Table 2.10 shows a typical RF transmitter link budget. The amplifier can be implemented as a switched gain amplifier or as a linear controlled gain amplifier. The minimum step size requirement is 1 dB. The advantage of a switched gain amplifier is that the transmitter is easier to calibrate to meet the various step and absolute gain requirements. The disadvantage of the switched gain amplifier is that the switching process can produce switching transients which will produce spectral components in the adjacent frequency band (channel).

An RF SAW bandpass filter is required to reduce out of band noise. The two key areas are the transmit noise in the receive band (2110 MHz to 2170 MHz) and the transmit noise in the GSM 1800 band (1805 MHz to 1880 MHz).

Table 2.10   Typical transmitter link budget for class 3 handset at
maximum power.

Circuit block	Gain in dB	Output level in dBm
Modulator	0	−9
RF gain controlled amplifier	10	1
RF SAW filter	−3	−2
Power amplifier	28	26
Isolator	−0.5	25.5
Duplexer	−1.5	24
Antenna	0	24

The power amplifier will typically have a fixed gain of between 25 dB and 30 dB depending on the number of gain stages. The amplifier has to maintain a good degree of linearity because of the nature of vector modulation. Generally, it will be a class AB amplifier with an efficiency of around 40 % at maximum output power.

The power amplifier is followed by an isolator. This protects the power amplifier from power reflected back from the antenna due to mismatches that regularly occur at a handset antenna. Reflected power will increase distortion in the power amplifier output stage, and lead to a failure of the ACLR specification.

The final component before the antenna is the duplexer. The duplex filter does little for the transmitter other than add loss and reduce efficiency. However, the filter does reduce the harmonics from the power amplifier and noise and local oscillator leakage.

The detector allows the transmitter power to be measured by the baseband so that the power output can be prevented from exceeding the maximum allowed output, thus it is fed from the PA output rather than the antenna port. This ensures the detector sees a stable impedance and will not be affected by any antenna mismatch. Also, it may be used for reporting transmit power level back to the Node B. Power control may be required over a range of 74 dB for a class 3 handset. It is very difficult to implement a detector accurately over this range, and so the detector is only used over the upper part of the transmitter power range where the absolute accuracy requirements are most stringent. In this case, the dynamic range requirement of the detector is 25 to 35 dB. For further discussion, see Section 2.6.7 for power control and reference [40] for ACLR measurements.

2.6.3 Transmitter IF Based up Conversion

The main advantage of an IF based up convertor is that most of the gain control can be implemented at the IF frequency. The main disadvantage is that the transmitter requires an additional up conversion stage. A typical IF based transmitter is shown in Figure 2.26.

In this example, the IF is set at 380 MHz and the RF synthesizer (LO) could be shared with the receiver if a superhet receiver design is used. It is likely that direct conversion receivers will become commonplace for UMTS, in which case, the RF synthesizers would likely be separate, and variable duplex operation would be supported.

The transmit I and Q signals are up converted to 380 MHz. As there is no baseband gain control, carrier leakage in the modulator is not as important as in the direct conversion case. A transmit carrier leakage of better than −20 dB is readily achievable. The 380 MHz IF filter can be implemented as an LC filter as the main requirement is to filter noise that will appear in the receive band after up conversion to the output frequency. If a narrowband SAW filter is used to filter the IF, then this would provide 15–20 dB of rejection of the adjacent channel noise and would largely remove the need for baseband filtering before the modulator.

For a typical transmitter, the control will be split so that the IF gain control amplifier will have a range of up to 60 dB while the RF gain control amplifier will have up to 35 dB of range. As previously noted, the power control range required for a class 3 handset is 74 dB. The transmitter is designed to have a larger range, up to 95 dB, to take into account temperature, frequency and process variations in the transmitter gain.

2.6.4 Transmitter Spurious and Noise Emissions

3GPP FDD is a full duplex system, where it is important that transmitter noise and spurious output are isolated from the receiver. This is largely achieved by the use of a SAW bandpass filter before the power amplifier and a duplexer filter. The SAW filter attenuates receive band noise generated in the RF gain controlled amplifier, modulator and local oscillator. The duplexer attenuates noise generated from the power amplifier. A typical power amplifier will have a gain of 28 dB and a noise figure of 6 dB and so isolation of greater than 34 dB is required between the transmit output and receive input to ensure interference is below the level of thermal noise. A typical SAW filter will have 30 dB of rejection in the receive band and this determines the maximum acceptable receive band noise generated by the RF gain controlled amplifier, modulator and local oscillator. The SAW filter also is needed to reduce noise in the *Digital Cellular Standard* (DCS) band (1805 MHz to 1880 MHz) also referred to as the GSM 1800 band. A typical ceramic duplexer has insufficient selectivity to reject noise in this band and only contributes a few dBs of noise reduction. The duplexer filter requires minimal rejection at the transmit harmonic frequencies as the power amplifier is operated in a linear mode and output harmonics are usually below the −30 dBm requirement without additional filtering.

2.6.5 Transmitter Distortion and ACLR

ACLR is determined largely in the modulator, RF gain control amplifier and power amplifier, and is directly related to the IP3 performance. The local oscillator phase noise typically has a minimal impact in the adjacent channel. Figure 2.27 shows a typical transmit output spectrum. The main channel power is shown in the central division with the adjacent channel distortion shown either side.

Improving ACLR invariably involves increasing current, i.e. power consumption, in the relevant stage. Since battery current must be kept to a minimum, it is important that the ACLR performance is optimized. Measurements on a typical power amplifier show that the ACLR degradation through the transmit chain may not be as severe as expected.

Table 2.11 shows the ACLR performance of a typical WCDMA power amplifier when fed with a WCDMA downlink signal with varying ACLR performance. The first column shows the power amplifier ACLR performance when it is fed with a signal with an ACLR > −55 dB (ideal signal). The remaining columns show the power amplifier performance when fed from signals which have an ACLR of −37 dB, −39 dB, −41 dB and −43 dB

Table 2.11 ACLR measurement with a third-order compressed signal.

	ACLR > −55 dB	ACLR > −37 dB	ACLR > −39 dB	ACLR > −41 dB	ACLR > −43 dB
Input level (dBm)	−1.7	1.0	1.0	1.5	1.5
Output power (dBm)	26.47	26.5	26.3	26.6	26.5
ACLR (dBc)	−41.0	−35.9	−37.9	−37.6	−38.9

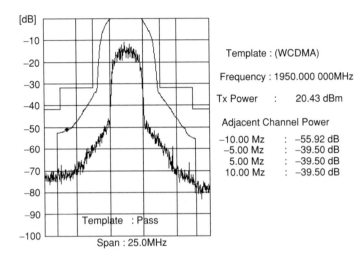

Figure 2.27. Typical transmitter output spectrum.

respectively. When the ACLR degradation of the ideal signal due to PA impairment is the same as the ACLR of signal input of the PA, then a further 6 dB degradation at the PA output would be expected. For example, compare the ACLR in column one and column four in Table 2.11. There is no 6 dB degradation but the PA's ACLR performance is about 2.5 dB better than expected from the simple power summation.

As well as exhibiting compression at high signal levels, a typical power amplifier characteristic will also have a region in which expansion of several decibels occurs. Figure 2.28 shows the effect of expansion in a typical power amplifier.

At low output power levels the gain is around 27 dB. Once the output signal reaches +10 dBm, the quiescent current starts to rise, as does the gain. A gain increase of around 1.5 dB is seen by the time the output power reaches +25 dBm. At higher input levels the power amplifier begins to compress and the gain reduces. It is important that these effects of expansion and compression are calibrated out so that the transmit step accuracy is not compromised.

2.6.6 Key Isolation Issues

When designing any RF transceiver there are generally a number of areas where care has to be taken with the design and layout to avoid interference problems. Two main areas of concern are highlighted next.

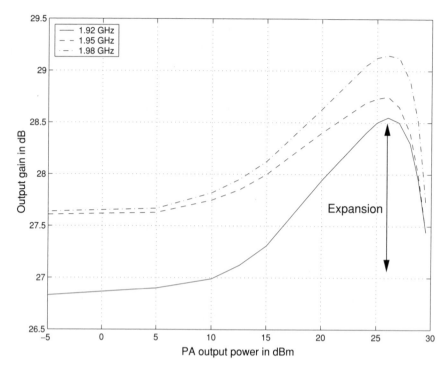

Figure 2.28. Output power versus output gain for a typical WCDMA power amplifier.

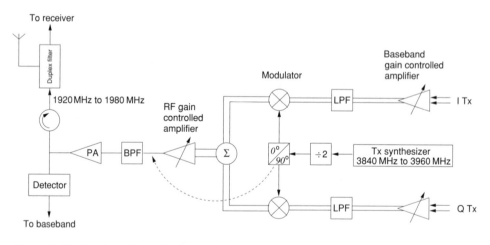

Figure 2.29. Local oscillator leakage path.

2.6.6.1 *Transmit Local Oscillator to RF Gain Control Amplifier*   This LO isolation issue as illustrated in Figure 2.29 only applies to the direct conversion transmitter and is worst at lowest output power, that is at −50 dBm.

Referring back to Table 2.10, the transmitter link budget, it can be seen that when the transmit output level at the antenna is −50 dBm, the level at the output from the RF gain

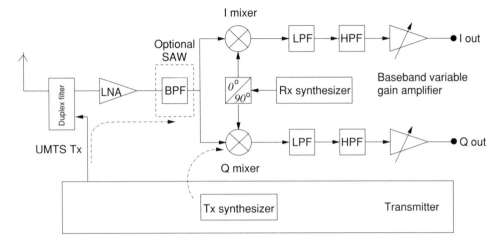

Figure 2.30. Transmit local oscillator to receive mixer isolation.

control amplifier is −73 dBm. Ideally, any local oscillator leakage should be at least 10 dB below the wanted signal. For a local oscillator power drain (to the mixer) of −10 dBm, this means that the isolation must be at least 73 dB. Great care is required in layout to achieve this amount of isolation at frequencies around 2 GHz. One way to mitigate this problem is to put some gain control in the power amplifier. If the power amplifier had 20 dB of gain control, the minimum level at the output of the RF gain controlled amplifier would be −53 dBm, thereby reducing the isolation requirement by 20 dB.

2.6.6.2 *Transmit Local Oscillator to Receive Mixer Isolation*   The transmit output signal will feed through the duplexer, LNA and BPF filter into the receive mixer as a blocking signal. Assuming a direct modulation transmitter, the transmit local oscillator will be at the same frequency as the transmit signal. If the transmit local oscillator leaks across to the receive mixer, it will mix with the transmit blocker to give interference in the receive baseband. This is illustrated in Figure 2.30.

This interference level is very dependent on whether or not a SAW filter is used in the receiver between LNA and mixer. Assuming a SAW filter is not used, the following calculation shows the level of isolation required. Assume that the transmit output signal is at maximum power for a class 3 UE (+24 dBm). The duplexer will attenuate the transmit signal by at least 45 dB in the receiver. Then the transmit signal at the receiver input is −21 dBm. An acceptable blocking level referred to the receiver input is about −100 dBm. Therefore, the transmit synthesizer level at receive mixer must be < −79 dBm. For a local oscillator power level of −10 dBm, this means that the isolation must be at least 69 dB. Once again, great care is required to achieve this isolation.

2.6.7 Transmitter Power Control and Calibration

Transmitter power control is complex and defined in several areas of the 3GPP specifications. Transmission power is controlled under both open and closed loop conditions [41]. The transmitter is only under open loop control when it first starts transmitting. Once the Node

B can receive and measure the transmitted signal, it sends commands to the handset to control its transmission power via a series of up or down commands. Once these power control commands from the base station are being acted upon by the handset, the handset is under closed loop power control.

When initially in open loop control, the handset has to estimate its required transmitter power based on a measurement of pathloss from the downlink CPICH *Received Signal Code Power* (RSCP). This is a measure of the signal strength received from a particular cell, and is measured in the receiver. The accuracy of this measurement depends on an accurate knowledge of the receiver gain. Thus the open loop power control accuracy is a function of receiver gain as well as transmitter gain accuracy. The requirements on the transmitter can be put into three categories:

Step accuracy. The smallest step is 1 dB with an accuracy of ± 0.5 dB. This constraint means that it is important that there are no discontinuities in the transmit gain control. The tightest tolerance comes with a 10 dB step which must be accurate to ± 2 dB.

Absolute accuracy. The maximum power level must be controlled to an accuracy of ± 2 dB. Under open loop power control the transmitter must be able to set its output power to an accuracy of ± 9 dB (± 12 dB under extreme conditions). As stated above, the open loop power accuracy is a function of both the receiver and transmitter. Therefore, it is sensible to split the open loop tolerance and allow for only a ± 4.5 dB (± 6 dB) error due to the transmitter gain.

Reported accuracy. This refers to the UE reported transmit power and is only specified from maximum power down to 10 dB below, and varies from ± 2 dB at maximum power to ± 4 dB at 10 dB below maximum.

The transmitter gain varies with frequency, temperature, supply voltage and from device to device. These variations are too large for the handset to meet the above requirements without some sort of correction mechanism. Two methods in common use are:

1. Calibrate each handset transmitter and generate a look up table to control the gain for varying frequencies and levels. It is not practical to calibrate each handset over temperature, so a small number are measured and an 'average' set of temperature offsets are added to each calibration table.

2. Use a detector to measure the actual transmitter output power and feed back the value to control the power level.

The problem with using a detector is that the dynamic range is large (74 dB) and it is very difficult to maintain accuracy down to such low levels. The calibration method can be made to work but can get complex and lead to a lengthy procedure which must be avoided when trying to simplify the mass production of handsets.

One method is to use a combination of both tables and detector. In this scheme a detector is used to limit the maximum power level and is also used to report the transmit power to at least 10 dB below maximum power. The transmitter chain needs to be calibrated to an absolute accuracy of ± 4.5 dB and a linearity of ± 20%. It may be possible to achieve this with one calibration table for all frequencies. However, the tables will have to be updated for temperature and possibly supply voltage variations. Figure 2.31 shows a plot of accuracy requirements from this calibration process.

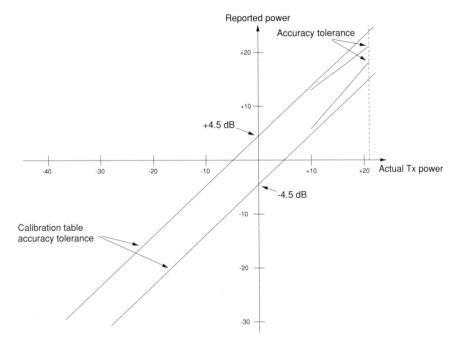

Figure 2.31. Calibration accuracy requirements for a class 4 UE.

2.6.8 The Power Amplifier

The output power amplifier (PA) presents a significant design challenge for UMTS terminal designer. A power amplifier will introduce distortion of the phase and amplitude of the modulated signal and the level of the distortion increases as the output power level approaches saturation. The amplitude component of the modulation will result in both amplitude and phase errors in the output signal, which will manifest as an increase in the modulation EVM and in unwanted power appearing in adjacent channels. The ultimate result is a reduction in the network capacity, therefore limits for adjacent channel power are imposed [9]. In practice, the limit for adjacent channel power defines the linearity requirement for the amplifier and a significant level of quiescent bias current is needed in the output amplifier to meet it. The efficiency suffers as a result, and at full power the efficiency of typical amplifiers is in the range of 35–40 % for a single DPDCH uplink signal, compared with 50–60 % for amplifiers used with the constant envelope GSM modulation, GMSK. For a standard voice call, the uplink transmits continuously and so the amplifier power consumption is a significant consideration with regard to battery drain and hence talk time.

Figure 2.32 shows the output stages between the RF up converter and the antenna. The integrated amplifier typically has two or three stages and a gain between 25–30 dB. The linearity of the power amplifier is very dependent upon the load impedance, and a ferrite isolator is required at the output of the amplifier to ensure consistent performance when connected to an antenna. To allow full duplex operation with a single antenna connection, a duplex filter is required between the isolator and antenna [11]. Typical filters have an insertion loss of 1.5 dB to 2 dB and, including the isolator and transmission lines, the total

Table 2.12   Typical figures for a multistage power amplifier
module with $V_{cc} = 3.4$ V and $f_c = 1950$ MHz.

N_{DPDCH}	PAPR at 0.001	P_{out}	η
1	3.4 dB	26.7 dBm	37.0 %
2	4.0 dB	26.4 dBm	35.5 %
4	6.2 dB	24.7 dBm	30.5 %
6	6.9 dB	23.9 dBm	27.5 %

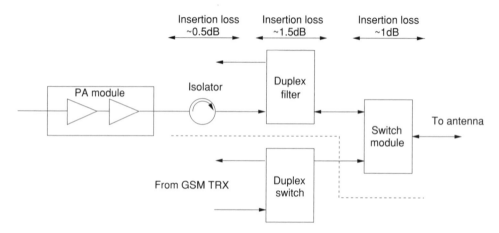

Figure 2.32. Typical RF output stages for a multimode receiver [11].

loss from the amplifier to the antenna is between 2–3 dB. For a single DPDCH uplink
signal with a continuous +21 dBm output power, the amplifier will have a power supply
consumption of approximately 0.6 W which is equivalent to the power consumption of a
GSM power amplifier with 2 W output for one active slot. As the number of uplink DPDCH
channels increases, the peak to mean power ratio also rises and the amplifier has to be
operated below the 1 dB compression point. Table 2.12 shows the maximum carrier output
power and the efficiency η at P_{out} that can be achieved with an *Adjacent Channel Power*
(ACP) of -38 dBc for a multistage amplifier module with different N_{DPDCH}, the number of
DPDCH channels.

2.6.9 Power Efficiency Enhancement

Reducing the output amplifier power consumption is an area that will receive considerable
attention as system deployment proceeds. During the initial stages and while the network is
sparse it is likely that terminals will spend a greater portion of the time transmitting near full
power. However, as the network matures the transmit power of a typical mobile will reduce,
and under these circumstances it is very wasteful to maintain the amplifier quiescent bias
at a level required for full output power. Several simple schemes are proposed to reduce
power consumption:

1. Reduce the DC bias current and leave the DC voltage constant when the amplifier is transmitting at lower power.

2. Reduce the bias current and switch the output matching network to make use of the full available voltage range at the lower current.

3. Vary the voltage using a DC switch mode converter.

In each case a control signal based on the power setting is required from the baseband processor, which can be as simple as a single bit used to switch between two different power ranges. In this case the switching point must be optimized based on network statistics to achieve the greatest overall efficiency. There are several practical difficulties associated with these schemes. The power control method must respond quickly enough to cope with fast power control. If the bias control method is very fast, then the bias can be changed at the slot boundary in a period of <50 µs at the same time the RF gain is changed. However, if there is a lag due to DC supply filtering then synchronization of the bias control with the power control becomes a problem. For switched bias control schemes, the gain of the output amplifier will usually change as the bias is adjusted. This may create difficulties achieving the power control step size accuracy, particularly if the step is large relative to the 1 dB minimum step size, or is variable with temperature and frequency. Finally, the bias control scheme must be able to cope with pulsed transmission. This will occur during the *Physical Random Access Channel* (PRACH) transmission prior to establishing a dedicated channel, but can also occur frequently during packet data transmission using the P-CPCH packet channel.

Figure 2.33 shows a graph of the operating current for a two stage integrated amplifier against output power. The upper line (diamonds) shows the case where the quiescent bias current is maintained at a constant level, in this case 80 mA. The amplifier operates in class A mode below 5 dBm output power and moves to class AB operation above this level as signal self biasing becomes significant. The lower line (squares) shows the operating current for the case where the bias current of the output stage is adjusted to maintain adjacent channel leakage at −40 dBc. Adjusting the quiescent bias current also changes the gain, however, by only adjusting output stage bias, the gain change is minimised whilst still achieving most of the available power saving.

Control of the power amplifier bias voltage whilst maintaining a constant quiescent current yields a greater power saving than bias current adjustment. The technique is widely used in CDMA handsets and has the additional advantage that the amplifier gain changes less with voltage than with quiescent current. However, the supply voltage must be adjusted in a period of less than 50 µs between slots to cope with the fast power control and the potential for an increase in output power. The variable supply voltage may be derived from a DC–DC converter and controlled from a DAC. However, it is difficult to achieve adequate filtering at the DC–DC converter output and achieve the required voltage switching times. The amplifier supply current is modulated by the presence of the AM component of the WCDMA modulation and this may reduce the stability of the DC–DC converter control loop. A possible alternative scheme provides a number of fixed voltage rails (power supplies), e.g. four, generated from a multiple output DC–DC converter. The optimum supply voltage for the present RF output power is selected using a fast low loss switch from the available supply voltages.

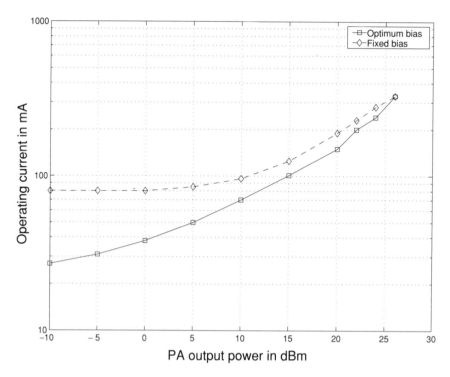

Figure 2.33. Relationship between current and power of a typical two stage PA.

In addition to the bias control schemes, a further set of techniques based on linearization is available for improving amplifier efficiency [42]. These techniques include feedforward and various predistortion schemes and have found most application in base station equipment [43–45]. The cost and complexity of these schemes have so far made them unattractive for implementation in a handset, and in many cases the power consumption of the additional hardware would be greater than the power saving for a handset amplifier. However, there may be some scope for using analogue predistortion schemes integrated with the power amplifier module as an additional method of increasing efficiency.

2.7 FUTURE TRENDS

The future of mobile communications is likely to see a progressive shift away from voice only services towards packet based mobile multimedia services. This will require greater flexibility in the transceiver design which will have to work with a variety of modes, bands and modulation formats. Future RF architectures and front end designs may consider software radio, see Section 8.4 in [8] for example, or other techniques which simplify multimode and multiband operation. Paths for evolution of the transceiver include the addition of HSDPA or multimode UMTS-GSM-802.11WLAN support. Support for *High Speed Data Packet Access* (HSDPA) requires that the receiver can demodulate 16 QAM modulation. The support for UMTS and *Wireless Local Area Network* (WLAN) offers the potential for terminals that can use the higher data rates offered by IEEE 802.11g of typically 20 Mbps where a hot spot is available, and UMTS in other areas to provide ubiquitous coverage with lower data rates. The 802.11g specification is licensed in the ISM band (allocated 2.4–2.4835 GHz) and is close enough to the UMTS Band 1 allocation to allow potential sharing of radio hardware. Support for 802.11g will require a transceiver that will operate with OFDM with each carrier using BPSK, QPSK, 16 QAM or 64 QAM and CSK modulation in addition to the requirements for UMTS.

Increased use of high data rate services will not be possible without the allocation of additional spectrum. It is likely that more bands will be added to the specifications as use of multimedia data services progresses. This will present a significant challenge to the RF designer as transceiver circuitry and filtering will have to cope with a wide range of frequencies. A shift towards broadband circuitry may be needed to simplify transceiver design but will create new challenges in terms of meeting performance without excessive power consumption. Further developments will be required in the design of efficient power amplifiers and on improved ceramics and filter technology.

Another rich field is the development of antennas for UEs, especially for mobile devices such as laptops [46], also for multiantenna systems, as introduced in Chapter 12, to enable increased data rates.

3

Physical Layer Chip Rate Processing

Thomas Keller and Morag Clark

3.1 INTRODUCTION

In this chapter we will discuss the principles of operation and some of the implementational issues of a WCDMA (UMTS-FDD) mobile terminal receiver.

3.1.1 Code Division Multiple Access

Wideband CDMA is an air interface technology based on artificially increasing the bandwidth of a signal by modulating each baseband symbol with a binary or quaternary signature of much higher rate than that of the original data symbol. Each sample of the signature is called a chip, and the number of chips that are modulated by a single data symbol is referred to as the *Spreading Factor* (SF) of the system in question. The process of modulating a high rate signature with lower rate data symbols is called 'spreading', as the resulting bandwidth is spread by a factor of SF relative to that of the original signal.

By raising the bandwidth of the signal, the potential diversity gain in dispersive propagation channels is increased. As the signal energy is spread over a larger bandwidth, the power spectral density of the transmitted signal is lower by a factor of SF than that of the original data signal. At the receiver, the signal is sampled at the appropriate rate, and the received samples are correlated with the known signature code. This correlation over SF chips results in significant inband noise and interference rejection, commonly labelled the 'processing gain' or 'spreading gain'. For a single path propagation channel, the processing

WCDMA – Requirements and Practical Design. Edited by R. Tanner and J. Woodard.
© 2004 John Wiley & Sons, Ltd. ISBN: 0-470-86177-0.

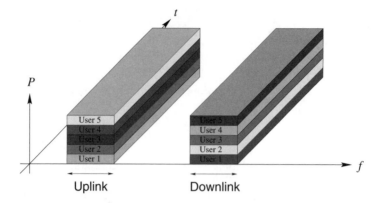

Figure 3.1. The principle of CDMA, where the user data is transmitted simultaneously in a wide spectrum at very low power.

gain is equal to the SF. This means that in the first approximation the overall performance of a CDMA system in Gaussian noise in a *Line of Sight* (LOS) channel is equal to that of a non CDMA system employing direct modulation of a carrier.

The benefits of *Wideband CDMA* (WCDMA) systems lie in the diversity gains that can be achieved over multipath propagation channels and the significant improvement of intercellular handovers. Also, WCDMA has resistance to narrowband jamming; the wideband signals can be hidden at very low power spectral density levels, which was of importance in the early military roots of CDMA technology.

Using appropriate signature codes, multiple information streams can share the same signal bandwidth by using different orthogonal or near orthogonal codes to spread the different data symbols. In the downlink (Node B to UE) of a cellular CDMA system, different scrambling codes are used to separate the different Node Bs because they use the same frequency band, and different spreading codes are employed to discriminate the different users. In the uplink, the UEs use different scrambling codes and may also use different spreading codes (see Figure 3.1). Reserved signatures are also used to transmit synchronization, signalling and broadcast channels alongside the user data.

We will not discuss the theory behind CDMA or all of its forms in this book; good starting points for study would be [7, 32, 47, 48].

3.1.2 The WCDMA Air Interface

The WCDMA air interface employs *Frequency Division Duplexing* (FDD), which means that the downlink employs a different carrier frequency than the uplink. The frequency range depends on local regulations, but a common spectral allocation is 1920–1980 MHz for the uplink and 2110–2170 MHz for the downlink. The uplink and downlink frequency bands are split into channels of 5 MHz width, and the frequency spacing between the uplink and the downlink channel is 190 MHz.

Because the uplink and the downlink transmissions employ different frequency bands, the instantaneous propagation channel fading conditions will typically be different: the short term channel quality of the downlink and that of the uplink are not correlated.

Both the uplink and downlink transmissions are organized in the time domain in frames, which have a duration of 10 ms. Given the chip rate of 3.84 Mchips per second (Mcps), a frame contains 38 400 chips. The frame is the fundamental time unit for the processing of channels within the WCDMA physical layer: channels are started and stopped at frame boundaries. The scrambling codes employed in both uplink and downlink repeat on a frame by frame basis. *Coded Composite Transport Channel* (CCTrCH) TTI lengths (see Chapter 4) are integer multiples of frames.

Each frame is further subdivided into 15 time slots, or power control periods, of length 2560 chips each. The slot is the time period within which all the L1 closed loop procedures take place: inner loop power control (see Section 3.6) and the closed loop transmit diversity procedures (see Section 3.8). Within each slot there is a time multiplexed structure to the signal, within which data symbols, physical layer signalling information and pilot symbols are multiplexed into a single symbol stream (see Section 3.3).

Spreading factors in the WCDMA air interface vary from 4 to 256 (512 for some special downlink signalling channels, see Section 3.3), allowing transmitted symbol rates between 960 ksymbols/s and 15 ksymbols/s on a single code (see Section 3.3). These symbol rates include time multiplexed pilots and physical layer signalling, so that the actual raw user data rate depends on the slot format. The modulation scheme is 4 PSK, which transmits two bits per symbol.

In 3GPP Release 99, WCDMA includes a number of advanced CDMA techniques that will be discussed in this chapter: fast closed loop power control (see Section 3.6), three different schemes for two antenna transmit diversity in the downlink (see Section 3.8) and fast downlink cell selection (see Section 3.7.4), all of which have great influence on the design of a WCDMA physical layer. It is beyond the scope of this chapter to reiterate all the information contained in the 3GPP specifications on the physical layer; instead, we will attempt to clarify the principles and to point out specific challenges for the design of a WCDMA physical layer. The authoritative references are, of course, the specification documents, and this chapter should be read in conjunction with the relevant ones.

The main 3GPP specifications describing the operation of the physical layer are:

- TS25.211 [49] and TS25.213 [29] describe the physical channels (see Sections 3.2 and 3.3);

- TS25.212 [50] describe channel coding and associated topics (see Chapter 4);

- TS25.214 [51] define the physical layer procedures (see Sections 3.5 to 3.10);

- TS25.133 [52] and TS25.215 [53] define the physical layer services to higher layers (including measurements);

- TS25.101 [9] and TS34.121 [10] define mobile terminal performance requirements.

3.1.3 Role of Chip Rate Processing

Within the physical layer, the role of *Chip Rate Processing* (CRP) is to receive the baseband signal from the output of the RF/IF subsystem and to deliver independent received data streams (*Physical Channels*, PhCHs) to CCTrCH processing (see Chapter 4). The CRP subsystem is also responsible for generating uplink chips to be transmitted given streams of encoded PhCH bits from BRP, and for managing the physical layer closed loops (power control, *Site Selection Diversity Transmit* (SSDT) and transmit diversity).

3.2 SPREADING AND SCRAMBLING

WCDMA uses a two level code system: orthogonal spreading codes and pseudo random scrambling codes. In order to support variable data rates, the WCDMA air interface allows for per-channel selectable spreading factors, and this family of spreading codes is called *Orthogonal Variable Spreading Factor* (OVSF) codes. The use of OVSF and scrambling codes is different in the downlink and uplink. In the downlink, the OVSF codes, also called channelization codes, are used to multiplex different channels transmitted in the same cell. In the uplink, the OVSF codes are used to separate data and control channels from a specific user. Scrambling, using pseudorandom sequences, is used in addition to spreading. In the downlink, different scrambling codes separate different cells, and in the uplink they separate different users. Figure 3.2 shows the process of spreading and combining in the UE, where $c_{d,1}$ denotes the OVSF code for the first data channel, β_d is a scaling factor, and $s_{dpch,n}$ denotes the scrambling code.

Spreading and scrambling are applied differently in the uplink and downlink as illustrated in Figure 3.2 and Figure 3.3. This will be discussed later in this chapter, see Section 3.3.

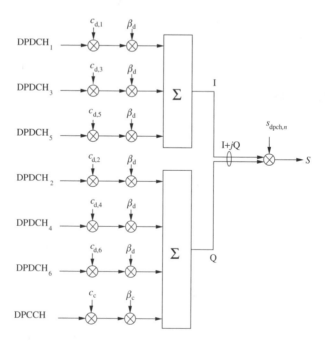

Figure 3.2. Spreading for UL dedicated channels.

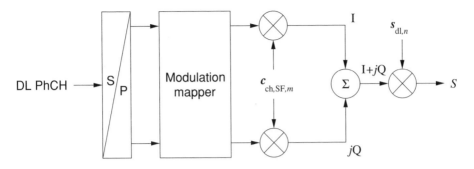

Figure 3.3. Spreading for all DL PhCH except synchronization channels.

3.2.1 Spreading

3.2.1.1 OVSF Codes We have seen in Section 3.1.1 that the ratio of the data symbol rate to the chip rate is the spreading factor SF. By keeping the chip rate constant, at 3.84 MHz for WCDMA, and by varying the spreading factor for a specific code in the range from 4 to 512, a wide range of data rates can be supported. The OVSF codes used for spreading are named $C_{sp,i,j}$, where i is the spreading factor, $i = 2^n$, $n \in \{2, 3, \ldots, 9\}$, and j is the index of the spreading factor $j \in \{0, 1, \ldots, (i-1)\}$. The $C_{sp,i,j}$ are perfect orthogonal sequences such that:

$$\sum_{n=0}^{i-1} C_{sp,i,u}(n) \times C^*_{sp,i,v}(n) = 0 \quad \text{for} \quad u, v \in \{0, 1, \ldots, (i-1)\} \quad \text{and} \quad u \neq v \quad (3.1)$$

The method of generation of OVSF codes $C_{ch,i,j}$ is defined as:

$$C_{ch,1,0} = 1 \quad (3.2)$$

$$\begin{bmatrix} C_{ch,2,0} \\ C_{ch,2,1} \end{bmatrix} = \begin{bmatrix} C_{ch,1,0} & C_{ch,1,0} \\ C_{ch,1,0} & -C_{ch,1,0} \end{bmatrix} \quad (3.3)$$

$$\begin{bmatrix} C_{ch,2^{n+1},0} \\ C_{ch,2^{n+1},1} \\ C_{ch,2^{n+1},2} \\ C_{ch,2^{n+1},3} \\ \vdots \\ C_{ch,2^{n+1},2^{n+1}-2} \\ C_{ch,2^{n+1},2^{n+1}-1} \end{bmatrix} = \begin{bmatrix} C_{ch,2^n,0} & C_{ch,2^n,0} \\ C_{ch,2^n,0} & -C_{ch,2^n,0} \\ C_{ch,2^n,1} & C_{ch,2^n,1} \\ C_{ch,2^n,1} & -C_{ch,2^n,1} \\ \vdots & \vdots \\ C_{ch,2^n,2^n-1} & C_{ch,2^n,2^n-1} \\ C_{ch,2^n,2^n-1} & -C_{ch,2^n,2^n-1} \end{bmatrix} \quad (3.4)$$

which results in the code tree shown in Figure 3.4. This code tree schematically depicts the relationship between the different OVSF codes that can be generated using the rules above; each node corresponds to an OVSF code $C_{ch,i,j}$, and the edges show the relationship between the codes. Two codes that lie on different subbranches on the code tree are orthogonal, and an OVSF code with a spreading factor of less than 256 interferes with all codes in the subtree to the right of itself. In Figure 3.5, the codes at the tips of the bold branches would be a selection that could be used simultaneously while maintaining

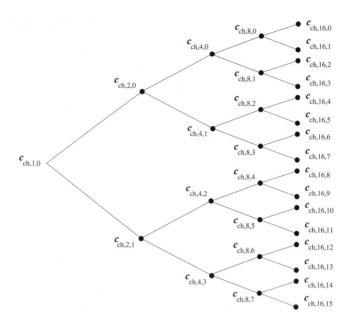

Figure 3.4. The OVSF (Walsh) code tree.

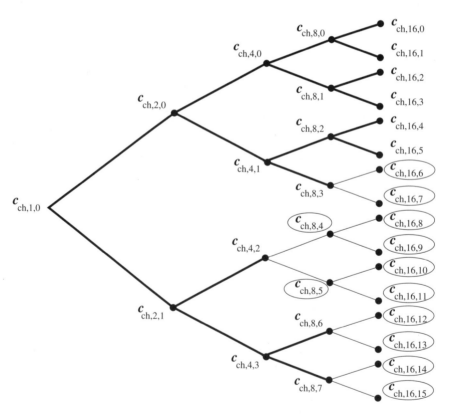

Figure 3.5. The codes on the bold branches can be used simultaneously, while adding any of the encircled codes would result in code clashing.

orthogonality. However, if an encircled code shown was also added, orthogonality would not be maintained.

The WCDMA air interface's OVSF codes allow implementation of flexible data rates: instead of receiving two codes of spreading factor 256 (multicode), a user is allocated a single code of factor 128. This single code covers the same space in the code tree that two neighbouring codes of length 256 would, and offers the same data rate as the two longer codes in parallel, but only a single code needs to be decoded at the receiver. For higher data rates, bundling multiple neighbouring codes into a single higher rate code gives an even more significant receiver simplification.

3.2.1.2 *OVSF Code Allocation*

In the uplink, OVSF codes are employed for multiplexing the data and the control channels, and in the downlink they are used to multiplex all the common and dedicated channels to different users within one cell. In the downlink, the physical layer is informed of the code allocation for different channels via higher layer signalling. Two common channels in the downlink always employ standardized OVSF codes: the P-CCPCH (which carries the *Broadcast Channel*, BCH) is on code $C_{ch,256,1}$, and the primary CPICH is on code $C_{ch,256,0}$. These channels are described in more detail in Section 3.3.

The fixed allocation of the two downlink broadcast channels on the first two codes in the OVSF code tree has effects on the codes available for other channels. All spreading codes $C_{ch,i,0}$ for all allowed values of i clash with the primary CPICH and the P-CCPCH and must not be allocated to other channels, see Figure 3.6.

In the uplink, allocation of channelization codes to physical channels is performed according to defined rules, which are the same for all mobile stations [29].

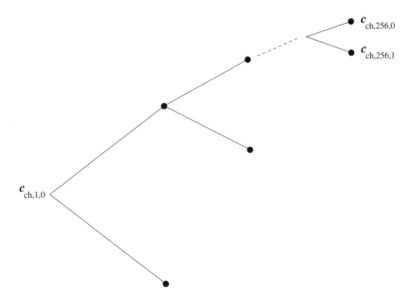

Figure 3.6. The effect of DL OVSF code allocation to P-CPICH and P-CCPCH.

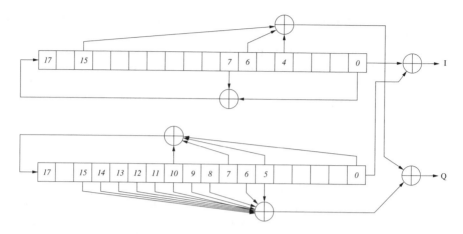

Figure 3.7. Downlink scrambling code generator.

3.2.1.3 Auto- and Crosscorrelation Properties of OVSF Codes   The OVSF codes are orthogonal Walsh codes, this class of code loses orthogonality if they are not time aligned. This poor crosscorrelation causes problems in multipath propagation channel conditions, where the echoes of the downlink signal would interfere significantly with each other. The necessary improvement in the signal's auto and crosscorrelation properties is achieved by the scrambling code, which is discussed in the next section.

3.2.2 Scrambling

Scrambling is applied on top of spreading and does not change the chip rate of the spread signal. In the uplink, scrambling is used for the separation of users and in the downlink different scrambling is used for the separation of cells. Section 3.4.3.2 discusses implementation details of scrambling code generation.

   Scrambling codes are 10 ms long codes formed using shift registers such as that shown in Figure 3.7 for generation of downlink scrambling codes. Unlike spreading codes, scrambling codes are not orthogonal sequences, but they have good auto- and crosscorrelation properties.

3.2.2.1 Scrambling in the Downlink   In the downlink, the shift register arrangement above allows $2^{18} - 1$ different scrambling codes to be generated. However, for ease of code acquisition and cell search, not all of these are used. The scrambling codes are arranged into 512 sets of primary scrambling codes, each with an associated 15 secondary codes. The 512 primary codes are further divided into 64 code groups of 16 codes. Each code has associated with it an alternative left and right associated scrambling code that can be used in compressed mode (see Section 3.11).

   Each cell is allocated exactly one primary scrambling code; the code number of this primary scrambling code is the cell's Cell ID, and the primary scrambling code is used to transmit the P-CCPCH, S-CCPCH, P-CPICH, PICH, AICH, AP-AICH, CD/CA-ICH and the CSICH. See Section 3.3 for details on these channels. The other physical channels can either be transmitted on the primary scrambling code or on one of the 15 secondary codes from the set associated with the primary code. The primary scrambling code is searched for by the mobile terminal during cell search (see Section 3.5).

The mobile terminal finds the primary scrambling code during the cell search procedure. Detection of the S-SCH gives the scrambling code group to which the primary code belongs and correlation of the CPICH with all of the codes in that group identifies the code (see Section 3.5).

3.2.2.2 *Scrambling in the Uplink*   In the uplink, each mobile terminal is allocated its own scrambling code by the network; this code must be unique in the immediate network neighbourhood as it is received by all the cells with which the mobile terminal is in soft handover (see Section 3.7). The scrambling code selection for common uplink channels like PRACH and PCPCH and their associated preambles are beyond the scope of this section and are detailed in the 3GPP specifications [29].

3.3 PHYSICAL CHANNELS

Physical Channels (PhCHs) are L1 communication streams characterized by a combination of scrambling code, OVSF code, spreading factor and format. In the downlink, physical channels fall into two groups: common and dedicated.

Common physical channels in the downlink carry cell broadcast data used for:

- L1 handshaking, synchronization and channel estimation (the *Primary Synchronization Channel* (P-SCH), the *Secondary Synchronization Channel* (S-SCH) and the *Common Pilot Channel* (CPICH));

- L3 information transmission (the *Broadcast Channel* (BCH) *Forward Access Channel* (FACH) and the *Paging Channel* (PCH));

- Non connection oriented user data (the FACH), and for shared time/code multiplexed resource (the *Downlink Shared Channel* (DSCH));

There is only one dedicated physical channel type, the *Dedicated Physical Channel* (DPCH), and it is used for user data.

Higher layer data streams [50] are organized as transport channels that are mapped to the physical channels in Layer 1. This mapping is shown in Figure 3.8. Also shown are those physical channels that are used only within the physical layer and do not have an equivalent transport channel in higher layers. These are indicator channels used in random access procedures to indicate resource availability to the mobile terminal (see Section 3.9).

The following sections describe the physical channels in terms of the type of information they carry (for example, data, control, preset sequence) and how this information is arranged in the frame and slot format of the channel. We will concentrate on the general principles of operation for each channel, additional information can be found in [49] and [6].

3.3.1 Synchronization and Channel Estimation Channels

The primary and secondary synchronization channels and the common pilot channel are downlink physical channels for the benefit of L1 alone; they carry no user data. The P-SCH

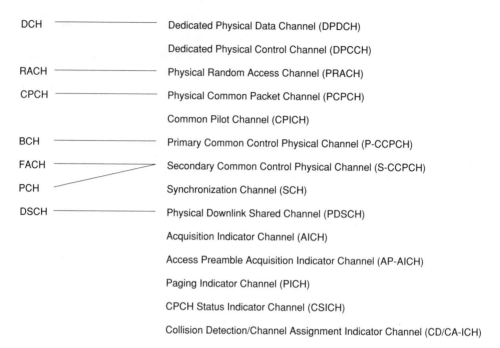

DCH ──────────── Dedicated Physical Data Channel (DPDCH)

Dedicated Physical Control Channel (DPCCH)

RACH ──────────── Physical Random Access Channel (PRACH)

CPCH ──────────── Physical Common Packet Channel (PCPCH)

Common Pilot Channel (CPICH)

BCH ──────────── Primary Common Control Physical Channel (P-CCPCH)

FACH ──────────── Secondary Common Control Physical Channel (S-CCPCH)

PCH ──────────── Synchronization Channel (SCH)

DSCH ──────────── Physical Downlink Shared Channel (PDSCH)

Acquisition Indicator Channel (AICH)

Access Preamble Acquisition Indicator Channel (AP-AICH)

Paging Indicator Channel (PICH)

CPCH Status Indicator Channel (CSICH)

Collision Detection/Channel Assignment Indicator Channel (CD/CA-ICH)

Figure 3.8. The mapping of the transport channels (left) onto the physical channels (right).

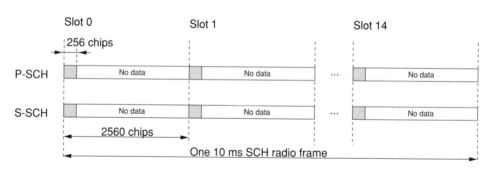

Figure 3.9. The structure of the SCH, showing P-SCH and S-SCH.

and the CPICH both consist of known sequences, and the S-SCH consists of one of a set of specified sequences. Because the sequences are known, these channels can be used by L1 for cell search and channel estimation (see Sections 3.4 and 3.5).

The *Synchronization Channel* (SCH) is a physical channel made up of two subchannels, the primary SCH (P-SCH) and secondary SCH (S-SCH). Both the P-SCH and the S-SCH are bursty and transmitted by the base station in the first 256 chips of every slot. Figure 3.9 schematically shows the frame format of the SCH, with the synchronization bursts aligned with the slot boundaries. The SCH transmits no power for the remainder of the slots.

The P-SCH is a fixed 256 chip sequence that is repeated at the start of every slot by every WCDMA cell; it is used by mobiles to detect the presence of a WCDMA cell at a specific

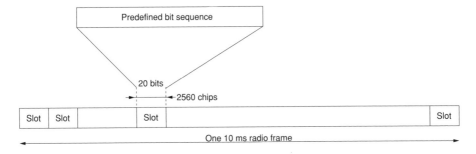

Figure 3.10. Frame structure for the CPICH comprising 15 slots.

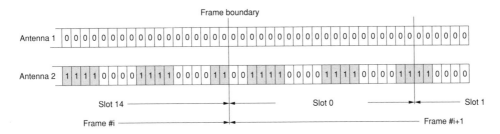

Figure 3.11. Modulation pattern for CPICH.

point in code space (see Section 3.5.1). Correlating the received signal with a copy of the P-SCH sequence yields a correlation peak indicating the time of a slot start for each cell within reach of the mobile.

The primary and secondary SCH differ in that the S-SCH encodes information by transmitting a series of fifteen 256 chip sequences per frame. This series of fifteen S-SCH bursts signals the cell's scrambling code group (see Section 3.5.2), and is repeated every frame. By virtue of the code used to signal the code group, the S-SCH can also be used to identify the frame start of the downlink transmission of the transmitting cell. Neither the P-SCH nor the S-SCH are scrambled with the cell's scrambling code.

The *Common Pilot Channel* (CPICH) is a fixed downlink physical channel transmitting a sequence of pilot symbols at 30 kbps (which is the raw data rate of a PhCH with SF = 256) with the frame and slot format shown in Figure 3.10. The physical layer uses the CPICH to gain an estimate of the downlink channel conditions as discussed in Section 3.4.

If transmit diversity (see Section 3.8) is employed in any downlink channel, then different CPICH sequences are transmitted from the two base station antennae. The pilot bit sequence transmitted from Antenna 1 is a fixed +1, while on Antenna 2 an orthogonal sequence is employed, as is shown in Figure 3.11. The use of orthogonal sequences allows the physical layer to separate the signals arriving from each antenna and therefore to estimate the independent propagation channels from each antenna as described in Section 3.8.

Each WCDMA cell transmits at least one CPICH, called the primary CPICH (P-CPICH). This P-CPICH is always carried on OVSF code $C_{256,0}$ on the cell's primary scrambling code, and it is guaranteed to be transmitted omnidirectionally over the whole cell; which means that no beam steering can be used at the Node B for the P-CPICH. The P-CPICH can be used as a phase reference for demodulating all broadcast channels. If a cell employs

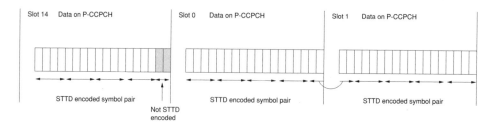

Figure 3.12. STTD encoding for the data bits of the PCCPCH.

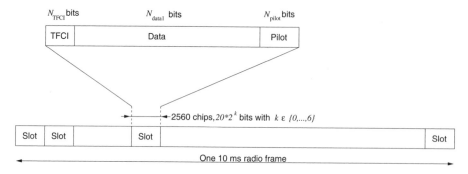

Figure 3.13. Frame structure for the S-CCPCH over 15 slots.

beam steering, then it may broadcast a secondary CPICH; this has the same characteristics as the primary CPICH but has a different OVSF and may have a different scrambling code. The mobile terminal is instructed by the network which CPICH to use for its dedicated channel.

3.3.2 Cell Broadcast Channels

The broadcast channels are carried on the two common control physical channels: the *Primary Common Control Physical Channel* (P-CCPCH) carries the BCH, and the *Secondary Common Control Physical Channel* (S-CCPCH) carries the FACH and the PCH.

The BCH, carried on the P-CCPCH is used to transmit system information and is broadcast over the entire cell. It is of fixed spreading factor (SF $= 256$), slot format, and OVSF code ($C_{256,1}$). Its slot format is unusual for a data bearing PhCH in that the P-CCPCH is not transmitted during the first 256 chips per slot, when the SCH is transmitted. The P-CCPCH can be STTD encoded (see Section 3.8) like other channels but because of the 256 chip gaps, the scheme is modified as shown in Figure 3.12.

An S-CCPCH can carry two different transport channels: the *Forward Access Channel* (FACH) and the *Paging Channel* (PCH). Apart from the paging procedure involving the *Paging Indication Channel* (PICH), which is discussed in Section 3.9, the FACH and PCH are indistinguishable in the physical layer as higher layer transport channels on an S-CCPCH. FACH and PCH can be multiplexed onto a single S-CCPCH or can be carried on independent S-CCPCHs. The FACH configuration is broadcast to the mobile over the BCH.

The S-CCPCH supports raw data rates between 30 and 1920 kbps (spreading factors range from 256 to 4) and the frame and slot structure is shown in Figure 3.13. The first block of bits transmitted in every slot carries TFCI information, this is used to signal the BRP settings for the current TTI (see Chapter 4). Following the TFCI field are the data bits, which will be passed to the receiver's BRP. The last field in each slot carries pilot bits, which are known sequences of bits that can be used at the mobile terminal to estimate the propagation channel's phase and magnitude (see Section 3.4.3.4), as well as the *Signal to Interference Ratio* (SIR) (see Section 3.6) and the channel quality measure (see Section 3.10).

There are 18 different slot formats defined for the S-CCPCH, offering a range of user data rates and L1 signalling bit numbers. The slot format is chosen by the network and signalled to the mobile station via the cell broadcast information on the BCH.

3.3.3 Dedicated Channels

Dedicated physical channels carry user data specific to a given user's mobile. There is only one type of dedicated physical channel, the DPCH. The DPCH carries user data in its data subchannel, the *Dedicated Physical Data Channel* (DPDCH), and L1 control information in its control subchannel, the *Dedicated Physical Control Channel* (DPCCH). The multiplexing of the DPDCH and the DPCCH differs between the downlink and uplink. In the downlink the two subchannels are time multiplexed to form the DPCH, while in the uplink they are code multiplexed and transmitted in parallel to form the DPCH (see Figure 3.2).

The physical layer control information carried on the DPCCH is made up of known pilot bits, feedback commands for closed loop power control in both the uplink and downlink, feedback commands for closed loop transmit diversity and SSDT (FBI bits, present only in the uplink DPCCH) and TFCI bits. Again, the pilot bits are a predefined sequence which can be used at the receiver for channel estimation. The operation of the power control loop is described in detail in Section 3.6, transmit diversity techniques are discussed in Section 3.8, and SSDT is discussed in Section 3.7.4.

3.3.3.1 *Downlink Dedicated Channel* The slot format of the downlink DPCH is shown in Figure 3.14. Depending on the data rate requirements, the structure of the slot can be chosen from specified slot formats shown in Table 3.1. The slot format used on a specific DPCH is configured (and reconfigured) by higher layers. Note that the slot formats

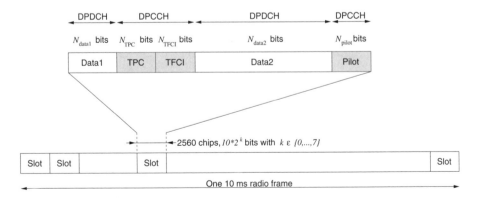

Figure 3.14. Frame structure for the downlink DPCH.

Table 3.1   Downlink DPCH slot formats, excerpt from Table 11 in [49].

Slot format #i	Channel bit rate (kbps)	Channel symbol rate (ksps)	SF	Bits/slot	DPDCH Bits/slot		DPCCH Bits/slot		
					N_{Data1}	N_{Data2}	N_{TPC}	N_{TFCI}	N_{Pilot}
0	15	7.5	512	10	0	4	2	0	4
1	15	7.5	512	10	0	2	2	2	4
2	30	15	256	20	2	14	2	0	2
3	30	15	256	20	2	12	2	2	2
4	30	15	256	20	2	12	2	0	4
5	30	15	256	20	2	10	2	2	4
6	30	15	256	20	2	8	2	0	8
7	30	15	256	20	2	6	2	2	8
8	60	30	128	40	6	28	2	0	4
9	60	30	128	40	6	26	2	2	4
10	60	30	128	40	6	24	2	0	8
11	60	30	128	40	6	22	2	2	8
12	120	60	64	80	12	48	4	8	8
13	240	120	32	160	28	112	4	8	8
14	480	240	16	320	56	232	8	8	16
15	960	480	8	640	120	488	8	8	16
16	1920	960	4	1280	248	1000	8	8	16

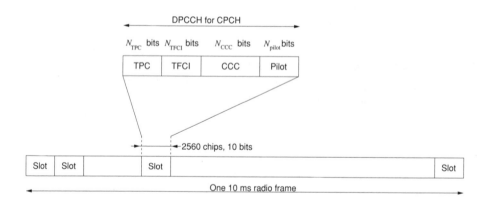

Figure 3.15. Frame structure for the downlink DPCCH for CPCH.

listed in Table 3.1 only include those that are used in noncompressed frames, for details of compressed mode handling refer to Section 3.11.

In addition to the DPCH slot formats described above, there is a special case of the DPCCH which is used to close the control loop for the UL common packet channel (DPCCH for CPCH). Its slot format is shown in Figure 3.15. The DL DPCCH for CPCH carries TPC, TFCI and Pilot bits as well as *CPCH Control Commands* (CCC) bits. The CCC bits are used to signal start of message and emergency stop indicators used for the CPCH access procedure described in detail in Section 3.9.

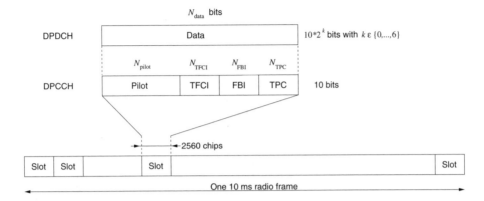

Figure 3.16. Frame structure for the uplink DPCH showing the DPDCH and the DPCCH.

Table 3.2   Uplink DPDCH slot formats, from Table 1 in [49].

Slot format #i	Channel bit rate (kbps)	Channel symbol rate (ksps)	SF	Bits/ frame	Bits/slot	N_{data}
0	15	15	256	150	10	10
1	30	30	128	300	20	20
2	60	60	64	600	40	40
3	120	120	32	1200	80	80
4	240	240	16	2400	160	160
5	480	480	8	4800	320	320
6	960	960	4	9600	640	640

3.3.3.2 *Uplink Dedicated Channel*   In the uplink DPCH the control and data information, the DPCCH and the DPDCH, are I/Q code multiplexed (see Figure 3.2) with a frame and slot structure shown in Figure 3.16. As in the downlink, there is a list of slot formats specified that define the number of bits per field as shown in Tables 3.2 and 3.3. Again, the slot formats for compressed mode have been omitted from these tables; compressed mode handling is discussed in detail in Section 3.11. Unlike in the downlink, where the slot format is a higher layer parameter passed to the physical layer, the slot format in the uplink is chosen dynamically by the physical layer depending on the number of bits to be transmitted and this may change on a frame by frame basis. This is possible because every mobile uses its own scrambling code, so that there is no need for centralized OVSF code assignment. In both the uplink and downlink there is always exactly one DPCCH but there may be none, one or several DPDCHs.

3.3.3.3 *DPCH Timing Relationships*   The timing relationship between the uplink and downlink dedicated channel is controlled by the mobile terminal, the time delay between the downlink DPCH frame boundary being received from a specific cell designated by the network as time reference (the 'reference cell') at the mobile and the start of the mobile's uplink transmitted DPCH frame boundary is nominally 1024 chips. In practice, because

Table 3.3   Uplink DPCCH slot formats, from Table 2 in [49].

Slot format #i	Channel bit rate (kbps)	Channel symbol rate (ksps)	SF	Bits/ frame	Bis/ slot	N_{pilot}	N_{TPC}	N_{TFCI}	N_{FBI}
0	15	15	256	150	10	6	2	2	0
1	15	15	256	150	10	8	2	0	0
2	15	15	256	150	10	5	2	2	1
3	15	15	256	150	10	7	2	0	1
4	15	15	256	150	10	6	2	0	2
5	15	15	256	150	10	5	1	2	2

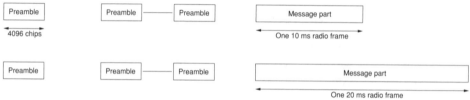

Figure 3.17. Structure of random access transmissions.

of the time varying nature of the propagation channel, the arrival time of the downlink channel is not constant and the uplink timing must be maintained by a tracking loop in the mobile's physical layer; because of restrictions on the speed at which the uplink channel timing can be adjusted, the instantaneous delay can vary significantly from the target delay of 1024 chips. This is discussed in Section 3.4.6.4.

3.3.4 Packet and Indicator Channels

There are three common packet channels: two on the uplink, the random access channel (PRACH) and the common packet channel (PCPCH) and one on the downlink, the down-link shared channel (PDSCH). The packet and indicator channels are all used in access procedures described in Section 3.9.1. In the current section, the format of the channels is covered briefly to aid understanding of these procedures.

The PRACH is used to carry the RACH transport channel and comprises a preamble and message part. Each preamble is 4096 chips long, made up of 16 repetitions of a 256 chip signature. The message part comprises data and control information that is transmitted in parallel. The RACH message part can be either 10 or 20 ms long. Figure 3.17 illustrates the two parts of the RACH.

In the 10 or 20 ms data bursts, data rates of between 15 and 120 kbps are supported in four defined slot formats, while the control information comprises a fixed 15 kbps rate with eight pilot bits and two TFCI bits. The frame and slot structures are shown in Figure 3.18.

Random access transmission is accomplished on the RACH in conjunction with the acquisition indicator channel (AICH), a downlink channel used to carry acquisition indicators (AI). The RACH procedure is described in Section 3.9.

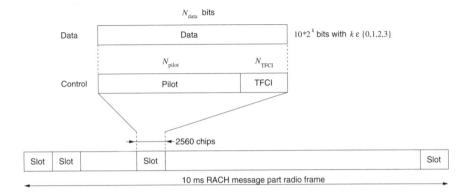

Figure 3.18. Frame structure of the random access message part.

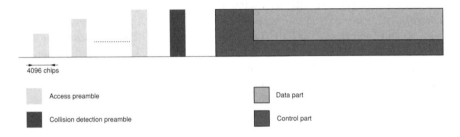

Figure 3.19. Structure of the CPCH access transmission.

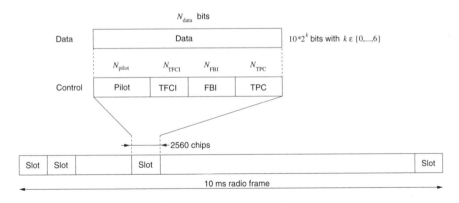

Figure 3.20. Frame structure for the uplink data and control parts of the PCPCH.

The PCPCH is used to carry the CPCH transport channel. It comprises an access preamble part, a collision detection part, a power control preamble part and a message part. These are illustrated in Figure 3.19.

There are one or more access preambles (AP) each of 4096 chips for which the RACH preamble signature sequences are used. The single collision detection access preamble (CDP) is also 4096 chips and again the RACH preamble signature sequences are used. There is one power control preamble part (PCP) which has the same slot format as the message part and a duration of between zero and eight slots as set by higher layers. The format of the message part (and the power control preamble) is shown in Figure 3.20.

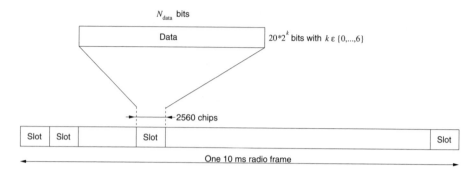

Figure 3.21. Frame structure of the PDSCH.

The data and control parts are transmitted in parallel. There are two defined formats for the control part, one with and one without FBI bits. The data part supports data rates of between 30 and 1920 kbps. The CPCH procedure uses the CPCH in conjuction with three downlink indicator channels: the CPCH access preamble acquisition indicator channel (AP-AICH), the CPCH collision detection/channel assignment indicator channel (CD/CA-ICH) and the CPCH status indicator channel (CSICH). The procedure is described in Section 3.9. The physical downlink shared channel is a variable rate channel supporting between 30 and 1920 kbps in seven defined slot formats and carrying only data bits. Its slot format is shown in Figure 3.21.

The PDSCH is always associated with a downlink DPCH and the TFCI bits in the DPCCH are used to inform the mobile terminal that there is data on the DSCH. Multiple PDSCHs can be transmitted in parallel, with the same spreading factor, to one mobile terminal. In this case, all PDSCHs have the same frame timing. However, the PDSCH(s) are not aligned with the associated DPCH: a DPCH frame is associated with a PDSCH frame if the PDSCH frame starts between 3 and 18 slots after the end of the associated DPCH frame. Closed loop transmit diversity [51] is used on the PDSCH if it is being used on the DPCH.

3.3.5 Overview of Physical Channel Timing

Physical channel timing uses the P-CCPCH, on which the system frame number is broadcast, as a reference. The P-CCPCH is frame aligned with the P-SCH, S-SCH, P-CPICH, S-CPICH and the PDSCH. The relative time offsets of the other downlink channels are shown in Figure 3.22. The relationship between the uplink and downlink timing at the mobile terminal is described in Section 3.3.3.3.

3.4 THE RECEIVER

3.4.1 Overview

The receiver is the heart of the chip rate processing block, it converts the oversampled signal from the CRP input into soft decisions that are passed to the bit rate processing

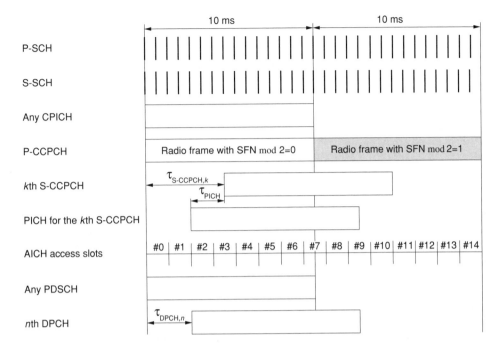

Figure 3.22. Radio frame timing and access slot timing of the downlink physical channels.

block discussed in Chapter 4. As we have discussed above in Section 3.2, the WCDMA air interface employs DS-CDMA with two overlaid codes to spread and scramble the data symbols for transmission over the air. We have seen in Section 3.3 that the physical channels are independent data streams received by the CRP, and that the PhCH's symbol stream includes time multiplexed data, control and pilot symbols. It is the role of the CRP to receive the physical channels, to demultiplex the data symbols and pass them on to the BRP, and to correctly act on the control information that is carried on the PhCH.

In this section we discuss the fundamental operation of the CRP's receiver block; the details of the receiver's operation depend on transmit diversity modes, power control modes and compressed mode parameters, which will be discussed in following sections. Here we will concentrate on the signal processing and control processes that are common to the reception in all air interface modes.

There are many different known receiver architectures for CDMA type air interfaces. The receiver architecture most commonly used for mobile CDMA terminals is the RAKE receiver [54, 55].

3.4.2 RAKE Receiver Overview

The RAKE architecture is a suboptimal but efficient implementation of a CDMA receiver [54, 56–58].

Figure 3.23 shows the components of a RAKE receiver: a set of fingers and a combiner block. Each single RAKE finger is an independent receiver for the signal from a specific cell on a specific propagation path. Multipath propagation channels and soft handover situations

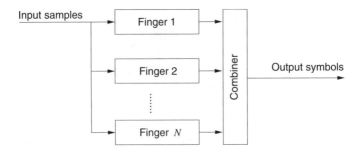

Figure 3.23. Key RAKE receiver building blocks.

are handled by employing multiple fingers, one for each propagation path in question. The output symbols of all the fingers are then combined coherently and synchronously by the RAKE receiver's combiner block to yield the received data symbols. We will discuss the operation of both in the following sections.

3.4.3 RAKE Fingers

Each RAKE finger independently receives the signal contribution of a single propagation path from a given cell; it performs this by correlating the received signal sample stream over a length of SF samples with a time aligned copy of the scrambling and spreading codes. At its input, the RAKE finger takes input samples at chip rate, and it outputs the correlation results at symbol rate.

The time and code selectivity of the fingers is achieved by exploiting the cross- and autocorrelation of the combined spreading and scrambling codes over SF samples. While the crosscorrelation between codes is nominally independent from their relative time delay, the correlation between the incoming samples and the local copy of the codes not only depends on the code's autocorrelation function, but also on the sampling position within the pulse shaping filter's impulse response, see Chapter 2: assuming a $\mathrm{sinc}(\pi t/T_c)$ shaped impulse response, a misalignment of $1/2$ chip translates into a drop of 3.9 dB in the correlation result. For this reason, the time alignment between the incoming signal path and the specific RAKE finger's signal sampling point must be better than one chip. Figure 3.24 shows the $\mathrm{sinc}(\pi t/T_c)$ shaped impulse response with sampling points at $1/2T_c$ and $1/4T_c$ misalignment from the ideal sampling point together with the loss in received signal power. The exact value of the signal loss due to the sampling point misalignment depends, of course, on the combined impulse response of the transmitter's and receiver's pulse shaping filtering. Also not taken into account in the numbers shown in the figure is the additional interchip interference caused by not sampling on the impulse response's zero crossings.

Practical RAKE receiver 'oversampling rates', i.e. the rate at which the input signal is sampled and presented to the input of the fingers, range from four to eight, and are not necessarily restricted to powers of two. A higher oversampling rate at the RAKE finger inputs means that more accurate alignment of the finger's sampling point with the received signal path can be achieved, at the cost of receiver complexity. Whatever oversampling rate is chosen, the correlation in the fingers still runs at chip rate.

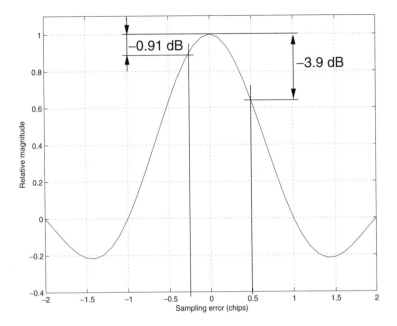

Figure 3.24. Loss of signal power as a function of the sampling point; assuming a sinc()
shaped filter impulse response.

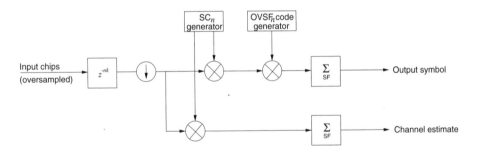

Figure 3.25. Simplified schematic diagram of a RAKE finger.

Figure 3.25 shows a simplified schematic of a RAKE finger, consisting of:

- a data correlator branch, consisting of two multipliers for applying the spreading and scrambling codes and an accumulator of length SF;

- a channel estimation branch receiving the CPICH, consisting of a scrambling code multiplier and an accumulator of length 256;

- an input sample delay memory block;

- a downsampling block, which reduces the rate of the input signal into the data and CPICH correlators to chip rate; and

- the scrambling and spreading code generators.

3.4.3.1 RAKE Finger Correlators   The RAKE fingers' correlators perform the despreading of the received symbol; this is achieved by correlating the input samples with a time aligned copy of the spreading and scrambling codes. The accumulation length of the correlator is the spreading factor of the received channel.

The set of RAKE finger correlators are the most computationally expensive part of a WCDMA receiver, which means that they are normally implemented in specialized hardware blocks. The multiplications in the signal path can be implemented as simple sign reversal operations, as the magnitude of the scrambling and spreading codes is constant.

3.4.3.2 Implementation of Code Generators   The downlink scrambling code generator is a set of shift registers as depicted in Figure 3.7, this is easy to implement in hardware or in software, and described in detail in the 3GPP specifications [29]. As discussed below in Section 3.4.5, the number of scrambling code generators required for the RAKE receiver depends on the receiver architecture chosen, as well as on the number of simultaneous downlink cells supported and the number of fingers.

Once seeded, the scrambling code generators will generate a scrambling code chip at each execution of the shift registers; as the scrambling code length is limited to a frame, the registers must be reset to their initial value on each frame boundary.

The main implementational problem is the generation of shift register seeds for starting fingers on non frame boundaries. Simple 'clocking through' of the shift registers until the required scrambling code offset within the frame is reached is slow and not energy efficient, and it is impractical to precompute lookup tables for every possible offset. Clearly, a sensible compromise between finger startup delay, memory and power consumption must be found; this depends to a large extent on the speed of the RAKE control discussed below and the acceptable delays in starting a new finger. Management of finger allocation and timing adjustments are discussed in Section 3.4.6 below. The spreading codes can easily be stored in a 256×256 bit lookup table and used by all correlators in the system.

3.4.3.3 Channel Estimation Using CPICH   The RAKE receiver operates by independently receiving signal contributions from different propagation paths in the fingers, and then combining these contributions coherently and synchronously in the combiner.

For coherent combination of the fingers' outputs, a phase estimate must be known to derotate each finger's output prior to combining. There are a range of sources for this reference phase, as discussed below, but the main source for phase estimation in the downlink is the CPICH.

As discussed earlier, at least one CPICH is always transmitted on every WCDMA cell, and this is called the primary CPICH, or P-CPICH. Its scrambling and spreading codes are known to be the cell's primary scrambling code and $C_{256,1}$, respectively. The received phase of the P-CPICH is valid as every channel's phase reference unless the mobile terminal is signalled otherwise by UTRAN. The main scenario in which the P-CPICH is not a valid phase reference is when the cell employs beam steering or directive aerials for dedicated channels only. If the P-CPICH is not available as phase reference for a specific channel, the mobile terminal will be notified by the network. In this case, a secondary CPICH, employing a different scrambling/OVSF code combination from the primary CPICH, may be transmitted in the cell. It is mandatory for the mobile terminal to support a secondary CPICH for receiving dedicated channels.

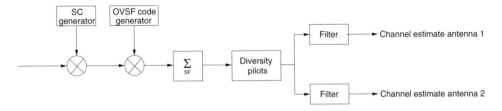

Figure 3.26. CPICH reception finger branch incorporating diversity pilot estimation and filtering.

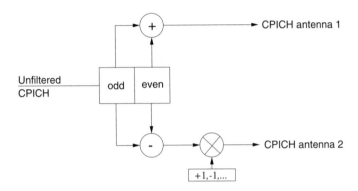

Figure 3.27. Diversity CPICH decoding.

The simplified schematic of a RAKE finger in Figure 3.25 does not support a secondary CPICH, an OVSF code generator is required in the CPICH correlator to receive a secondary CPICH instead of the primary CPICH. Also, in order to achieve the performance targets defined in [9] adequate channel estimation filtering is required to reduce the noise in the channel estimation. Lastly, in order to support the transmit diversity schemes discussed below (see Section 3.8) the CPICH reception branch needs to support CPICH with transmit diversity.

Figure 3.26 shows the CPICH correlation branch of a finger supporting all the required functionality in more detail. Diversity channel estimations are calculated by correlating the received CPICH symbols with the antenna specific two symbol diversity modulation sequences applied at the transmitter [49]. Figure 3.27 shows the CPICH diversity decoder block used to separate the channel estimates from the two transmit antennas. This is achieved by combining two consecutive CPICH symbols. To generate the channel estimate from the first antenna, the two received CPICH symbols are added, while for the second antenna's channel estimate, the two received CPICH symbols are subtracted. The inversion operation in the second antenna's channel estimation stems from the CPICH modulation pattern described in [49] and must be reset to +1 at the beginning of each frame.

The channel estimation filters following the CPICH diversity decoder have a major impact on system performance [59, 60]. Too little filtering leads to too much noise on the channel estimates, especially in slowly fading multipath channel conditions [9]. Too much filtering introduces significant group delay onto the channel estimation values, which, if not compensated for by delaying the data symbols for the same amount of time prior to combining, decreases combiner performance.

3.4.3.4 Other Sources of Phase Reference
The CPICH may not always be available as phase reference for combining and demodulation, depending on the downlink transmit diversity settings (see Section 3.8), and on whether the cell employs anisotropic antenna patterns for dedicated channels. In the latter case, the mobile terminal would be informed that the P-CPICH is not a valid phase reference for the dedicated channel.

In either case, if no CPICH is available as a phase reference, then the pilot symbols time multiplexed within the slots are available for channel estimation. The number of the pilot symbols depends on the slot format chosen for the channel and it is the responsibility of the network to select a slot format with a sensible number of pilot symbols if no CPICH is available.

Using time multiplexed pilot symbols as the downlink phase reference adds complexity to the RAKE receiver's data flow. The time multiplexed pilot symbols must be demultiplexed from the received symbols based on the slot format, the received pilot symbols must be correlated with the known pilot symbol sequence for the slot in question, and the resulting channel estimate must be filtered.

3.4.4 The Combiner

We have discussed above that a RAKE receiver employs multiple correlators (fingers) to independently receive the signal contribution from specific propagation paths. For each of these contributions a channel estimation value, giving the reference phase and relative magnitude, is available from the channel estimation branches of the fingers.

The role of the combiner block is to combine the independent signal contributions from the fingers coherently and synchronously, such that the signal to noise ratio of the resulting combined signal is maximized. The optimal coherent combining strategy is *Maximum Ratio Combining* (MRC) [47, 61, 62].

Figure 3.28 schematically shows an MRC combiner: the signal contribution from each finger is derotated and scaled with estimated phase and magnitude of the propagation path. In numerical terms, this is achieved by simply multiplying the received data symbol with the conjugate complex of the channel estimation.

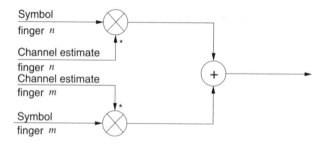

Figure 3.28. Simplest combining strategy of the finger outputs for non transmit diversity channels.

As can be seen in Figure 3.28, the optimal combining strategy is simple to implement in principle, but the problem is that the channel estimation is noisy. This leads to a bias in the channel magnitude estimation which reduces the performance of the receiver. In practical terms, this must be managed by RAKE control (see Section 3.4.6); only 'good' propagation paths should be allocated fingers, and the strategy with which path selection is achieved has a major effect on the receiver's performance.

3.4.5 RAKE Architectures

We have seen above that the alignment of the RAKE fingers' correlators with the received propagation paths is crucial for the operation of the receiver, and we have assumed in Figure 3.25 that the alignment between signal and finger is achieved by delaying the input samples; this concept is demonstrated in Figure 3.29(a) for a RAKE receiver consisting of multiple fingers. It can be seen that in this configuration, the delay memory at the input of the fingers takes the shape of a common delay line from which a tap feeds delayed input symbols to each finger at chip rate. The delay line operates at oversampling rate and the total length of the delay line depends on the total delay spread between the earliest and the latest path in the combined impulse response from all cells. This receiver architecture is called the 'front end model', and is conceptually the simplest way of building a RAKE receiver.

Because all the relative delays between the different propagation paths (and cells) are taken up in the front end delay memory, all the fingers operate synchronously: the correlators all output a symbol at the same time, which simplifies the finger to combiner interface. Another benefit is that, as all the fingers' combiners run synchronously, only a single scrambling and OVSF code generator is required for each scrambling and OVSF code in the downlink.

The main disadvantage of this RAKE architecture is the large block of memory at the receiver input: the delay line operates at oversample rate, and the maximal delays the memory must be able to accommodate are long (hundreds of chips). Memory access bandwidth might be an issue if all fingers' delays happen to be chip spaced and therefore all the reads from the delay memory coincide on a single oversample. It is difficult to switch off parts of the memory block if a short delay spread is observed.

An alternative RAKE architecture is demonstrated in Figure 3.29(b). In this configuration, dubbed the 'back end memory model', no delay is positioned at the input of the

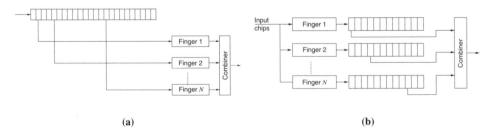

(a)   (b)

Figure 3.29. RAKE receiver architectures: front end (a) and back end (b) memory architectures in comparison.

fingers. Instead, each finger's timing is aligned with its input signal and all fingers operate asynchronously from each other. A block of delay memories at the output of the fingers serves as a buffer to allow synchronous combining of the fingers' outputs at the combiner.

Benefits of the back end memory architecture are that the total amount of memory is typically smaller than that for the front end memory: the signal delay buffers between the fingers and the combiner operate at symbol rate, not oversample rate, which reduces the size of each individual block by at least a factor of 16 (assuming four times oversampling and a minimal supported SF of four). The memory blocks at the output of unused fingers can be switched off to minimize receiver power consumption, and the memory access timing requirements are a lot more relaxed.

The downside of this architecture is the loss of synchronization between the fingers. This translates into more complicated finger control and typically into more complicated finger resource: multiple scrambling and spreading code generators are required, normally one per finger.

3.4.6 RAKE Control

We have discussed above that the RAKE receiver works by independently receiving signal contributions from different propagation paths and by combining those contributions. We have seen that each RAKE finger requires accurate knowledge of its corresponding propagation path's timing, and that combining the output of weak fingers decreases the total signal quality.

The role of the RAKE control is therefore to supply up-to-date timing information and resource allocation to the RAKE receiver. This entails:

- identifying new propagation paths from cells in the active set;
- identifying paths that have disappeared from view;
- tracking the temporal position of propagation paths;
- assigning RAKE resources (fingers) to paths such that the overall SIR is maximized.

At the core of the RAKE control's operation is the estimation of the propagation channels' impulse responses. On the basis of these, finger allocation and deallocation decisions can be made, maximizing the expected SIR. The estimated propagation path delays can be used to set up RAKE fingers on the chosen paths.

The RAKE control functionality can be broken down into the following blocks:

- the path searcher;
- the path tracker;
- finger manager.

3.4.6.1 Path Searcher The aim of the path searcher [63] is to estimate the impulse response of the propagation channel between a known cell and the mobile. From the RAKE receiver's point of view, the goal is to identify changes in the impulse response that, in turn, drive the reallocation of the RAKE fingers. If a new strong path is detected, then it makes sense to use a RAKE finger resource to receive its signal contribution. If a propagation path which is currently received by a RAKE finger becomes weak, then its RAKE finger should be stopped and the resource should be employed on a stronger path. This operation is performed by the finger manager (see Section 3.4.6.3). The path searcher can also be used to estimate the received power of cells that are not currently received by the RAKE receiver; this is discussed in Section 3.10.

One possible implementation of a path searcher uses correlators that receive the CPICH of the target cell over a window of possible path offsets around the known rough timing of the cell. The offset employed for the correlation is swept over a search window and the correlation output is recorded for each offset value. The resultant vector of correlation values gives a good estimate of the impulse response. The 'sampling rate' of this impulse response estimation is the value by which the correlator's time offset is changed between correlations. Typically a time resolution of 1/2 chip gives acceptable performance to the RAKE manager.

It is important to note that searchers based on the CPICH can only find paths for cells that are known to the mobile, i.e. paths for which the primary scrambling code and the position of the search window are known. This limitation means that a second searcher, used for detecting and identifying cells, must be present in the system. This process is discussed in Section 3.5.

As discussed in Section 3.5, CPICH based searchers offer better performance and built in cell selectivity compared to cell searchers. The performance is better because the CPICH is typically transmitted with more power than the SCH (see [9]) and the nonbursty nature of the CPICH allows coherent averaging over more than 256 chips, improving the processing gain.

The search window size covered by the path searcher must be chosen such that it covers the expected delay spread of the channel around the known cell position. Typically, this is in the order of $\pm$ 80 chips around the nominal cell position, which covers the worst case 'birth death' channel model in the 3GPP specifications [49].

3.4.6.2 *Path Tracker*

The path tracker is a control loop that adjusts the finger position in code space to track the downlink propagation path delay in time. It refines the path timing information gained from the path searcher and tracks propagation paths with time varying propagation delays. Path tracking is generally performed by means of a *Delay Locked Loop* (DLL). This DLL operates by running two correlators with small time offsets relative to the CPICH correlator, and by comparing the magnitude of the early and late correlations. This is possible as the signal's autocorrelation function is smooth, symmetrical and monotonous around the ideal sampling point for a single specular propagation path. A common time offset between the early and late correlators and the (on time) CPICH correlator in the finger is 1/2 chip, but other values are possible.

The DLL measures the asymmetry in magnitude between the early and late correlation result. The three possible outcomes are shown in Figure 3.30, assuming a spacing of 1/2 chip between the on time and the early and late correlator positions. If the early and the late correlation result are approximately the same, as depicted in Figure 3.30 (b), then the finger position is correctly centred on the true path position. If the early correlation is significantly less than the late one, as shown in Figure 3.30 (a), then the finger position is too early, and should be delayed.

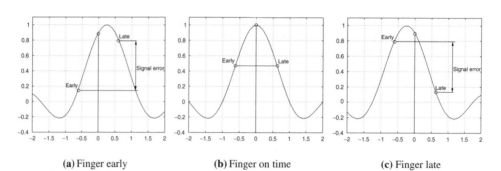

(a) Finger early   (b) Finger on time   (c) Finger late

Figure 3.30. Early/late sampling around the finger position.

Clearly, as the correlation results are noisy, some form of filtering and thresholding should be included in the finger position adaptation algorithm. This is unproblematic, as the propagation path delays change slowly over time. If the impulse response is not composed of single rays separated by more than a chip, then the impulse responses of the different rays overlap. This scenario is shown in Figure 3.31. The overlapping rays lead to a continuous, time varying and nonsymmetrical impulse response when sampled by the DLL correlators. The simple path timing adjustment rules described above would lead to all fingers converging towards the same position. This would mean that the noise samples on the fingers become correlated and therefore cause loss of diversity gain from multiple fingers and reduce the overall signal to noise ratio.

In the example given in Figure 3.31, two fingers are allocated to the cluster of closely spaced propagation paths. Although both finger 1 and finger 2 are positioned on local maxima of the impulse response, the DLL process for finger 2 would advance its position towards finger 1 ($e_2 > l_2$). This can be avoided if finger positions are not adjusted independently, for example if minimal allowed distances between finger positions are defined which constrain the finger position adjustment loop. The minimal allowable finger position would normally be chosen to be in the order of the reciprocal of the pulse shaping filter's bandwidth, i.e. about 1 chip.

3.4.6.3 *Finger Manager*   The finger manager controls the allocation and deallocation of finger resources to propagation paths. It uses the estimated impulse responses for all the cells in the active set and calculates the best use of its RAKE fingers.

The rules applied by the finger manager have a profound effect on the system performance: not allocating a RAKE finger to a 'good' propagation path decreases the overall received signal quality because not all available signal energy is collected. Conversely, allocating RAKE fingers to 'poor' paths will inject additional noise because the combiner cannot perform perfect MRC (see Section 3.4.4). Fast response to changes in the impulse responses can improve system performance in some (birth death channel conformance tests [9]), but harm it in other operating conditions (multipath Rayleigh fading channels).

3.4.6.4 *Uplink Timing Control Loop*   We have seen in Section 3.3.3.3 that the nominal timing difference between the downlink reception and the uplink transmission at the

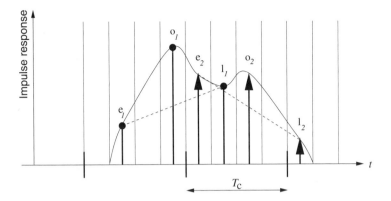

Figure 3.31. DLL operation in a closely spaced multipath channel. The DLL correlations are e = early, o = ontime and l = late. Two rays are shown: subscripts 1 and 2.

mobile is 1024 chips. This time delay is measured between the arrival time of the downlink frame boundary and the transmission time of the uplink frame boundary. The reference point for this time difference measurement is the mobile's antenna connector[1]. The downlink frame boundary arrival time is defined as the arrival time at the mobile of the first chip of the frame on the propagation channel's first propagation path. In soft handover situations, one cell is nominated to be the 'reference cell' by higher layers, and the first path from this cell is the downlink reference time.

The arrival time of this first path is not static, but changes over time as the mobile moves and as the downlink propagation channels' impulse responses change due to fading. It is the uplink timing control loop's responsibility to adjust the uplink transmission's timing such that the uplink delay stays as close to 1024 chips as possible. This is achieved by adjusting the uplink timing gradually, with a slew rate of up to 1/4 chip per 20 frames.

3.5 CELL SEARCH

Cell search is the process by which the mobile terminal finds the timing, cell ID (primary scrambling code) and BCH STTD status of cells in its neighbourhood. Cell searches are scheduled by higher layers either explicitly, e.g. after a power on, to detect cells to camp on, or implicitly, by adding unknown cells to the monitored set (see Section 3.10). In this section we will examine how the mobile terminal can detect cells it has no information about. In Section 3.10 we will discuss simplifications of this process if some a-priori knowledge about the cell (e.g. cell ID or some timing information) is available.

The cell search procedure is based on the three fundamental physical layer signalling channels available in every cell: the primary and secondary synchronization channels (P-SCH and S-SCH), and the primary CPICH (P-CPICH).

---

[1] The filter delays in the downlink and uplink data paths must be taken into account when controlling this delay.

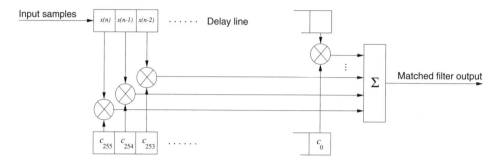

Figure 3.32. P-SCH matched filter, where c_n is short for $C_p(n)$ in Equation (3.7).

3.5.1 P-SCH Detection

The P-SCH is a burst which is transmitted by every WCDMA cell during the first 256 chips of every slot. It is used as a sounding sequence to detect the presence of cells and to give information about each cell's slot timing. The P-SCH burst $C_p(n)$ is defined as a hierarchical sequence with:

$$a(j) = [1, 1, 1, 1, 1, 1, -1, -1, 1, -1, 1, -1, 1, -1, -1, 1] \qquad (3.5)$$

$$b(k) = [1, 1, 1, -1, -1, 1, -1, -1, 1, 1, 1, -1, 1, -1, 1, 1] \qquad (3.6)$$

$$C_p(n) = b(\lfloor n/16 \rfloor) \times a(n \bmod 16) \qquad (3.7)$$

Because the timing of the cells is completely unknown at this stage, the mobile terminal must receive the P-SCH using a matched filter rather than a time aligned correlator. The simplest implementation of a P-SCH matched filter is shown in Figure 3.32.

Given the hierarchical properties of the P-SCH sequence, a more efficient hierarchical correlator can be employed to detect the P-SCH; this is shown in Figure 3.33. The top half of the figure shows a short 16-tap FIR filter matched to the sequence $a(j)$, and the bottom half is a 16-tap FIR filter matched to the sequence $b(k)$. The two implementations of the P-SCH matched filter shown here are equivalent, but the hierarchical correlator is significantly less power hungry than the 256-tap FIR implementation.

Figures 3.32 and 3.33 assume an input sample rate to the P-SCH detector of one per chip. This is not sufficient for reliably detecting synchronization bursts when the burst arrival time does not coincide with the sampling time at the receiver, and so typically the P-SCH receiver runs at least at twice chip rate; in the figures this would correspond to a delay tap between each of the FIR taps in the first filter, and clocking the detector at twice chip rate. This increase in temporal resolution improves performance, but increases complexity two fold.

3.5.2 S-SCH Detection

The secondary synchronization channel serves as a means of detecting the frame start timing of cells detected by the P-SCH correlator and as an aid to cell ID detection. Each cell transmits this channel alongside the P-SCH and, like the P-SCH, the S-SCH consists of fifteen bursts of 256 chips length. Unlike the P-SCH, different S-SCH bursts are transmitted

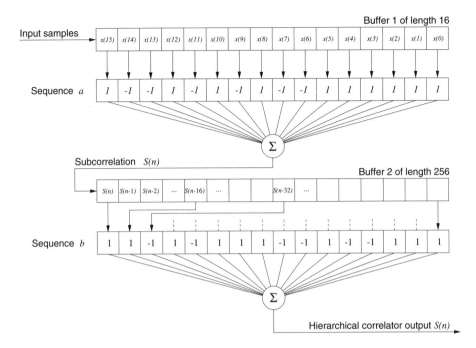

Figure 3.33. P-SCH matched hierarchical correlator.

in each slot, and the sequence of these bursts depends on the scrambling code group the cell belongs to (see [29]). Each cell repeats its sequence of S-SCH bursts in every frame.

The S-SCH bursts are, like the P-SCH, hierarchical codes, with a constituent inner code length of 16 (see [53]) and a higher level length 16 Hadamard code, leading to 16 different S-SCH bursts. Therefore, each S-SCH burst can be decoded with a 16 chip correlation followed by a 16 point Hadamard transform.

The position in the frame and the cell's code group are encoded in the sequence of S-SCH bursts transmitted from the cell; the sequences of bursts are Reed–Solomon (RS) codes, so that an RS decoder can be employed to find both the position and the sequence from the identified S-SCH burst sequence.

3.5.3 Cell ID Detection

Once the cell's frame timing and the scrambling code group are known from the S-SCH detection process, the cell's primary scrambling code can be determined by tentatively receiving the primary CPICH on all eight primary codes in the code group. The primary code which gives the strongest correlation on the OVSF code $C_{\mathrm{sp},256,0}$ is detected as the cell ID.

3.5.4 P-CCPCH Transmit Diversity Status Identification

The last step of the cell search is to detect whether the cell's P-CCPCH is open loop transmit diversity (space time transmit diversity, STTD) encoded; this information is required to receive the BCH. Information about the STTD status of the P-CCPCH is encoded in the relative phase between the P-CPICH and the synchronization channels; see [49] for details. Detecting the relative phase between the SCH and the P-CPICH requires receiving both in parallel, a P-SCH and a CPICH correlator are required. Note that because of the *Time Switched Transmit Diversity* (TSTD) scheme operating on the synchronization channels, this phase comparison should be done between the P-CPICH received from the first antenna and the P-SCH in even slots only. In odd slots, it is unknown at this stage from which antenna the synchronization channels have been transmitted, and it is therefore unclear which P-CPICH phase reference to use for phase comparison: if STTD is not used on the BCH, then the SCHs do not employ TSTD, and therefore the synchronization bursts are always transmitted from the first base station antenna. If STTD is used on the BCH, then TSTD is employed, and the SCH are transmitted from the second antenna in the odd numbered slots, and from the first antenna in the even slots.

3.6 POWER CONTROL

The capacity of CDMA systems is generally interference limited in both the uplink and the downlink. In the downlink, there is interference between different users, as well as between the different propagation paths of the same user, depending on the impulse response of the propagation channel. In addition, the downlink signals from different cells interfere with each other. In the uplink, as every mobile terminal uses a different nonorthogonal scrambling code, each user's received uplink signal suffers from interference between all the uplink channels in the cell, in addition to intercell interference from users in the neighbouring cells. This interference stems from the cross- and autocorrelation properties of the scrambling and spreading codes, as discussed in Section 3.2.

The interference problem is most troublesome in the uplink: the distance of mobiles from the base station varies greatly, so that uplink signals from different users arrive at the base station receivers with widely different power levels. The interference from strong users, close to the base station, can completely drown the signal from mobiles far from the base station. This scenario is sometimes referred to as the 'near far problem' [48].

The solution to this is closed loop power control. Both the base station and the mobile terminal continually monitor the received signal quality and send feedback power control commands to each other on the reverse links. In WCDMA, the signal quality is estimated every slot, and the *Transmit Power Control* (TPC) feedback command is sent in each slot. The update rate of the power control is therefore 1500 Hz. We will see that this high rate puts significant constraints on the physical layer system design.

Closed loop power control can only operate if a feedback channel is available and if the transmit power of a channel can be adjusted to suit a specific mobile's requirements. This means that only the dedicated channel (DPCH) and the common packet channel (CPCH) are power controlled with closed loop updates. The common channels, in contrast, must be transmitted at a level that is appropriate for all the intended recipients in the cell.

In this section, we will describe the power control mechanisms for dedicated channels from the mobile terminal's point of view; the faster, inner power control loop residing in the physical layer, as well as the outer loop power control in higher layers.

3.6.1 Inner Loop Power Control

Inner loop power control consists of two control loops that drive the SIR estimated at the UL and DL receivers towards their respective SIR targets. The basic mechanism is the same for UL and DL: the estimated SIR of the given link is compared with a target value, which can be different for UL and DL, and a step up/down command is sent on the reverse link to the transmitting device.

From the mobile terminal's point of view, inner loop power control can be broken down into the following tasks:

- receive UL power control bits in the downlink TPC field of the DPCCH and apply the resulting power step command at the transmitter;

- estimate the DL channel SIR and decide on the next downlink power control command;

- send the resulting DL power control commands to the Node B in the uplink.

We will discuss these tasks in turn.

3.6.1.1 Power Control Timing at the Mobile Station The main problem with implementing power control functionality on a mobile is the timing constraints. Figure 3.34 shows the relative timing of the downlink and uplink channels as experienced by the mobile terminal; the two time intervals important for closed loop power control are the delay between the end of the DL DPCCH pilot field and the first chip of the uplink TPC field in the same slot, and the delay between the end of the TPC field in the downlink and the slot boundary of the next uplink slot. Both these delays are nominally 512 chips, but this is

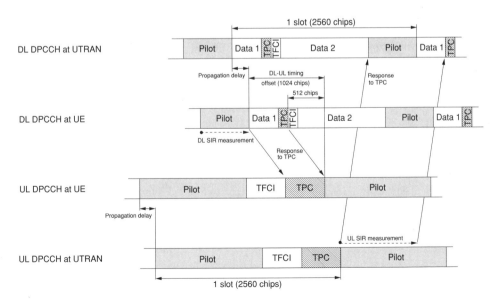

Figure 3.34. Power control timings.

reduced by the uplink timing inaccuracy that stems from time varying impulse responses in the downlink propagation channels (see Section 3.4.6.4) and by the downlink delays between cells in soft handover (see Section 3.7). Realistically, the receiver has a safe window size of about 256 chips within which it needs to close the two power control loops.

3.6.1.2 UL Power Control Commands

A new power control command is sent to the mobile terminal via the TPC bits of the downlink DPCCH once per slot; the mobile terminal receives the TPC bits and adjusts its transmit power by a defined step for the first chip of the UL slot following the TPC command, up to an upper power limit communicated by the network.

The method by which the mobile terminal derives the step from the TPC bits depends on several factors. 3GPP defines two algorithms for uplink power control: using the first algorithm, the mobile terminal derives and applies the requested step every slot, and using the second algorithm, the mobile terminal uses five slots' worth of TPC bits to derive the step, which is then valid for the next five slots. Since there is a defined minimum step size, the second algorithm makes it possible to emulate smaller step sizes than algorithm 1, or even to effectively turn off uplink power control by transmitting an alternating series of TPC commands. These algorithms are complicated by the fact that there is no guarantee that all Node Bs will request a power step in the same direction. The 3GPP specifications [51] therefore describe in detail how TPC commands from different radio link sets must be combined. Compressed mode and radio link initialization procedures also have defined effects on the derivation of the power control step from the TPC bits.

Because the timing to close the power control loop is critical, the processing of the TPC bits to obtain the power control step is usually implemented in hardware: efficient DSP type block processing is very difficult because of timing constraints.

3.6.1.3 Estimating the DL SIR

The DL SIR estimation is the critical part of the mobile terminal's power control tasks; from an implementational point the main problem is tight timing. The Node B will update the downlink power at the first chip of the pilot field, controlled by the last TPC bits transmitted in the uplink. The next uplink TPC bits are due to be transmitted nominally 512 chips after the end of the pilot field, see Figure 3.34.

As it is not guaranteed that the last TPC bits were received correctly by the Node B, or that the Node B honoured the power control command, the SIR estimation has to be achieved in this time frame.

In practice, SIR estimation can be split up into two parts: interference plus noise estimation, which is slowly time varying, and signal power estimation, which is changing with power control commands as well as with the time variant propagation channel characteristics.

3.6.1.4 DL TPC Generation

The mobile terminal requests a power up or power down command from the Node B via the TPC command to be transmitted with the next UL slot. This TPC command is generated by comparing the DL SIR estimate with an SIR target, which is chosen to achieve the specific QoS performance of the DL channel. If the estimated

SIR is worse than the target, then a power up command is sent; otherwise, a power down command is sent. As for uplink power control, there are two modes (see [51] for details): in the first mode, DPC mode 0, the mobile terminal sends a unique TPC command in each slot, and in DPC mode 1 the mobile terminal sends the same command for three slots. The behaviour of downlink power control TPC generation is also affected by compressed mode in a manner specified by 3GPP in [51].

3.6.2 Outer Loop Power Control

The SIR target for the inner loop power control must be chosen by the RRC, see Chapter 10, such that the link's QoS requirements are satisfied without requiring too much transmitted power from the Node B. In practice, this is achieved by defining an upper and a lower threshold for the coded block error rate (see Chapter 4), and by employing a control loop adjusting the SIR target dynamically to keep the *Block Error Rate* (BLER) in that range. This is a nontrivial exercise, as measuring BLER in the range of 10^{-2} to 10^{-3} is a slow process. Further physical layer measurements other than the BLER alone are generally used to achieve faster convergence.

3.6.3 Other Power Control Mechanisms

Other than the steady state power control for DCH discussed here, there are further power control mechanisms defined for WCDMA: open loop power control on the access preambles for RACH and CPCH (see Section 3.9), and initial convergence mode for closed loop power control on the CPCH message part, or on the DCH (see [51]).

3.7 HANDOVER

3.7.1 Introduction

All cellular module telecommunications systems implement some support for handing a user from a serving cell to another during a call. Changing the active connection from one cell to another once the link quality to the alternative cell has become better than that of the serving cell is called 'hard handover', and this is the only mechanism supported for non CDMA systems like GSM. CDMA systems allow a different form of handover, where the mobile is in communication to multiple cells at the same time; this mechanism is called soft handover. WCDMA allows three handover schemes:

- *Soft Handover* (SHO), between different cells employing the same Radio Access Technology (RAT) and downlink frequency;

- *Softer Handover* (SOHO), between different sectors of the same cell; and

- *Hard Handover* (HHO), which is handover between noncompatible RATs (e.g. WCDMA to GSM) or across different downlink frequencies.

We will discuss the principles of these, as well as the fundamental impacts of handover on a WCDMA mobile implementation, in this section.

Any kind of handover is only supported for the dedicated channels: the uplink and downlink DPCH. The PDSCH, being a common channel, is never in handover, although the supporting DPCH may be.

3.7.2 Soft and Softer Handover

From the physical layer's perspective, soft and softer handovers are treated in the same way: the downlink DPCH is transmitted from a set of different cells (or 'radio links' or RL), and it is guaranteed that the data bits from all the cells are the same. As we have seen in Section 3.4, the RAKE receiver combines these signal contributions in exactly the same way as it combines the different propagation paths from a single cell. The set of cells that the mobile is in soft handover with is called the 'active set', and it is not under the mobile's control: a change in soft handover configuration is performed by the network, via higher layers, modifying the current active set in the physical layer.

3.7.2.1 *Channel Timing in SHO/SOHO*   In soft handover, the delay between signal contributions from different radio links must be compensated for by the finger delay memory so that the signal contributions can be combined synchronously and coherently (see Section 3.4).

The finger delay memory is limited and the delays are potentially long, therefore the downlink must be synchronised between different cells. This is done by the network which adjusts $\tau_{DPCH,n}$ in Figure 3.22 for each cell in the active set. $\tau_{DPCH,n}$ is quantized to 256 chips (required to preserve orthogonality between OVSF codes). The adjustment is done by signalling timing measurements between different cells (specifically SFN–CFN, see [53]) back to the network.

One cell, the 'reference cell', provides the time reference for the uplink and $\tau_{DPCH,n}$ can be reconfigured by the network if the relative cell timing drifts to more than 128 chips from the 'reference cell'.

The 'reference cell' is replaced by another if it is taken out of the active set, which is signalled by higher layers. This leads to a potentially large error between uplink and downlink timing (up to 128 chips) that is slowly corrected by the UL timing control loop (see Section 3.4.6.4).

3.7.2.2 *Connection Frame Number*   The *Connection Frame Number* (CFN) is a DPCH specific time base (frame counter) that is used to signal actions on the DPCH (e.g. modification, release), and is maintained during handovers. The SFN, on the other hand, is cell specific and is not synchronized between cells.

3.7.2.3 *Downlink TPC and TFCI Bits in SHO/SOHO*   While the downlink data and pilot bits transmitted by all cells in the active set are the same, the TPC bits signalling the next uplink power control step and potentially the TFCI bits for PDSCH may be different from the different cells. The physical layer is informed by higher layers whether these are guaranteed to be the same or not, and has to perform the appropriate combining method. For more details on TPC combining, see Section 3.6 and the specifications ([51]), where the different TFCI signalling schemes for PDSCH are also defined.

3.7.2.4 *Effects on Receiver Implementation*   Soft handover support puts restrictions on the design of a WCDMA RAKE receiver:

1. The RAKE memory length depends on the maximum delays between the downlink paths to be combined. Because $\tau_{DPCH,n}$ is quantized to 256 chips there will be delays of up to 128 chips plus propagation delays between the arrival times of paths from different cells.

2. Because there is more than one cell's downlink propagation channel to be estimated, the multipath searcher must be faster than it would have to be in a single downlink radio link receiver.

3. The design of the multipath searcher must permit measurements on all cells in the active set in a sensible time frame.

4. The different treatment for TPC and potentially TFCI bits complicates data flow in the receiver and requires special treatment instead of standard RAKE combining. See Section 3.6 for more details.

3.7.3 Hard Handover

In hard handover scenarios, the mobile stops communicating with its serving cell(s) at a predefined time and, after a synchronization procedure, resumes the link with a different set of serving cells; the synchronization procedure is defined in the specifications [51]. Hard handover is employed whenever soft handover is impossible: the main scenario is interfrequency handover, where the carrier frequency of the link changes with the handover. Note that the active sets before and after the hard handover may contain more than one cell, so that a mobile can be instructed by the network to perform hard handover from one soft handover situation into another.

3.7.4 SSDT

Site Selection Diversity Transmission (SSDT) is a special mode of soft handover where the mobile terminal selects one of the serving cells as 'primary' by signalling an identification word to the Node B. The primary cell transmits the full DPCH, and all other cells stop transmitting the DPDCH, transmitting only the DPCCH. This aims to reduce interference caused by multiple transmissions in soft handover, and to provide fast site selection.

The mobile terminal selects the strongest cell as primary and it reselects the primary cell with the next signalled ID word. SSDT only works in slowly fading environments since the signalling delays are multiple slots.

Feedback errors in SSDT are problematic since the situation could arise where no Node B is transmitting. Each Node B decides independently whether it is primary or not: the received feedback bits must fulfil a quality check for a Node B to assume the secondary state. To minimize the likelihood of no Node B transmitting, the quality check is such that the Node B obeys, 'if in doubt, transmit'. It is therefore likely that more than one Node B will transmit the DPDCH.

In the RAKE receiver, the fingers and finger control are not affected by SSDT since the CPICH and DPCCH still transmit. The RAKE receiver combiner needs to maximize the

SIR and it can combine all contributions from the primary cell and from those cells that have wrongly assumed primary status. To avoid adding unwanted noise, it may exclude fingers from secondary cells for the DPDCH.

3.8 TRANSMIT DIVERSITY IN THE DOWNLINK

3.8.1 Background

As we have discussed in the introduction to this chapter, wideband systems are inherently superior to narrowband mobile communications systems in receiving signals in multipath propagation conditions. The higher temporal resolution of wideband receivers allows these to coherently combine paths that would add incoherently for narrower band systems. Incoherent addition of propagation paths manifests itself as multipath interference, and the effect of this is fast fading, i.e. chaotic constructive and destructive interference between propagation paths of different phases, but with the same arrival time within the receiver's temporal resolution.

A WCDMA receiver, with a chip rate of 3.84 MHz, can resolve multipaths down to about one chip duration delay difference (see Section 3.2). This corresponds to a path difference of approximately 78 m. Propagation paths that arrive with a path length difference of less than the receiver's resolution limit cannot be combined coherently, and this leads to fading of the paths observed by the band limited WCDMA receiver. Still, the increased multipath resolution ability of a wideband CDMA system gives a significant performance gain in multipath propagation conditions, as the observed paths fade independently from each other, and the RAKE receiver is able to combine the time variant set of 'strong' paths. This is called 'multipath diversity', or 'path diversity'.

This advantage is lost, however, if the total impulse response length is short, for example in very small cells, or in situations where there are reflectors only in the immediate vicinity of the mobile. In these cases, the bandwidth of the system is not enough to resolve the multipaths, and no path diversity gains are achievable. To make things worse, environments with low path diversity typically also imply low vehicular speeds, so that time diversity is poor (see Section 3.1). An example of this is an indoor office environment with small cells and low mobility.

The WCDMA specifications allow network designers to overcome this problem by deliberately introducing new propagation paths that exhibit independent fading and that are resolvable by the mobile receiver. This technique is called 'downlink transmit diversity'.

WCDMA downlink transmit diversity is based on the base station employing two transmit antennas for the downlink. These antennas are spaced far enough apart from each other to make the fading on the propagation paths to the mobile receivers uncorrelated (more than a carrier wavelength apart), but close enough so that the arrival time difference between the signals from the two antennas is much less than one chip. This translates to transmit antenna distances of between 0.3 and 78 m, well suited for a base station installation[2].

In order to make the signal from the two antennas usable by the mobile receiver, the combining of the independently fading signal must be made controllable by the receiver.

---

[2]Clearly, employing antenna diversity on a mobile is a lot more difficult: antennas must normally be spaced at least half a wavelength apart to get significant diversity gains, this translates to about 7.5 cm.

Just transmitting the same signal on both antennas would result in more fading at the mobile receiver.

There are two downlink transmit diversity types defined in the 3GPP specifications: open loop transmit diversity, where the Node B employs space time coding across the antennas, and closed loop transmit diversity, which is applicable for dedicated channels only and which is based on a specific mobile sending feedback information in the uplink to instruct the set of Node Bs to adjust their downlink antenna usage.

Both modes are based on the mobile's ability to estimate the propagation channels from each antenna independently for a specific path delay. See Sections 3.3.1 and 3.4.3 for details on transmit diversity handling for the CPICH.

3.8.2 Open Loop Transmit Diversity

Space Time Transmit Diversity (STTD) is a space time coding scheme that employs the encoder shown in Figure 3.35(a) at the base station. It is an open loop transmit diversity scheme, operating without a feedback channel from the downlink receiver to the base station. Pairs of symbols (of two bits each) are encoded differently for the two antennas prior to transmission at the base station. This implies that the received symbols need to be decoded at the mobile's receiver prior to combining. A schematic of an STTD decoder is shown in Figure 3.35(b). This decoder takes the place of the derotation of finger output symbols in the combiner (as shown in Figure 3.28) and operates on pairs of input symbols.

Because the STTD coding scheme does not require a feedback channel, it can be applied for all data bearing downlink physical channels including the common physical channels P-CCPCH and S-CCPCH. The use of STTD for the different channels is broadcast as part of the cell information on the BCH. To decode the P-CCPCH itself, its STTD status must be known; this is signalled by the relative phase of the synchronization channels (P-SCH and S-SCH) to the cell's primary CPICH (see [49]). For a dedicated channel, the STTD

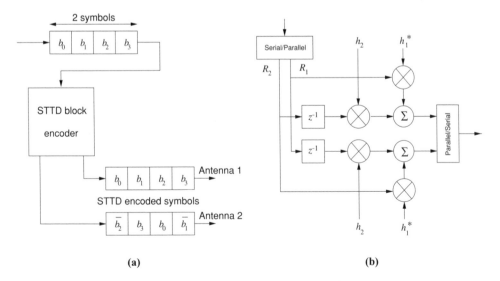

(a)   (b)

Figure 3.35. Schematic diagrams of (a) STTD encoder and (b) decoder.

status is signalled to the mobile at channel setup for each cell in the active set. Some of the cells serving the DPCH in a soft handover situation may employ STTD, while others may not use downlink transmit diversity.

We have seen that the STTD scheme operates on pairs of symbols. If a physical channel carries an odd number of symbols per radio frame, then different STTD rules are applied to make up for the missing symbol at the end of the frame; physical channels with odd numbers of symbols are the P-CCPCH (which carries nine symbols per slot, see Section 3.3.2) and channels with a spreading factor of 512. The rules for applying STTD encoding in these cases are given in the specifications [49]. Other exceptions to the STTD encoding scheme exist for the time multiplexed pilot symbols in certain channel configurations [49].

Support for STTD is mandatory for all mobile stations and does not introduce significant complexity into the downlink receiver. The main problem from an implementation point of view is the handling of concurrent STTD and nonSTTD encoded radio links in the combiner for a single soft handover DPCH. If STTD encoding is used, then a CPICH will be available as phase reference; this means that the encoded time multiplexed pilots are not necessarily used for channel estimation.

3.8.3 Closed Loop Transmit Diversity

In closed loop transmit diversity, the mobile terminal can steer the Node B's second antenna to optimize the signal it receives. The Node B applies complex weights for the DPCH (and PDSCH) on both antennas, as directed by mobile terminal feedback bits. There are two closed loop transmit diversity modes. In mode 1, the complex weight applied to the DPCH at antenna 2 is chosen from one of four phasors. In mode 2, there are eight phase settings selectable for the antenna 2 weight, and there is an additional amplitude weighting between the two antennas (see Figure 3.36).

Feedback generation at the mobile terminal involves selection of the antenna weights to maximize the signal strength at the mobile terminal. To determine the optimal weights at the Node B, the mobile terminal performs calculations to maximize the sum of pilot powers for all possible combinations of antenna weights. The mobile terminal then signals

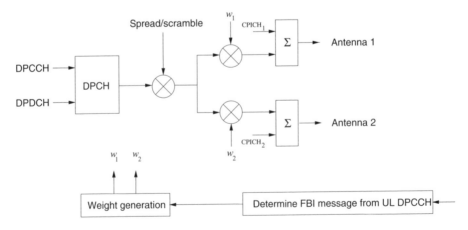

Figure 3.36. Closed loop transmit diversity.

the corresponding bits to the Node B on the uplink DPCCH. Calculation of the antenna weights is not very time critical and can either be done each signalling interval (one to four slots, depending on slot format and transmit diversity mode) or once per slot, based on antenna verification results (see below for antenna verification).

Closed loop transmit diversity adds complexity to the RAKE receiver. The RAKE receiver requires a coherent reference for each finger (see Section 3.4). For the DPCH in closed loop transmit diversity, the mobile terminal has a choice of coherent reference sources:

- time multiplexed pilot symbols embedded in the DPCCH;

- CPICH, combined using a priori knowledge of weights;

- CPICH, combined using a priori and antenna verification knowledge.

Use of the time multiplexed pilots requires no additional calculations in the RAKE and is performed as described in Section 3.4. However, the CPICH is not controlled by closed loop transmit diversity, so if the CPICH is used as the coherent reference for a DPCH that is using closed loop transmit diversity, then it must be adjusted to take into account the steering that has been applied to the DPCH. The CPICH coherent reference can be combined from diversity channel estimates and antenna weights for each finger, i.e. using a priori knowledge of the weights. The a priori weights used to correct the CPICH may not in fact be those requested by the mobile terminal; there may be transmission errors in the feedback bits transmitted by the mobile terminal and the feedback bits may be corrupted. Antenna verification compares the time multiplexed pilots with the rotated CPICH to determine what weights have in fact been applied by the Node B (see Figure 3.36). Since each Node B handles feedback independently, the calculations must be performed for each radio link separately. The CPICH phase reference is obtained using these weights, rather than the weights that the mobile terminal requested. Antenna verification is time critical, and because of the complexity of the hypothesis checking for mode 2, antenna verification is only often used for closed loop transmit diversity mode 1.

3.9 PHYSICAL LAYER PROCEDURES

3.9.1 RACH Procedure

The RACH procedure is a contention based procedure initiated by the mobile terminal in the uplink. It is fully defined in the 3GPP specifications ([51]).

The RACH procedure is controlled by a set of cell specific parameters, which the mobile terminal receives from the cell's broadcast information and which are passed to L1 from higher layers: the preamble scrambling code, the message length, information on the available signatures and subchannels, the preamble power step, the number of preamble tries, the initial preamble power, the offset between the last preamble power and the message power, transport format parameters, the access service class of the transmission and the data to be transmitted. The mobile terminal starts the RACH procedure by sending a preamble. There is collision detection at the network and the network responds to the preamble on the AICH with either a negative or positive acknowledgement. If there is a collision, then the network sends a negative acknowledgement and the mobile terminal tries again at a later time. A positive acknowledgement means that the mobile terminal should send its

10 or 20 ms data packet. The mobile terminal sends the data packet at a defined time after receiving the positive acknowledgement. If no acknowledgement is received, the mobile terminal increases the preamble power and tries again until it receives an acknowledgement or until it has reached the maximum allowed number of tries.

The random element of the procedure is in choosing the access slot and signature, which are chosen at random from the set of available access slots and signatures for the given access service class.

The AICH is not straightforward for the mobile terminal to detect because it is not a biphase signal. To make it easier, the power ratio between the CPICH and the AICH at the Node B is known and the mobile terminal uses thresholding to detect the presence/absence of the AICH.

3.9.2 CPCH Procedure

While the RACH can only carry short packets of 10 or 20 ms air time, the CPCH is a channel that allows for longer packet transmission from the mobile terminal to the network. The CPCH procedure is defined in the 3GPP specifications ([51]).

The access part of this procedure is similar to that of the RACH but with a second collision detection and preamble stage. The CPCH supports closed loop power control, and during its setup phase there is a power control preamble that aids fast convergence of the power control loops. To close the power control loop, there is a special instance of the downlink DPCCH, called 'DPCCH for CPCH' which is an SF 512 channel carrying the power control bits and additional CPCH control command bits that are used to signal an emergency transmission stop command to the mobile.

The CPCH procedure uses several downlink physical signalling channels in addition to the CPCH: the CSICH carries information on the availability of each PCPCH or, if *Channel Assignment* (CA) is active, on the maximum available data rate with the availability of each PCPCH. The AP-AICH is used to acknowledge the mobile terminal's *Access Preamble* (AP) and the CD/CA-ICH acknowledges reception of the mobile's *Collision Detection* (CD) preamble.

Before the CPCH procedure begins, higher layers set up the parameters L1 requires. Initially, the mobile terminal monitors the CSICH and determines which CPCHs will support the required transport format, or transport format and data rate if CA is active. The mobile terminal selects one CPCH at random and, providing the CSICH indicates that it is still available, transmits a CPCH-AP with a signature chosen at random from those available in an access slot. If no acknowledgement is received and the chosen CPCH is still available then the mobile terminal retransmits the AP with an increased power for a specified number of tries. If the mobile terminal detects a negative AP-AICH then it aborts the access attempt. If, however, the mobile terminal detects a positive AP-AICH then the contention resolution part of the procedure begins. The mobile terminal randomly selects a CD signature and access slot subchannel and transmits the CD preamble. If the mobile terminal does not receive the expected CD/CA-ICH then it aborts the procedure. If it does receive the expected CD/CA-ICH then transmission of the CPCH data packet begins.

The mobile terminal monitors the DL-DPCCH for CPCH during the first frames of the CPCH data packet transmission and if it does not detect a start of message indicator then it aborts the access attempt. After successful channel initialization, the mobile terminal monitors the DL-DPCCH control bits for an emergency stop command. Detection of an emergency stop will cause the uplink transmission to be aborted. During transmission of the CPCH packet data, the mobile terminal and UTRAN perform inner loop power control on the CPCH and the DL-DPCCH.

3.9.3 PICH/PCH Procedure

The PICH is a downlink physical layer signalling channel instructing the mobile terminal to receive an associated *Paging Channel* (PCH). Each mobile is allocated a paging indicator (PI) on a PICH, and the paging indicators are repeated in a defined pattern. This pattern is set up such that it allows the monitoring mobiles to turn off reception between paging indicators to preserve power.

The PICH is a simple channel to receive with no error correction coding. The number of paging indicators sent in a PICH frame is 18, 36, 72 or 144 and the mobile terminal listens to only one specific indicator. The formula that defines which indicator to listen to depends on the total number of indicators being sent on the PICH, on the SFN and on the index given to the mobile terminal. The PCH follows its associated PICH by three slots, thus giving the mobile terminal time to wake up processing resources if required.

3.9.4 DSCH Procedure

This is a downlink data only channel, similar to the uplink CPCH. It is used to support variable data rates.

The DSCH is not a standalone channel and is always associated with a DPCH. The TFCI of the associated DPCCH contains information on the SF, OVSF code and whether the DSCH is allocated to a specific mobile terminal. The DPCH frame with this information precedes the DSCH frame by between 3 and 18 slots and the DSCH and DPCH are not necessarily frame aligned. The delay between the DPCH and the DSCH gives the mobile terminal time to reconfigure and prepare to receive the DSCH. Power control and transmit diversity for the DPCH are also applied to the DSCH associated with that DPCH. This is awkward in closed loop transmit diversity since the weights change at different times for the DPCH and the DSCH.

The DSCH is allocated to a mobile terminal on a frame by frame basis and the network may allocate multiple users of DSCH by time multiplexing code tree resources (see Section 3.2).

3.10 MEASUREMENTS

3.10.1 Overview

In this section we will discuss the measurements that need to be reported by the physical layer to higher layers (see [53] and [52]). These need to be performed in addition to those required internally by Layer 1.

The set of measurements that are required by higher layers can be split into three groups:

- cell measurements;
- physical channel measurements;
- transport channel measurements.

Each of these groups represents a set of measurements that aid in some of the higher layers' functionality. We will discuss each group of measurements in turn.

3.10.2 Cell Measurements

Cell measurements are required by the RRC for cell selection and reselection procedures. This set of measurements gives the RRC (and the network) visibility of the mobile's radio neighbourhood. Cell measurement reports consist of power and timing measurements of neighbouring, as well as serving, cells.

3.10.2.1 *The Monitored Set*   The RRC controls cell measurements in Layer 1 by configuring a set of 'monitored' cells. This set is a list of cells for which cell measurements are explicitly requested by the RRC. It is the responsibility of Layer 1 to detect the timing and to measure the power of these cells. Obviously, the RRC can only add cells to the monitored set which it knows to be present: this information can either stem from earlier measurements or be signalled by UTRAN.

For each cell in the monitored set, the RRC signals to Layer 1 the available cell timing information as well as a measure of reliability for this timing information. This reliability measure determines the method by which Layer 1 can perform the measurements. If the cell timing is known to be accurate to a few chips, then the search window size of the multipath searcher resource may be big enough to estimate the cell's impulse response (see Section 3.4.6.1). If the accuracy of the timing information supplied by the RRC is too poor for the multipath searcher to reliably detect all significant paths within its search window, then the cell must be detected with a cell search (see Section 3.5).

Five different reliability classes for cell timings in the monitored set are defined:

- 1 chip: the cell timing is known perfectly. This is indicative that the cell has already been detected by Layer 1;

- 40 chips: the cell timing is supplied by UTRAN with timing accuracy of 40 chips;

- 256 chips: the cell timing is supplied by UTRAN with timing accuracy of 256 chips;

- 2560 chips: same as above, with an accuracy of 2560 chips;

- Timing unknown: a cell for which no information on the cell timing is provided by the UTRAN.

The cell measurements listed below generally translate to a multipath searcher run for Layer 1; the timing information is the time of the earliest propagation path from the cell in question, and the received CPICH code power is the power of the estimated impulse response (assuming the impulse response estimation is performed on the cell's primary CPICH as outlined in Section 3.4.6.1).

Clearly, the accuracy of the timing information for each cell will determine whether a multipath search can be run for a cell in the monitored set. If the timing accuracy is such that the impulse response is likely to be within the searcher's window, then a multipath search is all that is required for cell timing and power estimation. This is typically the case for the 1 and 40 chip accuracy classes, but depends on the window size of the multipath searcher.

For accuracy classes of 256 or 2560 chips, on the other hand, a multipath search with a typical search window is normally not sufficient. A cell search procedure, ideally taking the timing hints from UTRAN into account, must normally be scheduled to gain a time estimate that is accurate enough for a multipath searcher run. It must be noted, however, that the detection probability for a P-SCH/S-SCH cell search is much lower than that for a CPICH based multipath search, and that therefore the CPICH search should be used wherever possible.

In addition to measurements for cells in the monitored set, the RRC can request Layer 1 to report other, 'unlisted' cells that may be detected by the mobile. If this flag is set, then Layer 1 is expected to perform a cell search and to report any detected cells to the RRC.

3.10.2.2 *Physical Layer Cell Measurements* Cell measurements are performed to gather information for cell selection/reselection. For each cell in the monitored set, this entails measuring:

- the received power of the primary CPICH (CPICH_RSCP);

- the arrival time of the first propagation path received.

The timing information reported by the physical layer measurement does not directly map to any of the timing measurements reported by the RRC to the network: the SFN–SFN and SFN–CFN measurements defined in [53] require knowledge of the different cells' asynchronous SFN values; this is beyond the capability of the physical layer, and must be controlled by the RRC. The physical layer's measurement task is to measure the subframe arrival time, i.e. the time the frame boundary is received from a specific cell, to within the accuracy specified in [53], on the first detected propagation path. This can be estimated as the time of the first detected path in the impulse response measured with a multipath search on the cell in question; to employ this, the cell's timing must already be known to sufficient accuracy to position the path searcher's search window (see Section 3.4.6.1). Once the path search has been run on the cell, the received CPICH power (called CPICH_RSCP in [53]) can be estimated by adding the power of all the detected rays. If the cell's timing is not known to a sufficient degree, then a cell search must be performed beforehand to find the cell timing.

In addition to frame timing and CPICH power measurements for each cell, each cell measurements report also contains an RSSI measurement, used as a reference point in determining signal quality. The CPICH_Ec/Io measurement listed in [53] is simply CPICH_RSCP divided by the RSSI at the receiver input.

In summary, if the timing of the cell to be measured is known to within the path searcher's window, then the cell's timing and power measurements are simple to perform. The only implementation problems are that the power measurements must be corrected for the total gain of the receiver chain (which depends on the current AGC setting), and the scheduling of measurements within Layer 1.

For cells that are not part of the monitored set, the following cell parameters are reported by Layer 1 to the RRC in addition to the measurements discussed above:

- the primary scrambling code index for the cell;

- the STTD status of the BCH.

We have discussed the BCH STTD status in Section 3.5.

3.10.3 Radio Link Measurements

Radio Link (RL) measurements include timing information for all cells in the active set, uplink transmit power measurements, current uplink timing and the current downlink SIR estimate. This information is used at the network for handover control and DPCCH timing updates (see Chapter 10). RL measurements typically need no special measurement support within Layer 1, as all the information can be easily extracted from the receiver's RAKE control, the power control loop and the uplink timing control loop. RL measurements are only applicable for dedicated channels.

3.10.4 Transport Channel Measurements

Transport Channel (TrCH) measurements are specific to channels with closed loop power control (DCH and DPCCH for CPCH), and serve to report the TrCH quality to the RRC. For the DCH, TrCH measurements consist of a short term block error rate estimate (CRC failure rate) gathered from the BRP (see Chapter 4). For the DL DPCCH for CPCH, the DPCCH bit error rate is the measurement reported. This bit error rate can be measured on the known dedicated pilot bits, which introduces additional complexity in the receiver.

3.11 COMPRESSED MODE

3.11.1 Introduction

We have seen in Section 3.10 that cell measurements are fundamentally important for the operation of the RRC; in connected mode (while a downlink channel is being received), a single radio receiver can only detect and monitor cells on the same frequency as the serving cell. This means that such a receiver could not perform measurements on different downlink frequencies or RATs.

Compressed mode is used to support measurements that cannot be performed by a single radio receiver while a downlink channel is in progress: interfrequency measurements (cell measurements) and inter-RAT measurements (UMTS-TDD, GSM). Compressed mode allows the mobile terminal to take these measurements in gaps in the downlink DPCH. These gaps are scheduled by the network and their location is signalled to the mobile terminal. Compressed mode is only available on the DPCH, although FACH measurement occasions are the equivalent measurement windows for the S-CCPCH.

Note that it is the network's responsibility to allocate compressed mode gaps to support the mobile's measurements; each mobile signals to the network whether it requires compressed mode gaps for the different measurement classes (intrafrequency, interfrequency and the different inter-RAT measurements) as part of its capability message. This request for transmission gaps can be for the downlink, uplink or both. Uplink compressed mode allows the mobile to switch off transmission during the measurements, if transmitting the uplink would interfere with the measurements.

In this section, we will discuss the impact of compressed mode support on the mobile CRP; we will concentrate on the effects compressed mode handling has on the receiver design. The full details of compressed mode handling, including all the different recovery

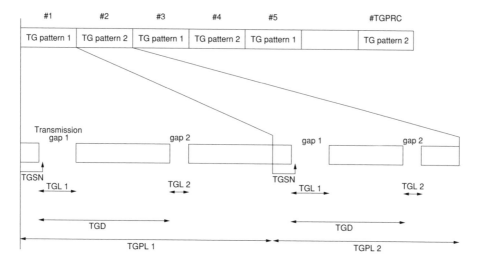

Figure 3.37. Transmission gap pattern sequence.

modes for transmit diversity and power control, are defined in the specifications [51]. Compressed mode support issues in the bit rate processing block are discussed in Chapter 4.

3.11.1.1 *Compressed Mode Configuration Parameters*   Compressed mode gap parameters are signalled to the mobile with a DCH configuration or reconfiguration; compressed mode gaps are set up in bursts:

- a transmission gap is a period of multiple slots length where the DL (or UL) DCH is compressed;

- a transmission gap pattern is a group of up to two compressed mode gaps in a repeatable pattern, and up to two patterns can be defined;

- a transmission gap pattern sequence is a sequence of transmission gap patterns.

The parameters used to describe compressed mode are illustrated in Figure 3.37 and described below:

Transmission Gap Starting Slot Number (TGSN). Slot number of the first transmission gap in the pattern counted from pattern start (frame boundary).

Transmission Gap Length 1 (TGL1). Duration of the first transmission gap in the pattern in slots.

Transmission Gap Length 2 (TGL2). Duration of the second transmission gap in the pattern in slots.

Transmission Gap Start Distance (TGD). Distance between the start of the first and the second transmission gap in slots.

Transmission Gap Pattern Length 1 (TGPL1). Length of the first transmission gap pattern in frames.

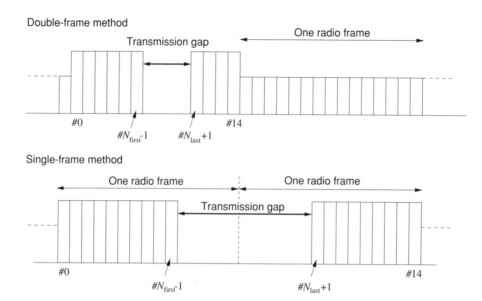

Figure 3.38. Positioning of compressed mode transmission gaps within radio frames.

Transmission Gap Pattern Length 2 (TGPL2). Length of the second transmission gap pattern in frames.

Transmission Gap Pattern Repetition Count (TGPRC). Number of transmission gap patterns to repeat in this sequence.

Transmission Gap Connection Frame Number (TGCFN). Start CFN of this transmission gap pattern sequence.

The maximum compressed mode gap length, TGL1 and TGL2, is seven slots, and the minimum gap length is three slots. Up to seven slots can be compressed in a single radio frame. However, the two gaps in the pattern can be lined up such that they span a frame boundary, giving a maximum total combined gap length of 14 slots. Figure 3.38 illustrates the positioning of compressed mode transmission gaps within a frame and spanning two consecutive frames.

In addition to the parameters describing the exact position of transmission gaps, the additional parameters supplied to the physical layer during compressed mode setup are:

- UL, DL, UL + DL compressed mode selection, i.e. flags to indicate whether the transmission gap pattern is for uplink only, downlink only, or uplink and downlink;

- Downlink compressed mode method;

- Uplink compressed mode method;

- Downlink frame type 'A' or 'B';

- Alternative scrambling code mode for compressed mode by *Spreading Factor Reduction* (SFR);

- RPP (recovery period power control) mode flag;

- ITP (initial transmit power control) mode flag.

These parameters are listed here for completeness, they are discussed in more detail below.

3.11.1.2 *Compressed Mode and the Physical Layer*   We have seen that the *raison d'être* for compressed mode is to supply time windows during which the mobile can perform measurements on frequencies or RATs other than the serving frequency during the lifetime of a dedicated channel. Compressed mode support in the physical layer is designed to handle these gaps in downlink transmission gracefully, minimizing their impact on both the existing dedicated link and the total system capacity.

Clearly, transmission gaps interrupt the continuous reception in downlink and uplink underlying the dedicated channel operation, which reduces the number of chips actually transmitted for some radio frames, and which interrupts the fast closed control loops in operation on a DPCH: power control (see Section 3.6), closed loop transmit diversity (see Section 3.8) and SSDT (see Section 3.7.4). The interruption of the control loops is handled with special recovery procedures, discussed in more detail below.

To cope with the reduced system throughput (in chips) during frames affected by compressed mode transmission gaps, there are two strategies:

- reduce the number of PhCH bits to fit the available PhCH bandwidth. Two methods for this are defined: 'compressed mode by puncturing' and 'compressed mode by higher layer scheduling';

- increase the throughput in the leftover slots in a frame with a transmission gap by reducing the spreading factor. This is called 'compressed mode by spreading factor reduction'.

The choice of the compressed mode scheme is done by the network and signalled to the mobile station.

Whichever of these methods is chosen by the network, frames incorporating transmission gaps are transmitted using a different slot format from noncompressed frames, to keep up the number of DPCCH bits even in the worst case transmission gaps. The different slot format variants will be discussed below.

3.11.2 Effects on CRP: an Overview

Compressed mode affects the CRP in the RAKE receiver and in the closed loop mode handling (power control, transmit diversity, SSDT). In the RAKE, both the fingers and the combiner need to handle the change in slot format when a compressed frame begins and ends, and there is the option of stopping the fingers and combiner in the transmission gaps.

3.11.3 Compressed Mode Support in RAKE Fingers

The DPCH slot format changes at the compressed frame boundary to use slot format variant 'A' or 'B' from the slot formats defined in [49] depending on whether compressed mode by

Figure 3.39. Compressed frame type 'A'.

Figure 3.40. Compressed frame type 'B'.

puncturing or higher layer scheduling (slot format variant 'A') is being used, or compressed mode by spreading factor reduction (slot format variant 'B'). This means that the fingers must be programmable with the new slot format at the correct time. The fingers must also be stopped at the start of the gap and restarted at the end of the gap.

Where exactly the fingers are stopped and started depends on the frame type being used. Although transmission gaps nominally start and end at slot boundaries, there are two variants: frame type 'A' and frame type 'B'. In frame type A, shown in Figure 3.39, the pilot bits of the last compressed slot are transmitted to allow use of time multiplexed pilots for demodulation of the following slot, to speed up power control reconvergence and to aid recovery of the closed loop modes.

In frame type B, shown in Figure 3.40, the pilot bits of the last compressed slot are transmitted (as in frame type A), and the TPC bits of the first compressed slot are transmitted to reduce the effects of compressed mode on the UL power control.

Note that slot format variants A and B are independent from the two compressed frame types, labelled A and B.

3.11.3.1 Slot Format A: Maintaining the Spreading Factor

Table 3.4 shows part of the DCH slot format table from [49] and lists some of the slot formats with the corresponding variants A and B for compressed mode. The first column in Table 3.4 lists the slot format number followed by the channel bit rate R_{ch}, the channel symbol rate R_s, the spreading factor and the number of bits per slot. Columns N_{Data1} and N_{Data2} represent the bits per slot of the DPDCH, and N_{TPC}, N_{TFCI} and N_{Pilot} the bits per slot of the DPCCH. The last column, N_{Tr}, denotes the transmitted slots per radio frame.

In slot format variant A, the spreading factor is preserved, and therefore the number of physical channel bits in the compressed frames is reduced; the number of physical channel bits lost depends on the length of the transmission gap in the frame. In order to preserve the number of TFCI bits in the radio frame, the slot format variant A contains a greater number of TFCI bits per transmitted slot than in noncompressed frames. This increase in TFCI bits per slot is at the expense of the data bit field length.

Because the number of data bits transmitted in compressed frames is reduced, the BRP must adjust the physical channel data streams to fit the reduced physical channel throughput. The compressed mode method employed in this case is either compressed mode by puncturing or by higher layer scheduling (both of these are discussed in Chapter 4). The choice of method makes no difference to the receiver's chip rate processing, but increased puncturing reduces the performance of the receiver's bit rate processing. To help counteract

Table 3.4   Subset of slot formats for DCH according to [49].

#i	R_{ch} #i (kbps)	R_s #i (ksps)	SF	Bits/ slot	N_{Data1}	N_{Data2}	N_{TPC}	N_{TFCI}	N_{Pilot}	N_{Tr}
0	15	7.5	512	10	0	4	2	0	4	15
0A	15	7.5	512	10	0	4	2	0	4	8–14
0B	30	15	256	20	0	8	4	0	8	8–14
1	15	7.5	512	10	0	2	2	2	4	15
1B	30	15	256	20	0	4	4	4	8	8–14
2	30	15	256	20	2	14	2	0	2	15
2A	30	15	256	20	2	14	2	0	2	8–14
2B	60	30	128	40	4	28	4	0	4	8–14
3	30	15	256	20	2	12	2	2	2	15
3A	30	15	256	20	2	10	2	4	2	8–14
3B	60	30	128	40	4	24	4	4	4	8–14

this, the transmit power and the corresponding power control target (see Section 3.11.5 below) are adjusted during compressed radio frames.

3.11.3.2 Slot Format B: Compressed Mode by Spreading Factor Reduction

In slot format B, the dedicated channel's spreading factor is halved, and the number of bits in all fields per slot is doubled, this method is called compressed mode by *Spreading Factor Reduction* (SFR). Compressed mode by SFR is not possible if the spreading factor in noncompressed frames is four. The total number of data bits per frame is increased, as less than seven slots are compressed, but all additional data bits are filled with DTX bits (see Chapter 4).

Halving the spreading factor guarantees constant physical channel throughput in compressed and noncompressed frames, but also reduces the receiver's noise rejection by 3 dB. To counteract this loss of spreading gain the transmit power is adjusted. When compressed mode by SFR is used, the fingers must be reconfigured to change spreading factors at frame boundaries, and the change in SF means that the OVSF code must also change.

3.11.3.3 Scrambling Codes for SFR

Halving the spreading factor during radio frames containing transmission gaps means that the channel will occupy twice the amount of code resource (see Section 3.2), which might lead to clashes with other channels in the downlink. This is avoided by the network signalling whether the same scrambling code, or an alternative scrambling code, is to be used for frames in compressed mode by SFR.

If the mobile terminal is instructed not to change scrambling code for compressed frames then the new OVSF code is $C_{ch,SF/2,\lfloor n/2 \rfloor}$, assuming an OVSF code of $C_{ch,SF,n}$ in noncompressed frames. The new OVSF code must not clash with any other user or common channels, and this is the responsibility of the network. In soft handover situations, the new OVSF codes are all calculated following the same rule.

In the case where the mobile terminal is instructed to use the alternative scrambling code, two alternative scrambling codes are associated with each ordinary scrambling code k: the left alternative scrambling code ID ($k + 8192$) and the right alternative scrambling code ID ($k + 16384$). The OVSF code $C_{\text{ch,SF},n}$ used for noncompressed frames determines the compressed frame codes. If $n < \text{SF}/2$ then the left alternative scrambling code is used, otherwise the right alternative scrambling code is used. In any case, OVSF code $C_{\text{ch,SF}/2,n\%\text{SF}/2}$ is used.

3.11.3.4 Regaining the Channel Estimate

On restarting a finger after a compressed mode gap the channel estimate will be poor and potentially out of date from the last received slot. There is very little time to fill the channel estimation filter with up-to-date data and there will be a performance loss during the recovery phase while this happens. This effect can be minimized by adjusting the channel estimation filter parameters appropriately during compressed mode recovery.

3.11.4 Effects of Compressed Mode on the Combiner

The slot format change at the compressed frame boundary also affects the combiner. The combiner must be able to change slot format and spreading factor at the start of the compressed frame. In addition, the combiner may have to deal with invalid data symbols output from the fingers. In the case where closed loop transmit diversity is being used in conjunction with compressed mode, there are recovery mode changes to the antenna weights.

3.11.5 Power Control in Compressed Mode

The changes in power control operation in compressed mode aim to conserve the performance of compressed frames as well as to enable fast reconvergence of the power control loop interrupted by transmission gaps in the uplink and downlink. The parameters that may change, depending on the compressed mode parameters, are:

- DL power control SIR target;
- DL power control step size;
- UL power control step size;
- UL power control target.

3.11.5.1 Uplink Power Control in Compressed Mode

The uplink power control target (at the Node B receiver) is increased in frames containing uplink transmission gaps:

$$\text{SIR}_{\text{cm\_target}} = \text{SIRtarget} + \Delta\text{SIR}_{\text{PILOT}} + \Delta\text{SIR1\_coding} + \Delta\text{SIR2\_coding}$$

The SIR target correction factors $\Delta \text{SIR}\{1,2\}\_$coding apply corrections for compressed mode frames: $\Delta \text{SIR1\_coding} = \text{DeltaSIR1}$ if the first transmission gap in the current pattern starts in the current frame, or DeltaSIR1After if the first transmission gap in the current pattern started in the last frame; the analogous rules apply for DeltaSIR2 and DeltaSIR2After. The factor $\Delta \text{SIR}_{\text{PILOT}}$ corrects for a change in the number of pilot bits:

$$\Delta\text{SIR}_{\text{PILOT}} = 10 \cdot \log_{10}(N_{\text{pilot},N}/N_{\text{pilot,curr\_frame}})$$

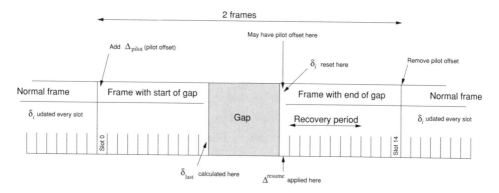

Figure 3.41. Compressed mode power control recovery.

where $N_{\text{pilot},N}$ is the number of pilot bits in an uncompressed frame, and $N_{\text{pilot,curr\_frame}}$ is the number of pilot bits in the current frame.

In parallel to the change in the uplink power control target, the mobile will adjust its transmit power each slot by Δ DPCCH:

$$\Delta \text{DPCCH} = \Delta \text{TPC} \cdot \text{TPC\_cmd} + \Delta_{\text{PILOT}}$$

where

$$\Delta_{\text{PILOT}} = 10 \cdot \log_{10}(N_{\text{pilot,prev}}/N_{\text{pilot,curr}})$$

$N_{\text{pilot,prev}}$ and $N_{\text{pilot,curr}}$ are the number of pilot bits in the previous and current uplink slot, respectively. This correction factor increases the uplink power in compressed frames, as the number of pilot bits is decreased for compressed frame slot formats (see [49] for details).

After the transmission gap, the power control recovery period begins, see Figure 3.41. Two compressed mode setup parameters influence the recovery period power control behaviour: the *initial transmit power* mode (ITP) setting determines the power step size in the first slot after a transmission gap in uplink or downlink (see [51]), and the *Recovery Period Power* (RPP) control mode parameter determines behaviour during the recovery period of length RPL:

- if RPP $= 0$, then step size is the usual Δ_{TPC};

- if RPP $= 1$, then step size is $\Delta_{\text{RP-TPC}}$.

The calculation of $\Delta_{\text{RP-TPC}}$ is defined in [51]. The length of the recovery period, RPL, is defined as RPL $= \min(\text{seven slots, gap length})$.

3.11.5.2 *Downlink Power Control in Compressed Mode*   For frames containing transmission gaps in the downlink, the SIR target and the downlink transmit power are adjusted in a similar manner to that in the uplink:

$$
\begin{aligned}
\text{SIRcm\_target} = \text{SIRtarget} + \\
\max(\Delta\text{SIR1\_compression}, \ldots, \Delta\text{SIR}n\_\text{compression}) + \\
\Delta\text{SIR1\_coding} + \Delta\text{SIR2\_coding}
\end{aligned}
\tag{3.8}
$$

The SIR target correction factors Δ SIR[1,2]_coding take into account compressed mode: Δ SIR1_coding = DeltaSIR1 if the first transmission gap in the current pattern starts in the current frame, and Δ SIR1_coding = DeltaSIR1After if the first transmission gap in the current pattern started in the last frame. The same applies for Δ SIR2_coding, DeltaSIR2 and DeltaSIR2After.

The Δ SIRn_compression (where n are all TTI lengths in the downlink) are defined as:

$$\Delta Pi\_compression = \begin{cases} 3 \text{ dB} & \text{for compressed mode by SFR} \\ 10 \cdot \log_{10}\left(15F_i/(15F_i - TGL_i)\right) & \text{for CM by puncturing} \\ 0 \text{ dB} & \text{for noncompressed frames} \end{cases}$$

The Node B transmitter automatically applies this power step at the frame start, in parallel to the mobile terminal increasing its power control target.

3.11.6 Transmit Diversity in Compressed Mode

Compressed mode transmission gaps interrupt the closed loop transmit diversity. Recovery modes are defined to enable fast reconvergence of the control loops after the transmission gaps. The handling of transmit diversity is defined in the 3GPP specification [51].

3.11.6.1 Transmit Diversity Mode 1   In downlink only compressed mode, the mobile continues to send feedback commands to the base station during the downlink transmit gaps. These feedback bits are based on the last available channel estimation values before the transmission gap started. After the end of the downlink transmission gap, once new channel estimation values are available, the operation resumes as normal, such that the mobile reselects the most appropriate antenna weight and transmits the appropriate feedback bits.

In uplink compressed mode, the base station freezes the current antenna weights until new feedback bits arrive in the uplink after the transmission gap. If closed loop transmit diversity initialization is to be achieved during downlink compressed mode, then the mobile will send '0' bits while no measurements are available, and go through the normal initialization routine after the gap.

3.11.6.2 Transmit Diversity Mode 2   In downlink only compressed mode the base station sets both antenna powers to 0.5, and uses the last antenna phase information before the gap. During recovery mode after the transmission gap, if the first feedback bit after the gap is not on a *Feedback* (FB) word boundary (of length four bits), then the mobile will send the most significant bit of the phase word until the next FB word boundary is reached. In this context, 0 means antiphase, and 1 means in-phase.

In uplink only compressed mode, the base station maintains the antenna weights until new feedback bits are received. These feedback bits are combined with bits received before the gap to generate new weight vectors, the uplink updates whichever feedback bit is due to be transmitted in the current uplink slot, and this bit overwrites the local copy at the base station.

3.11.7   Measurements in Compressed Mode

The measurements performed in compressed mode transmission gaps are those that cannot be done while an active downlink or uplink channel is running: interfrequency WCDMA measurements and interRAT measurements (see [52]).

To make the best use of the transmission gaps, the scheduler must decide which interfrequency and interRAT measurements to do in which transmission gap and control the measurement resources efficiently. It must control carrier frequency changes for interfrequency FDD measurements and manage multimode functionality for interRAT measurements.

Intrafrequency measurements (on the serving frequency), including the channel estimation for RAKE control, must be scheduled around the downlink transmission gaps so the measurements scheduler must be aware of pending transmission gaps. This introduces strong coupling between RAKE control, intrafrequency measurements management and interfrequency/inter-RAT measurement management.

4

Physical Layer Bit Rate Processing

Jason Woodard, Chris Hayler and Rudolf Tanner

4.1 INTRODUCTION

In this chapter we describe the so called 'Bit Rate' Processing (BRP) used in the FDD mode of UMTS. By 'bit rate' we mean operations that happen at, or close to, the rate of the data source being transmitted/received, as opposed to the 'chip rate' operations that happen at, or close to, the transmission rate of 3.84 million chips per second used in UMTS.

In the transmitter these bit rate operations, specified in [50], happen prior to the chip rate operations described in Chapter 3. They take the 'raw' data from *Medium Access Control* (MAC) (see Chapter 6) and encode and interleave this data so that the inevitable errors produced by its transmission between the UE and the Node B can (at least in the majority of cases) be corrected or detected. Typically a single user will use several transmission channels simultaneously. These different channels will usually have different *Quality of Service* (QoS) requirements. By QoS we mean the delay and error rate a service can tolerate. The bit rate processing section of the physical layer is responsible for multiplexing these different channels together, and for matching their different QoS requirements. Finally, it matches the extremely flexible raw data rates offered to the various data sources, to the much less flexible data rates that are available at the interface to the chip rate processing.

We begin, in Section 4.2, by describing the interface offered by the physical layer to MAC and the RRC, and introducing the concepts of *Transport Channels* (TrCHs), *Coded Composite Transport Channels* (CCTrCHs), *Transport Formats* (TFs) and *Transport Format Combinations* (TFCs). We then give an overview of the chain of bit rate operations used

---

WCDMA – Requirements and Practical Design. Edited by R. Tanner and J. Woodard.

in both the uplink and downlink in Section 4.3, and describe the error detection and inter-leaving operations used. Sections 4.5 and 4.6 describe the convolutional and turbo error correction schemes used in UMTS, the theory behind these codes, how to decode them and how well they perform. Then in Section 4.7, we describe how the receiver is able to decide which TFC has been sent, using either blind TF detection or TFCI decoding. Section 4.8 describes compressed mode and its effect on bit rate processing. Finally, we describe the BRP limitations on different TrCHs and CCTrCHs, before we conclude in Section 4.10.

Before proceeding, we give a list of symbols frequently used in this chapter. These symbols and their meanings are all introduced later – they are listed here together as a convenient reference point.

4.1.1 Frequently Used Symbols

The following symbols are frequently used throughout this chapter, with the meanings given below:

i transport channel label.

l transport format label.

j transport format combination label.

A_i number of data bits per transport block for TrCH$_i$.

L_i number of CRC bits per transport block for TrCH$_i$.

C_i number of code blocks in one TTI for TrCH$_i$.

K_i number of bits per code block for TrCH$_i$.

Z maximum code block size. $Z = 504$ for convolutional coding and $Z = 5114$ for turbo coding.

Y_i number of coded bits per code block for TrCH$_i$.

u_k encoder input bit.

$\hat{u}_k$ decoded bit corresponding to encoder input bit u_k.

y_{kl} encoder output bits ($l = 1, 2$ or $l = 1, 2, 3$) corresponding to input bit u_k.

$\hat{y}_{kl}$ received version of encoder output bit y_{kl}.

4.2 TRANSPORT CHANNELS, FORMATS AND COMBINATIONS

In this section we describe the interface between the physical layer and the higher layers in the protocol stack, and define the terms used in this chapter to detail the operation of the bit rate processing section of the physical layer.

The physical layer, or Layer 1 (L1), interfaces with both Layer 2 and Layer 3, as shown in Figure 4.1. The medium access controller (MAC) in Layer 2 passes information to L1 for transmission across the air interface, and similarly receives decoded information, on

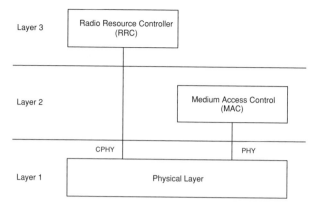

Figure 4.1. Interface between Layer 1 and Layers 2 and 3.

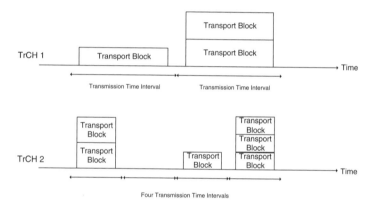

Figure 4.2. Exchange of data between the MAC and L1 for one CCTrCH.

the *PHYsical layer* (PHY) interface. Also, information is passed to Layer 2 from Layer 1 about the status of L1. The *Radio Resource Controller* (RRC) passes information to L1 that controls the configuration of the physical layer, and receives measurement information from L1, on the *PHYsical layer Control* (CPHY) interface.

Layer 1 offers several *transport channels* to the MAC in Layer 2. These transport channels are used to transfer data from the mobile to the Node B, or vice versa. This exchange of data can be modelled as shown in Figure 4.2. For each transport channel one or several *transport blocks* are passed per *Time Transmission Interval* between the MAC and Layer 1. The terms above in *italics*, and other related terms, are defined below.

Transport CHannel (TrCH). The channels offered by the physical layer to Layer 2 for data transport. Different types of transport channels are defined by how and with which characteristics data is transferred on the physical layer. So, for example, Figure 4.2 shows two TrCHs carrying data. The MAC uses these TrCHs to transport different logical channels, one or more logical channel can be mapped to a given TrCH by the MAC. See Chapter 6 for details of logical channels and this mapping.

Transport Block. The basic unit of data passed down to L1 from the MAC, for L1 processing. It consists of a group of bits to be transferred by Layer 1. Each transport block can be given its own *Cyclic Redundancy Check* (CRC) for error detection (see Section 4.3.3) by Layer 1.

Transport Block Size. The number of bits in a transport block. This parameter is extremely flexible and can take any integer value from 0 up to some maximum (which depends on the UE's capability). In Figure 4.2 this transport block size is represented by the size of the block drawn for each transport block.

Transport Block Set. A set of transport blocks that is passed to L1 from the MAC at the same time instance using the same transport channel. All the transport blocks within a given transport block set will be the same size. Figure 4.2 shows varying numbers of transport blocks being carried by the two TrCHs at different times. At any time it can be seen that all the transport blocks on a given TrCH are the same size, but from one time instant to another the size of the transport blocks on a given TrCH may vary.

Transport Block Set Size. The total number of bits in a transport block set, i.e. the number of transport blocks in the set multiplied by the transport block size.

Transmission Time Interval (TTI). The time it takes to transmit a transport block set. The TTI will be 10, 20, 40 or 80 ms (i.e. one, two, four, or eight radio frames), and as shown in Figure 4.2 different transport channels may have different TTIs.

Transport Format (TF). A format applied by L1 to a transport block set on a given transport channel for a given TTI. This 'format' controls how much data is transferred on the TrCH in that TTI, and how that data is coded etc. by the physical layer. A TF is composed of two parts, one dynamic and one semi-static. The semi-static attributes of a TF change relatively slowly (if at all) on instructions from the RRC. These attributes are:

- the TTI for the transport channel;
- details of the error detection scheme (CRCs of 0, 8, 12, 16 or 24 bits, see Section 4.3.3.);
- details of the error correction scheme (half or one third rate convolutional coding or turbo coding, see Section 4.3.4.);
- rate matching details, see Section 4.4.

The dynamic attributes of a TF may change from one TTI to the next, and at the transmitter are signalled to L1 from the MAC with each transport block set. These parameters are:

- the transport block size;
- the transport block set size.

Note that, as all the transport blocks in a given set must be the same size, we can easily infer the number of transport blocks within a set from the block size and the block set size.

Transport Format Set (TFS). A set of transport formats that can be used by a given TrCH. For example, a TrCH used to carry a variable rate data channel will have a TFS with one TF for each rate.

Coded Composite Transport Channel (CCTrCH). A data stream given by encoding and multiplexing one or several transport channels together to be transmitted together on one, or a set of, *Physical Channels* (PhCHs). Transport channels are multiplexed together into CCTrCHs like this so that the physical resource can be used as efficiently and flexibly as possible. Note, however, that there are restrictions on the types of TrCH that can be multiplexed onto a single CCTrCH. These restrictions are detailed in Section 4.9. Figure 4.2 shows the flow of data for a particular CCTrCH with two TrCHs.

Transport Format Combination (TFC). A combination of TFs, specifying the TF used on each TrCH multiplexed onto a CCTrCH.

Transport Format Combination Set (TFCS). The set of TFCs possible on a CCTrCH. Note that although each transport channel has a transport format set, not all combinations of different transport formats on different transport channels are necessarily allowed. The TFCS gives the permissible set of combinations of transport formats on different transport channels that can be used on a particular CCTrCH.

Transport Format Combination Indicator (TFCI). A representation of the current transport format combination used on the CCTrCH, that may be used at the receiver to determine which TFC was transmitted. See Section 4.7.1. For the *Downlink Shared Channel* (DSCH) the TFCI can also be used to represent which OVSF code is used. See Chapter 3 for details of OVSF codes.

These definitions and abbreviations are used throughout this chapter. To help clarify the meaning of these terms we now give an example of a transport format combination set.

4.2.1 Example TFCS

Tables 4.1 and 4.2 give an example of a simple downlink CCTrCH with two TrCHs. Each TrCH has a transport format set (TFS) shown in Table 4.1, and the CCTrCH has a set of allowed combinations of these TFs. This transport format combination set (TFCS) is shown

Table 4.1 Examples of transport format sets (TFSs) associated with two TrCHs.

Transport format	Dynamic part (Num Bits, Num TrBks)	Semi-static part
TrCH$_1$		
TF$_{1,1}$	40, 0	TTI = 40 ms
TF$_{1,2}$	40, 1	1/2 conv coding
TF$_{1,3}$	40, 4	12 bit CRC
		$RM_1 = 100$
TrCH$_2$		
TF$_{2,1}$	0, 1	TTI = 10 ms
TF$_{2,2}$	150, 1	1/3 conv coding
		24 bit CRC
		$RM_2 = 120$

Table 4.2 Example of a transport format combination set (TFCS) associated with a CCTrCH.

Transport format combination (TFC)	TFC indicator (TFCI)	TF for TrCH$_1$	TF for TrCH$_2$
TFC$_1$	0	TF$_{1,1}$	TF$_{2,1}$
TFC$_2$	1	TF$_{1,2}$	TF$_{2,1}$
TFC$_3$	2	TF$_{1,3}$	TF$_{2,1}$
TFC$_4$	3	TF$_{1,1}$	TF$_{2,2}$
TFC$_5$	4	TF$_{1,2}$	TF$_{2,2}$

in Table 4.2. Through a description of these tables, and what the parameters in them mean, we hope to clarify the concepts introduced above.

Table 4.1 details the TFS associated with each TrCH in our example. The first TrCH, denoted TrCH$_1$, has three TFs allowing three data rates. All three TFs carry transport blocks of 40 bits each, so the transport block size is 40 bits for all TFs. However, the first TF carries no transport blocks, so no data bits will be transmitted when this TF is selected. The second and third TFs carry one or four transport blocks, giving either 40 or 160 transmitted data bits. This gives the MAC three bit rates to choose from for this TrCH: off (TF$_{1,1}$), low rate (TF$_{1,2}$) or high rate (TF$_{1,3}$). For any of these rates, bits on this TrCH are transmitted over a 40 ms interval using half rate convolutional coding and a 12 bit CRC for error detection. Note that, as described in Section 4.3.3, one CRC is attached to each transport block. So for TF$_{1,3}$ a total of four CRCs, of 12 bits each, will be transmitted.

The second TrCH has only two TFs in its TFS. The first TF gives one transport block of zero bits. Like TF$_{1,1}$ for the first transport channel, this TF gives no transmitted data bits. However, as will be explained in Section 4.3.3, there is a difference between sending no transport blocks (as in TF$_{1,1}$), and sending one or more transport blocks of zero size (as in TF$_{2,1}$). The second TF of this TrCH (TF$_{2,2}$) carries one transport block of 150 bits. So for this TrCH we have two rates to choose from, on (TF$_{2,2}$) or off (TF$_{2,1}$). For either of these rates the bits are transmitted over 10 ms using one third rate convolutional coding and a 24 bit CRC.

Notice that the highest rate TF of TrCH$_1$ carries 160 bits, whereas the highest rate of TrCH$_2$ carries 150 bits. So it might be thought that TrCH$_1$ has a slightly higher maximum bit rate than TrCH$_2$. However, this ignores the different TTIs of the two TrCHs. TrCH$_1$ has a maximum rate of 160 bits every 40 ms TTI. So its maximum bit rate is given by 160/40 ms = 4 kbits/s. TrCH$_2$ can carry up to 150 bits every 10 ms, so its maximum bit rate is 150/10 ms = 15 kbits/s.

Another difference to note between the two TrCHs in our example is the channel coding used. TrCH$_1$ uses half rate convolutional coding, whereas TrCH$_2$ uses one third rate convolutional coding. As will be seen in Sections 4.3.4 and 4.5 this means the bits on TrCH$_2$ are given more protection against errors than those on TrCH$_1$. Similarly TrCH$_1$ uses a 12 bit CRC, whereas TrCH$_2$ uses a 24 bit CRC. So, as described in Section 4.3.3, we will be much more likely to detect any transport blocks containing errors on TrCH$_2$ than on TrCH$_1$.

Finally, note the parameters RM_1 and RM_2 in Table 4.1. These give what is called the rate matching attribute for each TrCH, and can take any integer value from 1 to 256. As described in Section 4.4 this parameter controls the relative amount of puncturing or repetition that is carried out on the different TrCHs in a CCTrCH. A higher value for RM means a TrCH will have more repetition of its bits, or less puncturing, when rate matching is carried out. So the coded bits of TrCH$_2$ will have a higher repetition ratio, or a lower puncturing ratio, than those of TrCH$_1$. It is even possible to choose values of RM such that the coded bits from one TrCH are repeated while those from another TrCH are punctured. Whereas the choice of channel coding gives a coarse means of controlling how much protection the bits of a TrCH are offered, the choice of the rate matching attribute RM gives a much finer means of controlling the different protection given to different TrCHs.

It should now be clear that, in our example, TrCH$_2$ has a much higher 'Quality of Service' than TrCH$_1$. Compared to TrCH$_1$ it offers a higher maximum data rate (15 kbits/s rather than 4 kbits/s), a lower delay (10 ms rather than 40 ms), better protection against errors (one third rather than half rate coding and a higher rate matching attribute), and better detection of errors (24 rather than 12 bit CRCs).

Table 4.2 shows the transport format combination set (TFCS) for our example CCTrCH. Each TFC gives a combination of TFs for the two TrCHs on the CCTrCH. Only combinations of TFs from within this set can be transmitted on the CCTrCH. With three TFs for TrCH$_1$ and two TFs for TrCH$_2$ there are a total of $3 \times 2 = 6$ possible unique combinations of these TFs. However, not all combinations are necessarily included in the TFCS; in our example the two TrCHs cannot both simultaneously use their maximum rate TFs.

As well as a TF for each TrCH, each TFC has an indicator, the transport format combination indicator (TFCI), that may be transmitted along with the coded data to allow the receiver to know which TFC was sent (see Section 4.7).

At this point it is useful to recap on some symbols that are used in this chapter and in the 3GPP specification defining bit rate processing [50]. We use the symbol i to indicate a TrCH number, the symbol l to indicate a TF number and the symbol j to represent a TFC number. So in our example above we have TrCHs TrCH$_i$ where $i = 1, 2$, TFCs TFC$_j$ where $j = 1, 2 \ldots 5$ and TFs TF$_{i,l}$ where $l = 1, 2, 3$ for $i = 1$ and $l = 1, 2$ for $i = 2$.

In this section we have described the interface between the physical layer and the RRC and MAC. In doing so we have introduced several key concepts, TrCHs, TFs, CCTrCHs and TFCs. In the next section we describe the different sequences of BRP operations that are carried out by the uplink and downlink transmitters, and the reasons for these differences.

4.3 OVERVIEW OF THE BIT RATE PROCESSING CHAIN

As mentioned in Section 4.1, the BRP takes several streams of data from the MAC and typically adds redundancy to this data to allow for error correction and detection at the receiver. It also multiplexes the different data streams together, balancing their quality of service requirements, and matches the extremely flexible rate at the input of the BRP to the restricted set of physical channel capacities available at the input to chip rate processing. In this section we give an overview of the bit rate processing operations that are used in UMTS to achieve these requirements.

Some BRP operations differ slightly, or are carried out in a different order, between the uplink and the downlink. These differences are due to the different 'chip rate' techniques (see Chapter 3) used in the uplink and downlink to separate different users' signals from each other, and also to simplify 'blind' transport format detection at the UE. Section 4.3.1 describes the uplink chain of operations, while Section 4.3.2 describes the downlink chain. We then give details of the individual blocks used in these chains in Sections 4.3.3 to 4.3.7. The details of rate matching, convolutional encoding and decoding and turbo encoding and decoding are described in Sections 4.4, 4.5 and 4.6 respectively.

4.3.1   BRP Chain in the Uplink

In this section we give an overview of the BRP operations used for the uplink transmitter. On the uplink the signals from different UEs arrive at the Node B receiver unsynchronized with each other. In contrast, on the downlink all signals from one Node B arrive at a given UE synchronized with each other. Because of this, as explained in Chapter 3, different UEs are separated on the uplink by giving each UE a different scrambling code. This gives better interference properties than if they shared a single scrambling code and were separated using different OVSF codes, as on the downlink.

This use of a different scrambling code for each user means that they can independently choose which OVSF code(s) to use on a frame by frame basis. Therefore, each UE has control over the bandwidth of its uplink PhCH. This is in contrast to the downlink, where the PhCH capacity is fixed for most PhCH types. This choice is carried out by the rate matching block, as described in Section 4.4.2.

The BRP operations carried out on the uplink are shown in Figure 4.3, and are listed below. Some of these operations are carried out individually for each TrCH, once per TTI of the TrCH. Others are carried out across all TrCHs once per radio frame.

- Individually for each TrCH, once per TTI, the following operations are performed:
  - CRC attachment to each transport block (for error detection, see Section 4.3.3);
  - channel coding (for error correction, see Section 4.3.4);
  - 'radio frame equalization'. This is just padding the data so that the number of bits is a multiple of the frames per TTI for this TrCH. These padding bits are concatenated at the end of the set of coded bits, and can be either logical 0s or 1s (it is not specified which to use). So, for example, if a TrCH with an 80 ms TTI had 70 bits after channel coding, two padding bits would be added to give a total of 72 bits which can later be split into eight sets (one per radio frame) of nine bits each;
  - TrCH, or '1st', interleaving across the whole TTI (see Section 4.3.5);
  - radio frame segmentation. The total number of bits is split into a set for each radio frame. In our 80 ms TTI example above, with 72 bits after radio frame equalization, the first nine interleaved bits will be transmitted in the first radio frame, the next nine bits in the second radio frame, and so on over all eight frames of the TTI.

- Across all TrCHs, every 10 ms frame, the following operations are performed:
  - rate matching. The total number of bits for a given radio frame from all TrCHs, after repetition or puncturing of the different TrCHs according to their rate matching attributes RM_i, is calculated. This number of bits is then used to choose the *Spreading Factor* (SF) and number of physical channels to use for this radio frame, as described in Section 4.4.2. Given this choice the coded bits from the different TrCHs will be repeated and/or punctured to fit onto the chosen physical channel(s). See Section 4.4 for details of rate matching;

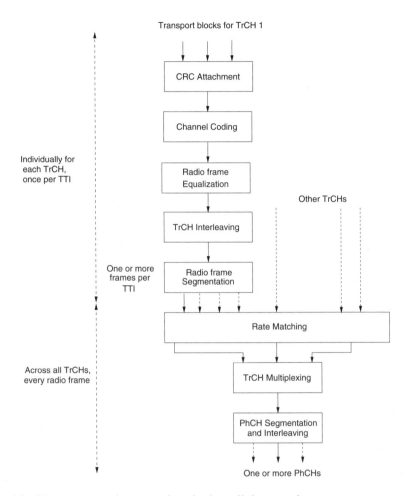

Figure 4.3. Bit rate processing operations in the uplink transmitter.

- TrCH multiplexing. The bits from different TrCHs are multiplexed together. See Section 4.3.6;
- physical channel segmentation and interleaving. The bits for this frame from all the TrCHs are interleaved. If we are using more than one PhCH, the bits for different PhCHs are segmented and then individually interleaved. See Section 4.3.7.

As mentioned earlier, the rate matching block selects which SF to use for each frame on the uplink, based on the requirements of the current TFC. In order to limit interference it is vital that each user transmits as little power as possible whilst still maintaining the quality of service he/she requires. An approximately constant amount of energy per *bit* is needed to give a certain error rate. So if the number of chips per bit increases, through either more repetition or a higher SF, the amount of energy per *chip* needed decreases. This means the power needed to give the QoS required changes as the SF and the amount of repetition or puncturing applied to the coded bits change. So, for example, a bit spread with SF = 256 will need much less power to receive it reliably than if it was spread with SF = 4.

So on the uplink, where the SF and the rate matching applied depend on the TFC used in each frame, the transmitted power also varies depending on the TFC used in each frame. Scaling factors, denoted β_d for the coded data and β_c for the L1 control information, are chosen each frame. These scaling factors control the power of the data channel relative to the L1 control channel. They are chosen, as described in Section 5.1.2.5 of [41], to take account of the amount of coded data each frame, and hence the amount of rate matching applied by the BRP and the SF used to transmit the data. For TFCs with a relatively high data rate we will have $\beta_d > \beta_c$ so that the the coded data is transmitted with a high power relative to the control information. Similarly, for TFCs with a relatively low data rate we will have $\beta_d < \beta_c$. The power of the control information stays constant (if we ignore power control) as the TFC changes, but the power of the coded data information relative to this varies with the β_d and β_c chosen each frame. Hence the power of transmitted coded data information varies as required with the TFC used.

We now discuss the BRP processing in the downlink transmitter, and the differences with the uplink processing described above.

4.3.2 BRP Chain in the Downlink

There are two fundamental differences between the uplink and downlink in WCDMA that affect the BRP operations:

1. A mode of rate matching that results in each TrCH having a fixed position within the CCTrCH, regardless of the TFC.

2. With the exception of the DSCH, a DL CCTrCH has a fixed PhCH allocation.

The first difference is used in order to simplify 'blind' detection by the UE of the TFC used on the downlink (called *Blind Transport Format Detection* or BTFD, see Section 4.7.2). On the uplink, the starting positions of the different TrCHs within the multiplexed CCTrCH data, at the point just after TrCH multiplexing, are flexible. This means that although the first TrCH always starts at the beginning of the multiplexed data, the starting positions of other TrCHs vary with the TFC used for a particular frame. This allows the rate matching block to divide the available PhCH capacity between different TrCHs depending on how many bits are needed by each for a particular TFC, and hence use the available capacity as efficiently as possible.

This mode of TrCH multiplexing can also be used on the downlink. However it significantly complicates the use of BTFD at the UE, as each TrCH may have several different possible starting positions as well as different possible numbers of coded bits. Therefore, on the downlink there is a second mode of rate matching and TrCH multiplexing which results in fixed starting positions for the TrCHs. In this mode a fixed proportion of the PhCH capacity is allocated to each TrCH depending on its maximum rate TF, and each TrCH is positioned assuming the other TrCHs have used their maximum rate TF. Hence the starting positions of TrCHs do not depend on the TFC used. These two different modes of rate matching calculation and TrCH multiplexing on the downlink are called flexible and fixed TrCH positions and are discussed in more detail in Section 4.4. See Figure 4.7 for an illustration of TrCH starting positions.

The other major difference to the uplink is that on the downlink different users' signals arrive at the UE receiver synchronized with each other. Because of this, on the downlink multiple users share a scrambling code and are separated by allocating them different OVSF codes. This separation of users by OVSF codes leads to better interference properties than if the users were separated using different scrambling codes, as on the uplink. Each channel for each user, (excluding the DSCH) has an OVSF code that is semi-statically allocated. By semi-statically allocated we mean that the code will not change on a frame by frame basis, but may be changed only relatively infrequently through higher layer signalling between the Node B and the UE. Hence, on the downlink a given user can be considered to have a fixed number of coded data bits that he/she can transmit every frame.

However, as on the uplink, the number of data bits to be transmitted every frame can vary dramatically from one frame to the next. On the uplink this is dealt with by having the ability to change the spreading factor used from one frame to the next. On the downlink we have a fixed spreading factor. The amount of puncturing or repetition to be applied to each TrCH is calculated so that either the maximum rate TF or TFC fills the available PhCH capacity. Then, for lower rate TFs or TFCs, approximately the same repetition or puncturing ratio (e.g. repeat every 4th bit, or puncture every 11th bit, etc.) is used. This maintains the QoS of each TrCH as its TF changes, but means that for lower rate TFs or TFCs there is an excess of PhCH capacity. This excess capacity is 'filled' using *Discontinuous Transmission* (DTX); for positions corresponding to excess capacity, no power is transmitted. These excess positions are marked with DTX indicators. So interference is limited on the downlink by transmitting for less of the time when we have a lower rate TFC (for flexible position) or TF (for fixed positions). So although, excluding the DTX positions, the transmitted *power* is not modified when the TFC changes, the total transmitted *energy* in a frame will change based on the TFC. Note that these power or energy changes based on the TFC, on both the uplink and downlink, are independent of the fast power control described in Chapter 3.

Note that as well as these DTX 'bits', we may also have notional 'p bits' on the downlink. These 'p bits' are a concept introduced in [50] to reserve space for a compressed mode gap created by extra puncturing (and/or less repetition) in the rate matching stage. This is called compressed mode by puncturing, see Section 4.8 for more details. Unlike the DTX indicators they are removed prior to the PhCH interleaver and form no part of the signal transmitted by the Node B or received by the UE. In effect, from the UE's point of view, they merely distort the pattern of the TrCH deinterleaver. However, they are useful conceptually as they preserve the number of 'bits' at various points in the BRP chain when there is a compressed mode gap by puncturing. Therefore we use the concept in our discussions.

These differences lead to the chain of BRP operations for the downlink shown in Figure 4.4 and listed below. Again these operations are divided into those that are carried out individually for each TrCH (once per TTI) and those that are carried out once per radio frame across all the TrCHs multiplexed into the CCTrCH.

- Individually for each TrCH, once per TTI, the following operations are performed:
  - CRC attachment to each transport block (for error detection, see Section 4.3.3);
  - channel coding (for error correction, see Section 4.3.4);
  - rate matching and first insertion of DTX. For fixed TrCH positions, coded bits are punctured or repeated so that the maximum rate TF fits into the PhCH capacity allocated for this TrCH. For flexible TrCH positions, each TrCH has coded bits repeated or punctured so that the maximum rate TFC fits into the total PhCH capacity for this CCTrCH. Then lower rate TFs or TFCs are punctured or repeated at

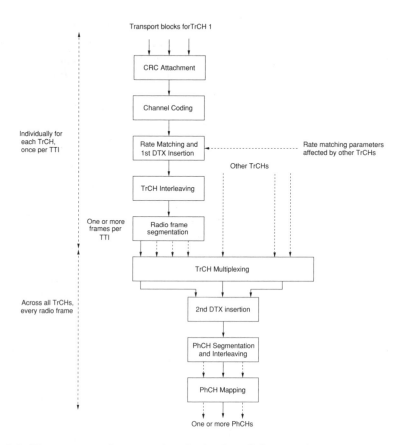

Figure 4.4. Bit rate processing operations in the downlink transmitter.

approximately the same rate so that the quality of service of the TrCH is maintained. For fixed TrCH positions, discontinuous transmission (DTX) indicators will also be inserted here if the current TF does not use all of the PhCH capacity that is reserved for this TrCH. See Section 4.4 for details;

- 'TrCH, or '1st' interleaving across the whole TTI (see Section 4.3.5);
- radio frame segmentation. The total number of bits is split equally into a set for each radio frame. Note there is no radio frame equalization stage in the downlink; the rate matching will have ensured that the number of bits, including any DTX 'bits' and notional 'p bits' is a multiple of the number of frames per TTI.

- Across all TrCHs, every 10 ms frame, the following operations are performed:
  - TrCH multiplexing. The bits from different TrCHs are multiplexed together. See Section 4.3.6;
  - second insertion of DTX. For flexible TrCH positions if the current TFC uses less of the PhCH capacity than the maximum, DTX indicators are added here to 'fill' the PhCH capacity;
  - physical channel segmentation and interleaving. If we are using more than one PhCH, the bits for different PhCHs are segmented and individually interleaved. See Section 4.3.7. Note the rate matching and DTX insertion blocks will ensure

that the number of bits (including DTX and 'p bits') will be an exact multiple of the number of PhCHs used;

– physical channel mapping. On the downlink the L1 control information (pilot and power control bits, see Chapter 3, and any coded TFCI bits, see Section 4.7.1) are multiplexed with the coded data bits before being spread and scrambled. In compressed mode by spreading factor reduction (see Section 4.8) extra DTX bits are also added at this stage to fill the increased physical channel capacity.

Having given a brief overview of the BRP operations in the uplink and downlink, we now describe each of these blocks in more detail.

4.3.3 Cyclic Redundancy Check Codes

Cyclic Redundancy Check (CRC) codes are used for error detection on decoded transport blocks. They are similar to the well known and very simple parity checking codes, in which a single check bit is added to a group of data bits in order to make the total number of 1s in the set of bits either even (for even parity codes) or odd. Then, using even parity codes as an example, if the number of 1s received in a block of data at the receiver is odd, the receiver will know there has been an error. So all single bit errors will be spotted by a parity check. However, if there is more than one bit error, there is a 50 % chance the error will be undetected at the receiver.

CRC codes are an extension of these parity checking codes. Instead of adding just one bit to a block of data, several bits are added. This gives the ability to reliably detect longer error *sequences*, and a lower chance of undetected errors when the error detection capability of the code is exceeded. Note the use of the term error *sequence* here. An error sequence of length y bits means that there is a set of y consecutive bits *some* of which are incorrect, including the first and last bits in the sequence. Note that usually not all the y bits will be incorrect.

In [50] the number of CRC bits applied to each block of data is denoted as L_i. We will use this notation here. In general, it can be proved (see [64]) that, with the correct choice of parameters, an L_i bit CRC will allow the detection of *all* error sequences of L_i bits or less, and longer error sequences have only a one in 2^{L_i} chance of escaping detection. Note that a CRC code's ability to detect all error sequences of less than a given length is useful because in communication systems, especially when channel coding is used, errors tend to occur in bursts.

As explained in Section 4.2, the length L_i of the CRC to add to each transport block in a given TrCH is one of the parameters given in the semi-static attributes of the TFs used for that TrCH. A CRC of the given length L_i is added independently to each transport block within the TrCH$_i$ before the transport blocks are concatenated and coded. Details of the available CRCs that can be chosen are shown in Table 4.3.

CRCs are typically implemented in hardware as a linear feedback shift register, with one element for each bit of the CRC, as shown in Figure 4.5. This figure shows the shift register used to generate the eight bit CRC from Table 4.3. Each connection in the register corresponds to a term in the generator polynomial given in Table 4.3. For each element in the shift register, if the corresponding power of D is present in the generator polynomial the connection to the next element is through an exclusive OR gate (XOR). Elements for which the corresponding power of D is not present are connected directly to the next element.

Table 4.3   Available CRCs.

Number of CRC bits	Generator polynomial
0	N/A
8	$D^8 + D^7 + D^4 + D^3 + D + 1$
12	$D^{12} + D^{11} + D^3 + D^2 + D + 1$
16	$D^{16} + D^{12} + D^5 + 1$
24	$D^{24} + D^{23} + D^6 + D^5 + D + 1$

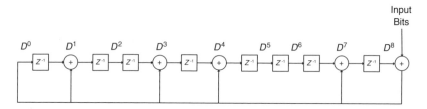

Figure 4.5.   Linear feedback shift register implementation of the eight bit CRC.

The D^{L_i} term and $D^0 = 1$ term correspond to the connections at the start and end of the shift register. For each transport block, initially the contents of the shift register are set to zero. Then each bit is shifted through from left to right. The contents of the shift register after all A_i bits of the transport block have been shifted through gives the CRC bits to be transmitted. These bits are transmitted in reverse order, i.e. in Figure 4.5 the CRC bit in the left most register is transmitted first.

A CRC check can be implemented by setting the initial state of the shift register to all 0, and then shifting the A_i received data bits through the register. The final state of the register will give the CRC bits that would have been sent if the received data bits were transmitted. These bits are compared to the received CRC bits; if the two sets of bits are the same then there have been no detected errors in the received bits. This is called a CRC pass. If they are different then there has been at least one error amongst the A_i data bits and/or the L_i CRC bits. We call this a CRC fail.

In software, CRCs can be implemented in a similar manner to the hardware implementation shown in Figure 4.5, using a shift register and a series of XOR operations. However, much more efficient software implementations are possible using a lookup table, see [65].

Note that although CRC checks, especially with the longer codes, provide very effective error detection capability, they are not perfect. Even with a CRC pass, we cannot be absolutely sure the A_i data bits have been received correctly; as noted earlier if there has been an error sequence of more than L_i bits there is a one in 2^{L_i} chance that this error sequence will escape detection. Also, note we may get a CRC fail even when the A_i data bits are received correctly because there have been errors in the L_i CRC bits. So neither a CRC pass nor a CRC fail can be treated with absolute confidence.

Table 4.4 Channel encoding options.

Option	Overview	Number of output bits Y_i (K_i input bits, one code block)	Maximum code block size Z
1/2 rate convolutional	$K = 9$, $G_0 = 561$ $G_1 = 753$ (octal)	$2 \times (K_i + 8)$	504
1/3 rate convolutional	$K = 9$, $G_0 = 557$ $G_1 = 663$, $G_1 = 711$	$3 \times (K_i + 8)$	504
Turbo	Parallel $K = 4$ half rate RSC codes	$3 \times K_i + 12$	5114

As stated earlier, a CRC of length L_i bits is added to each transport block for each TrCH$_i$. If there are no transport blocks then no CRC bits are added. However, in the special case of transport blocks of size zero bits, i.e. $A_i = 0$, a CRC code of L_i bits is still added to each transport block. This allows for monitoring of the quality of the radio link even when there are no data bits to be transmitted. It is also a necessary condition for blind transport format detection that for some TrCHs all TFs have CRC bits transmitted, even when no data bits are transmitted. See Section 4.7.2 for details. In this special case of $A_i = 0$ the L_i CRC bits added to each transport block will be all zero.

After CRC bits are added to each transport block, the transport blocks are concatenated before channel coding is applied.

4.3.4 Channel Coding and Code Block Segmentation

Three types of channel coding are specified for UMTS, either half or one third rate convolutional encoding or one third rate turbo encoding.[1] In this section we give a brief overview of these encoding options, and describe code block segmentation and concatenation. The coding options are summarized in Table 4.4, and more details of the encoders used are given in Sections 4.5.1 and 4.6.1. Note that, for clarity, the numbers of input and output bits shown in Table 4.4 do not take account of the code block segmentation detailed below.

Convolutional codes, which are typically decoded with a Viterbi decoder [66], have been used for many years, and are used in second generation systems such as GSM and IS-95. The codes specified for UMTS, have constraint length $K = 9$ compared to $K = 5$ or $K = 7$ in GSM. This increase in constraint length gives an improvement in performance, but also an increase in the decoder complexity.

The other coding option, turbo coding, is a relatively recent innovation [67], and gives an improvement in performance over convolutional coding, especially for long block lengths. Two half rate, constraint length four, *Recursive Systematic Convolutional* (RSC) codes are used in parallel with an interleaver between them. The data bits and the parity bits from both RSC codes are transmitted, giving a one third rate code.

---

[1] In earlier versions of [50] a fourth option of no channel encoding was also specified, but this has now been removed from the specifications.

Support of convolutional encoding and decoding is mandatory in UMTS terminals, whereas support of turbo encoding and decoding is an option. Typically, convolutional encoding is used for control channels and low rate channels such as AMR speech channels, and turbo encoding is used for high rate channels in higher capability terminals.

After the transport blocks in TrCH$_i$ have had CRCs added and been concatenated, we have X_i bits to be channel encoded. The number of bits X_i is limited only by the capabilities of the UE (how many bits and how many transport blocks it can handle per 10 ms), and so may be very large. For example, for a UE supporting the 2048 kbits/s capability class in [68] X_i may be as large as 22 784 bits (20 480 data bits split between 96 transport blocks with a 24 bit CRC each, so a total of $20\,480 + (96 \times 24) = 22\,784$ bits). As will be explained in Sections 4.5.6 and 4.6.7, the implementation of convolutional and especially turbo decoders can be significantly simplified if the number of bits to be decoded in one block is limited. For this reason, before channel encoding, the X_i data and CRC bits may be segmented into several (C_i) equal sized blocks, called code blocks, each of size K_i. Each of these code blocks is encoded separately, and so can be decoded separately.

Depending on whether turbo or convolutional coding is used, a maximum code block size Z is defined. The number of code blocks C_i is then given by

$$C_i = \left\lceil \frac{X_i}{Z} \right\rceil$$

i.e. the result of X_i divided by Z rounded up to the nearest integer. The size K_i of each of the C_i code blocks is then given by

$$K_i = \left\lceil \frac{X_i}{C_i} \right\rceil$$

As all the code blocks are of equal size, there may not be enough data and CRC bits to fill all the code blocks. In this case extra bits, called filler bits, are added to the start of the first transport block. These filler bits are set to logical 0s. For the special case of $X_i = 0$ (i.e. no data or CRC bits) there are no code blocks and $C_i = K_i = 0$. Also, for turbo coded TrCHs, if X_i is less than 40 (but greater than 0), $40 - X_i$ filler bits are added to the start of these X_i data and CRC bits to give one code block of 40 bits.

The performance of turbo coding tends to increase as the block size increases. However, this increase in performance levels off as the block size increases beyond a few thousand bits. Therefore the maximum code block size Z is set to 5114 for turbo coded channels. For convolutional coded channels there is little difference in performance as the number of coded bits increases. However, the overhead of termination bits (the $+8$ terms in the equations for the number of coded bits in Table 4.4) is obviously higher for very small code blocks. So the maximum code block size Z is set to 504 bits (or 512 bits including the eight termination bits) for TrCHs using convolutional codes.

After the C_i code blocks are separately encoded, they are concatenated to give a total of $E_i = C_i \times Y_i$ coded bits, where Y_i is the number of coded bits per code block shown in Table 4.4.

4.3.5 TrCH Interleaving

Interleaving is used to split up the errors at the receiver which, because of the nature of mobile radio channels [47], tend to occur in bursts. At the receiver, the deinterleaving splits

Table 4.5 TrCH interleaver column permutations.

TTI	Number of columns, F_i	Inter column permutation patterns $< P1_{F_i}(0), P1_{F_i}(1), \ldots, P1_{F_i}(F_i - 1) >$
10 ms	1	$< 0 >$
20 ms	2	$< 0, 1 >$
40 ms	4	$< 0, 2, 1, 3 >$
80 ms	8	$< 0, 4, 2, 6, 1, 5, 3, 7 >$

up these bursts of errors, which improves the performance of channel decoding. Typically, best performance is achieved by interleaving over as long a period as possible, in order to split up errors as widely as possible. So each TrCH is interleaved over its whole TTI. Also, bits from different TrCHs should be interleaved amongst each other, as each TrCH is individually decoded. For these reasons, two stages of interleaving are used in UMTS. The first, which we call the TrCH interleaver, operates individually on each TrCH over a whole TTI. The second, which we call the PhCH interleaver, operates over a frame on the data for that frame from all TrCHs.

The TrCH interleaver is a block interleaver with inter column permutations that processes a whole TTI of data from a given TrCH. The number of interleaver columns is equal to the number of radio frames per TTI, F_i, and each column in the block corresponds to a radio frame. The number of rows is given by $R1 = N/F_i$, where N is the number of 'bits' input to the interleaver. On the downlink with fixed TrCH positions these N 'bits' include DTX and 'p' bit [2] indicators. Note that N is always a multiple of F_i, so that $R1$ is always a whole number. On the uplink, radio frame size equalization ensures this is the case, and on the downlink, rate matching and DTX insertion ensure this is the case.

The interleaver operates as follows:

- the input bits are entered into the interleaver row by row;

- the columns are reordered according to the inter column permutation function $P1_{F_i}(j)$, shown in Table 4.5. If a column has position j after reordering, its position prior to reordering is given by $P1_{F_i}(j)$;

- the output bits are read out column by column. Each column contains the $R1$ bits from this TrCH for one radio frame.

Note that the pattern of the interleaver on the downlink is modified in compressed mode, for compressed mode by puncturing only. Notional 'p bits' are inserted at the start of the column corresponding to any frame with a gap for compressed mode by puncturing. The number of 'p bits' inserted is dependent on the size of the gap. As explained in Section 4.8, these 'p bits' are removed prior to transmission, and so effectively all they do is change the pattern of the TrCH interleaver.

To illustrate the operation of the TrCH interleaver, we now discuss an example for the case of a TrCH with a 40 ms TTI and $N = 28$. The number of columns $F_i = 4$, and the number of rows is given by $R1 = 28/4 = 7$. The input data x_k, where $k = 1, 2, \ldots 28$, is

---

[2] Used to mark gaps for compressed mode by puncturing, see Section 4.8

first written row by row into the interleaver matrix as shown below:

$$\begin{bmatrix} x_1 & x_2 & x_3 & x_4 \\ x_5 & x_6 & x_7 & x_8 \\ x_9 & x_{10} & x_{11} & x_{12} \\ x_{13} & x_{14} & x_{15} & x_{16} \\ x_{17} & x_{18} & x_{19} & x_{20} \\ x_{21} & x_{22} & x_{23} & x_{24} \\ x_{25} & x_{26} & x_{27} & x_{28} \end{bmatrix}$$

The columns are then shuffled according to the inter column permutation function $P1_4(j)$, as shown below:

$$\begin{bmatrix} x_1 & x_3 & x_2 & x_4 \\ x_5 & x_7 & x_6 & x_8 \\ x_9 & x_{11} & x_{10} & x_{12} \\ x_{13} & x_{15} & x_{14} & x_{16} \\ x_{17} & x_{19} & x_{18} & x_{20} \\ x_{21} & x_{23} & x_{22} & x_{24} \\ x_{25} & x_{27} & x_{26} & x_{28} \end{bmatrix}$$

Finally, the data for each frame is read out column by column as follows:

Frame 1: $x_1, x_5, x_9, x_{13}, x_{17}, x_{21}, x_{25}$
Frame 2: $x_3, x_7, x_{11}, x_{15}, x_{19}, x_{23}, x_{27}$
Frame 3: $x_2, x_6, x_{10}, x_{14}, x_{18}, x_{22}, x_{26}$
Frame 4: $x_4, x_8, x_{12}, x_{16}, x_{20}, x_{24}, x_{28}$

One can see that the interleaver distributes bits from any group of F_i consecutive bits to different radio frames. The inter column permutation distributes consecutive bits to nonconsecutive frames, where this is possible (i.e. for 40 ms or 80 ms TTIs). Notice that for a one frame TTI the interleaver is transparent – the output bits are in the same order as the input bits. In this case all interleaving for the TrCH is carried out by the PhCH interleaver – see Section 4.3.7. Also, for a two frame TTI the inter column permutation has no effect – the interleaver merely splits adjacent bits between the two frames.

4.3.6 TrCH Multiplexing and Second DTX Insertion

TrCH multiplexing simply involves concatenating the data from each TrCH together. The TrCHs are concatenated in order, i.e. TrCH$_1$ then TrCH$_2$, etc. On the downlink, when using flexible TrCH positions, additional DTX may also be inserted at this stage as described below.

On the uplink, after rate matching the number of coded bits will always exactly match the bits available on the selected set of PhCHs. Similarly, on the downlink for fixed TrCH positions, the number of coded bits, DTX indicators and notional 'p bit' indicators will match the space available on the PhCH(s) being used. Hence, DTX insertion is not needed at this stage for either the uplink or the downlink with fixed TrCH positions. However, on the downlink with flexible TrCH positions, for lower rate TFCs, the multiplexed TrCH bits will not fill the available PhCH bits. This shortfall is made up by appending DTX indicators to the end of the multiplexed TrCH data. See Figure 4.7 in Section 4.4.3 for an illustration of TrCH multiplexing and DTX insertion for fixed and flexible TrCH positions on the downlink.

The amount of DTX inserted at this stage on the downlink is reduced for frames containing a gap for compressed mode by higher layer scheduling, in order to make room for the gap. See Section 4.8 for details. As DTX is only inserted at this stage if flexible TrCH positions are used, on the downlink this form of compressed mode can only be used with flexible TrCH positions.

This insertion of DTX in the downlink at this stage for flexible TrCH positions is called second DTX insertion in [50]. Note that the PhCH interleaver will distribute these DTX indicators over all transmitted slots.

4.3.7 PhCH Segmentation and Interleaving

As described earlier, two stages of interleaving are used in UMTS. The first, known as TrCH or first interleaving, was described in Section 4.3.5. This interleaver distributes the bits from each TrCH as widely as possible amongst the number of frames per TTI for the TrCH. The second interleaving stage, which we call the PhCH interleaver, operates on the multiplexed bits from all TrCHs, one frame at a time, and is described in this section.

If more than one PhCH is used then the bits to be transmitted this frame are split equally between the PhCHs. If the total number of bits[3] is X, and the number of PhCHs to be used is P, then bits are split into P segments of length U each, where

$$U = \frac{X}{P}$$

Note that rate matching and DTX insertion will ensure that the number of bits matches the PhCH capacity exactly, and so U will be a whole number and will match the number of data bits available on the slot format to be used. The first U bits are transmitted on the first PhCH, the next U bits on the next PhCH, etc. Each set of U bits is interleaved separately with a PhCH interleaver.

Like the TrCH interleaver, the PhCH interleaver uses a block interleaver with inter column permutations. The number of columns is denoted by $C2$ and is fixed at $C2 = 30$. The number of rows $R2$ is given by:

$$R2 = \left\lceil \frac{U}{C2} \right\rceil$$

Note that, except for compressed mode, the number of bits per frame U will be a multiple of 30 and so the division above will give an exact match, i.e. we will have $C2 \times R2 = U$. The PhCH interleaver then operates as follows:

1. The input bits are entered into the interleaver row by row. If $(R2 \times C2) < U$, which happens only for frames with compressed mode gaps, then padding bits are used to fill the last row.

2. The columns are reordered according to the inter column permutation function $P2(j)$, shown in Table 4.6. $P2(j)$ gives the column position prior to reordering, of the jth column after reordering.

---

[3] In this section we include DTX and 'p bit' indicators in the term 'bits'

Table 4.6   PhCH interleaver inter column permutation pattern.

Number of columns $C2$	Inter column permutation pattern $< P2(0), P2(1), \ldots, P2(30) >$
30	$< 0, 20, 10, 5, 15, 25, 3, 13, 23, 8, 18, 28, 1, 11, 21,$ $6, 16, 26, 4, 14, 24, 19, 9, 29, 12, 2, 7, 22, 27, 17 >$

3. The output bits are read out column by column. If any padding bits were used in the last row they are removed at this stage.

This process is very similar to the TrCH interleaver example discussed in Section 4.3.5, with 30 columns rather than the four columns in this example.

4.4   RATE MATCHING

Rate matching provides a mechanism to map the encoded data bits from each TrCH on to the available PhCH resource. At the encoder this means either puncturing or repeating some encoded bits, depending on whether the number of bits available on the PhCH is greater or less than the number of coded bits. At the decoder these operations are reversed. For bits that have been repeated, the received soft decisions are combined to give a more reliable indication of the value of the bit. For bits that have been punctured at the encoder, soft decisions that indicate we have no information regarding the value of the bit are inserted in their place at the decoder.

In this section we give details of exactly how rate matching patterns (i.e. which bits are punctured or repeated) are determined, and how the parameters (i.e. how many bits are punctured or repeated) are calculated. It will be valuable for readers who wish to understand the detailed operation of rate matching. However, the basic ideas have been explained in Section 4.3, and so this section can be omitted without loss of continuity of the chapter.

Although rate matching is similar for both uplink and downlink, there are several key differences, which are listed below.

- On the uplink variations in the TFC are dealt with by varying the spreading factor (SF), and perhaps the number of PhCHs, on a frame by frame basis depending on the TFC. The uplink PhCH rate tends to be reduced for lower rate TFCs, which means the amount of repetition or puncturing applied to a TrCH varies as the TFC varies. However, the quality of service may be kept approximately constant for each TrCH by varying the transmitted power in line with the relative amount of puncturing or repetition done from one TFC to another.

- On the downlink the SF and number of PhCHs is constant[4], and variations in the TFC are dealt with using discontinuous transmission (DTX). DTX insertion reduces the effective PhCH rate, and is used to maintain approximately the same puncturing or repetition rate for all TFs on a given downlink TrCH.

[4]Except for the DSCH

- There is no concept of fixed TrCH positions on the uplink. However, on the downlink, fixed TrCH positions are required to simplify blind TF detection (BTFD) at the UE.

- On the uplink, rate matching is performed once a frame across all TrCHs, and after TrCH interleaving and radio frame segmentation. In contrast, on the downlink, rate matching is performed once a TTI for each TrCH and prior to TrCH interleaving.

Each TrCH in a CCTrCH is assigned a rate matching attribute, RM_i. The rate matching attributes are used to control the amount of puncturing, or repetition, on each TrCH. The ratio of the number of rate matching output bits to the number of input bits is approximately proportional to the rate matching attribute RM_i divided by the sum of these attributes over all the TrCHs in the CCTrCH. Hence, the ratio of these rate matching attributes between different TrCHs determines the relative amount of puncturing or repetition for each TrCH. Note, the rate matching attributes can also be set such that puncturing occurs on one TrCH, and repetition on another.

In the following sections we detail rate matching on the uplink and then the downlink. However, first we discuss the generation of a rate matching pattern, i.e. which bits are punctured or repeated, as this is common for both links.

4.4.1 Rate Matching Patterns

The rate matching pattern is determined by parameters X_i, ΔN_i, e_{ini}, e_{plus}, and e_{minus}. X_i gives the number of bits input to a given rate matching 'process'. The bits for each TrCH are rate matched separately, either one frame (for the uplink) or one TTI (for the downlink) at a time. Note however that, as explained below, there may be two rate matching processes for one set of input bits from a given TrCH.

The sign of ΔN_i determines whether repetition or puncturing is used. For the uplink and the downlink with flexible TrCH positions, the magnitude of ΔN_i gives the number of bits added or removed by rate matching. For the downlink with fixed TrCH positions, the magnitude of ΔN_i gives the number of bits added or removed by rate matching for the maximum rate TF. If ΔN_i is negative, puncturing is performed, if ΔN_i is positive, repetition is performed, and if ΔN_i is zero, then no rate matching is required.

The other parameters e_{ini}, e_{plus}, and e_{minus} control the pattern of bits that are punctured or repeated. The first, e_{ini}, is used to control the position of the first punctured or repeated bit, whereas e_{plus} and e_{minus} are set based on the number of bits to be punctured or repeated. Sections 4.4.2 and 4.4.3 detail how all these parameters are determined for both the uplink and the downlink. In this section we detail how they are used.

The following pseudocode details the algorithms for puncturing and repetition. For puncturing ($\Delta N_i < 0$) the following procedure is followed:

```
e = e_ini
for bit m = 1 to X_i
        e = e - e_minus
        if e ≤ 0 then
                puncture mth bit
                e = e + e_plus
        end if
end for
```

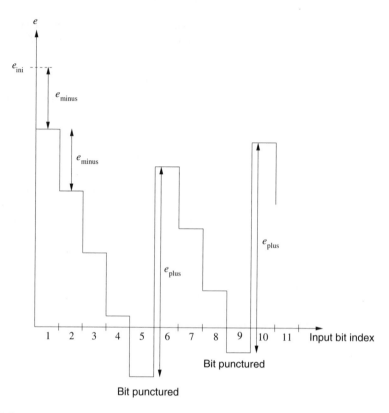

Figure 4.6. Variation of e as puncturing procedure processes input bits.

The effect of this can be seen in Figure 4.6, which shows how the variable e changes as the puncturing procedure moves through the input bits. It starts at $e = e_{\text{ini}}$, and is reduced by e_{minus} for each input bit. When e falls below zero, the corresponding input bit is punctured and e is increased by e_{plus}. So, in Figure 4.6, the input bits with indices 5 and 9 would be punctured.

For repetition ($\Delta N_i > 0$) a similar procedure is followed, modified to allow a given input bit to be repeated several times:

```
e = e_ini
for bit m = 1 to X_i
        e = e − e_minus
        do while e ≤ 0
                repeat mth bit
                e = e + e_plus
        end do
end for
```

For ($e_{\text{plus}} \geq e_{\text{minus}}$), the algorithms result in every $\lceil e_{\text{plus}}/e_{\text{minus}} \rceil$th, or $\lfloor e_{\text{plus}}/e_{\text{minus}} \rfloor$th, bit being punctured or repeated. For ($e_{\text{plus}} < e_{\text{minus}}$), which is only allowed for repetition, the algorithm results in repeating each bit $\lceil e_{\text{minus}}/e_{\text{plus}} \rceil$, or $\lfloor e_{\text{minus}}/e_{\text{plus}} \rfloor$, times. One can also see how the value of e_{ini} determines the position of the first bit that is repeated or punctured. Finally, notice that although the sign of ΔN_i is used to determine whether puncturing or repetition should be applied, the magnitude of ΔN_i is not directly used within the repetition or puncturing procedures above. Instead it is used to calculate e_{plus} which, together with e_{minus}, determines the repetition or puncturing rate.

For convolutionally encoded TrCHs, and for TrCHs where repetition is used, a single rate matching process is used for all data bits. However, for turbo encoded TrCHs where puncturing is employed, the systematic, parity 1 and parity 2 bits are split into three separate data sequences and puncturing is applied only to the two parity bit sequences. This is because the performance of turbo codes suffers more when systematic bits are punctured than when parity bits are punctured. This operation is known as bit separation, and when it is used the puncturing algorithm above is run separately for the two parity bit sequences. This gives two rate matching processes for a single TrCH, and requires two sets of rate matching parameters. After bit separation and puncturing, the systematic and remaining parity 1 and parity 2 bits are reassembled in the same order as they were before bit separation.

As will be described in Section 4.6.1, the output from the turbo encoder is ordered systematic, parity 1, parity 2, systematic, parity 1, parity 2, etc. On the downlink, rate matching is carried out immediately after channel encoding, and so bit separation is trivial and the three sequences are formed by taking every third input bit, offset by 0, 1 and 2 bits respectively. Notice that this results in each sequence containing both systematic bits and parity bits from the trellis termination stage. See Section 4.6.1 for details.

On the uplink, bit separation is complicated by performing rate matching after TrCH interleaving, radio frame equalization and segmentation. Fortunately, because of the design of the TrCH interleaver, after this interleaving the bits are left in the order systematic, parity 1 and parity 2, or systematic, parity 2 and parity 1, but for each frame there is an offset into this sequence. Hence, on the uplink each data sequence is again formed by taking every third input bit, but the offset for each bit type depends upon the number of frames per TTI and the frame number within a TTI.

Having described how the rate matching parameters X_i, ΔN_i, e_{ini}, e_{plus} and e_{minus} are used to perform the puncturing or repetition, we now describe how these parameters are calculated for the uplink and downlink.

4.4.2 Uplink Rate Matching

Uplink rate matching is performed every frame, with a frames worth of data input from each TrCH. The bits from different TrCHs are rate matched separately, but the controlling parameters e_{ini} etc. are calculated for all TrCHs based on the TFC for this frame. Note that for a given TrCH the TFC may change during a TTI, and so the amount of puncturing or repetition on different frames within a single TTI may vary.

The uplink rate matching is broken into the following stages:

1. Select the PhCH capacity required for this TFC;

2. Split the available PhCH capacity between each of the TrCHs;

3. Calculate the rate matching parameters for each TrCH to determine which bits are punctured or repeated this frame.

In the following sections the nomenclature below is used:

$N_{PhCH,j}$ number of PhCH bits per frame selected for TFC j;

RM_i the rate matching attribute for TrCH i;

$N_{i,j}$ number of input bits per frame from TrCH i, for TFC j;

$\Delta N_{i,j}$ difference in the number of bits per radio frame before and after rate matching on TrCH i, for TFC j, i.e. $|\Delta N_{i,j}|$ gives the number of bits that are repeated or punctured.

4.4.2.1 Uplink PhCH Rate Selection

When rate matching an uplink frame, the first stage is to determine the SF and number of PhCHs required. We refer to this as the PhCH combination to be used. At any time the list of SFs available to a UE, and the number of physical channels it can use, is set by the combination of:

1. The UE's capability (only high capability UEs will be able to transmit at the highest rate possible in the system);

2. The minimum SF and maximum number of physical channels for this UE allowed by the network. This is signalled to the UE when the call is set up. This allows the network to limit the interference generated between different UEs;

3. Restrictions due to the CCTrCH type. Although a DCH may use the full set of PhCH rates, a RACH may only use PhCHs with SFs of 32 or greater, and a CPCH is limited to a single PhCH.

Generally, the uplink rate matching will choose a PhCH combination from the set available to it so that all TrCHs use repetition rather than puncturing. However, in certain cases it will choose a PhCH combination that needs puncturing on some, or all, of the TrCHs. The amount of puncturing that can be performed is limited by the *Puncturing Limit*, PL, which is signalled to the UE by the network. Rate matching will only entail puncturing in the following cases:

- the number of PhCHs required is reduced by applying puncturing;

- insufficient PhCH capacity is available, since the PhCH rate is limited by one of the three factors above;

- insufficient PhCH capacity is available due to a transmission gap in a compressed mode frame (this applies to compressed mode by higher layer scheduling only, see Section 4.8).

Each frame, the UL rate matching control selects the PhCH capacities, from the set available to it, that would give no puncturing. This is referred to as SET1 in [50]. If SET1 is not empty, and includes a member with only one PhCH (i.e. it is possible to fit all the bits onto a single PhCH without puncturing) then the minimum PhCH capacity from this set is chosen. Otherwise, a second set of PhCH capacities, referred to as SET2 in [50], is prepared. This is the set of PhCH capacities that fit the bits to be transmitted without excessive puncturing on the TrCH with the minimum rate matching attribute RM_i. If SET2 contains any members using only one PhCH, then the largest such choice is taken. Otherwise, the minimum member of SET2 is chosen.

This choice of a PhCH combination from SET2 means that the combination is chosen so as to minimize the amount of puncturing, unless this would result in using more than one PhCH. If more than one PhCH is needed, the combination is chosen so as to minimize the number of PhCHs without exceeding the puncturing limit.

As discussed earlier, the uplink transmitted power is also varied on a frame by frame basis to compensate for changes in SF and repetition or puncturing rates.

Note that for frames containing an uplink compressed mode gap created by spreading factor reduction, the PhCH rate chosen at this stage will be modified to allow for the reduction in SF and the gap. See Section 4.8 for details.

4.4.2.2 Uplink PhCH Allocation

This section details how the PhCH bits for a frame are allocated between the TrCHs in the CCTrCH.

The number of PhCH bits allocated to TrCHs 1 to i, for TFC j, is denoted as $Z_{i,j}$, given by:

$$Z_{0,j} = 0$$

$$Z_{i,j} = \left\lfloor \frac{\left(\sum_{m=1}^{i} RM_m \times N_{m,j}\right) \times N_{\text{PhCH},j}}{\sum_{m=1}^{I} RM_m \times N_{m,j}} \right\rfloor \quad \text{for } i = 1 \text{ to } I \quad (4.1)$$

This process just allocates the $N_{\text{PhCH},j}$ bits available on the PhCH this frame between the TrCHs is proportion to the product of their rate matching attribute RM_i and the number of bits per frame $N_{i,j}$. The number of bits on the PhCH this frame allocated to TrCH i is then given by $Z_{i,j} - Z_{(i-1),j}$, and the number of bits to puncture or repeat this frame is given by:

$$\Delta N_{i,j} = Z_{i,j} - Z_{(i-1),j} - N_{i,j}. \quad (4.2)$$

As an example, consider a CCTrCH with two TrCHs, each having the same number of input bits per frame and the same rate matching parameters. In this case the available PhCH capacity is shared equally between the TrCHs.

$$RM_1 = 1, \; N_{1,j} = 100, \; RM_2 = 1, \; N_{2,j} = 100$$
$$Z_{1,j} = \lfloor 100 \times N_{\text{PhCH},j}/200 \rfloor$$
$$= \lfloor N_{\text{PhCH},j}/2 \rfloor$$

If the rate matching attribute of the second TrCH is doubled, it is allocated twice the PhCH capacity allocated to the first TrCH.

$$RM_1 = 1, \; N_{1,j} = 100, \; RM_2 = 2, \; N_{2,j} = 100$$
$$Z_{1,j} = \lfloor 100 \times N_{\text{PhCH},j}/300 \rfloor$$
$$= \lfloor N_{\text{PhCH},j}/3 \rfloor$$

If we now double the number of bits on the first TrCH, the PhCH capacity is again shared equally between the TrCHs.

$$RM_1 = 1, \; N_{1,j} = 200, \; RM_2 = 2, \; N_{2,j} = 100$$
$$Z_{1,j} = \lfloor 200 \times N_{\text{PhCH},j}/400 \rfloor$$
$$= \lfloor N_{\text{PhCH},j}/2 \rfloor$$

4.4.2.3 *Uplink Rate Matching Parameter Calculation*   As described earlier, the rate matching pattern is determined by the parameters X_i, ΔN_i, e_{ini}, e_{plus} and e_{minus}. Given these values the algorithm described in Section 4.4.1 determines which bits will be punctured or repeated. In a similar fashion to interleaving, the performance of the decoder is improved if the distance between repeated or punctured bits is maximized. Over a given block of rate matched data, the algorithm in Section 4.4.1 will achieve this. However, on the uplink, a TTI is split up into frames and the data from each frame is rate matched separately. Also, the TrCH interleaver operates between the channel encoder and the rate matching block. As the TrCH interleaver sends consecutive encoder output bits in different frames, the rate matching patterns for different frames in a TTI are offset with respect to each other using different values of e_{ini}. This offset aims to maximize the distance between punctured/repeated bits at the decoder input. The value of e_{ini} is therefore a function of the frame number within the TTI, the TrCH interleaver inter column permutation, and the average puncturing or repetition distance.

The other parameters are given as follows. For all TrCHs except turbo coded TrCHs with puncturing, we have:

$$X_i = N_{i,j}$$
$$\Delta N_i = \Delta N_{i,j}$$
$$e_{plus} = 2 \times N_{i,j}$$
$$e_{minus} = 2 \times |\Delta N_{i,j}|$$

For TrCHs using turbo coding for which puncturing is needed, as explained earlier, the bits are split into three sequences, two of which have rate matching applied. For these two parity sequences the rate matching parameters are given by:

$$X_i = \lfloor N_{i,j}/3 \rfloor$$
$$\Delta N_i = \lfloor \Delta N_{i,j}/2 \rfloor \text{ for parity 1 bit sequence}$$
$$\Delta N_i = \lceil \Delta N_{i,j}/2 \rceil \text{ for parity 2 bit sequence}$$
$$e_{plus} = a \times X_i$$
$$e_{minus} = a \times |\Delta N_i|$$

where $a = 2$ for the parity 1 bit sequence and $a = 1$ for the parity 2 bit sequence. These calculations serve to split the bits to be punctured approximately equally between the two parity bit sequences, and leave the systematic bit sequence unpunctured.

4.4.3 Downlink Rate Matching

On the downlink, the PhCH capacity is fixed for all TFCs in the TFC set, with the exception of a DSCH CCTrCH. In the case of a DSCH, different PhCH capacities may be configured for each TFC in the TFCS, and rate matching is performed on subsets of TFCs that use the same PhCH capacity. Therefore, unlike the uplink, the PhCH selection is not calculated by Layer 1 but is provided by higher layers. From this PhCH selection, Layer 1 calculates how the capacity is split between the TrCHs, and hence the rate matching parameters X_i, ΔN_i, e_{ini}, e_{plus}, and e_{minus} for each TrCH.

The TrCH positions on a downlink CCTrCH may be fixed or flexible, and the allocation of the available PhCH capacity depends upon this. The amount of puncturing or repetition to be applied to each TrCH is calculated so that either

Table 4.7   Coded bits per frame
for each TFC.

TFC	TrCH$_1$	TrCH$_2$
1	100	0
2	0	100
3	50	100

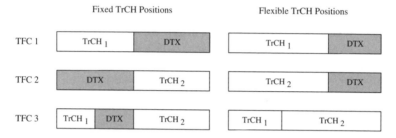

Figure 4.7. TrCH multiplexing for fixed and flexible positions.

- for fixed TrCH positions: at the maximum rate TF for a given TrCH the bits from that TrCH exactly fill the PhCH capacity allocated to the TrCH; or

- for flexible TrCH positions: at the maximum rate TFC for the CCTrCH, the bits from all TrCHs exactly fill the PhCH capacity available to the whole CCTrCH.

In either case, excess PhCH capacity is filled using discontinuous transmission (DTX) indicators when a lower rate TF (for fixed TrCH positions) or TFC (for flexible TrCH positions) is used. No power is transmitted from the Node B for this CCTrCH for DTX indicators, and the UE will receive effectively only noise for these bits. For fixed TrCH positions, the DTX indicators are inserted immediately after the rate matching to fill up the PhCH capacity for this TrCH. This is referred to as first DTX insertion in [50]. For flexible TrCH positions DTX is inserted after the TrCHs have been multiplexed, as described in Section 4.3.6, to fill the PhCH capacity for all TrCHs.

These ideas are illustrated above with an example CCTrCH. This CCTrCH has two TrCHs, and for simplicity both TrCHs have the same rate matching attribute RM_i. The first TrCH has three TFs, the second TrCH has two TFs, and the CCTrCH has three TFCs. Table 4.7 shows the number of coded bits per frame for each TrCH for these three TFCs, while Figure 4.7 shows the positions of the two TrCHs and the DTX for each TFC. The positions are shown for both fixed and flexible TrCH positions.

For fixed TrCH positions, TrCH$_1$ (or its DTX) always uses the first half of the PhCH capacity, and TrCH$_2$ uses the second half. As no TFC uses the maximum rate TF from both TrCHs, the entire PhCH capacity is never filled for fixed TrCH positions. In contrast, with flexible TrCH positions, the maximum rate TFC, (TFC 3) uses all of the PhCH capacity. Also note that for all TFCs, using flexible TrCH positions gives more capacity to the TrCHs, so that more repetition (or less puncturing) will be performed.

Fixed TrCH positions are used when UE is required to perform BTFD, see Section 4.7.2. In addition to each TrCH having the same position within the CCTrCH, the rate matching pattern, i.e. which bits are punctured or repeated, is the same for all TFs on a given TrCH. This means that at the receiver, the TrCHs can be derate matched without knowing the TF. As discussed in Section 4.7.2 this significantly reduces the complexity of BTFD.

For flexible TrCH positions, the position of a given TrCH depends upon the TFC, and the rate matching pattern for a given TrCH can vary depending on the TF. If there is more than one TFC when using flexible TrCH positions, TFCI detection is used to determine the TFC. See Section 4.7.1 for details of TFCI detection.

In the case of the DSCH, flexible TrCH positions are always used. For other types of CCTrCH, the TrCH positions are signalled from higher layers.

The following sections detail the calculations for PhCH capacity allocation between TrCHs for fixed and flexible TrCH positions. This results in a value for the number of bits to be repeated or punctured per TTI for each TrCH. We then detail how the rate matching parameters X_i, ΔN_i, e_{ini}, e_{plus}, and e_{minus} are calculated from this number of bits to be punctured or repeated.

In the following sections the nomenclature shown below is used:

$N_{i,l}^{TTI}$ number of input bits per TTI from TrCH$_i$, for TF l;

$N_{i,max}^{TTI}$ maximum number of input bits per TTI from TrCH i, across all TFs l;

N_{PhCH} number of PhCH bits per radio frame;

F_i number of frames per TTI for TrCH i;

RM_i rate match attribute for TrCH i;

I number of TrCHs in the CCTrCH.

4.4.3.1 Downlink PhCH Allocation for Fixed TrCH Positions

For fixed TrCH positions, the PhCH capacity allocated to a given TrCH depends upon the TF from the TFS which gives the greatest number of bits input to rate matching. The amount of the PhCH allocated to each TrCH is calculated based on this maximum rate TF, and then the rate matching parameters for the TrCH are calculated so that this maximum rate TF fits into the allocation. Other, lower rate, TFs are rate matched with approximately the same puncturing or repetition ratio, so as to keep the error rate approximately constant over all the TFs of the TrCH. Note that although the maximum rate TF will be in the TFS, it may not necessarily be included in any TFC. In this case, the PhCH allocation for this TrCH will not be entirely used by any TFC.

In a similar fashion to the calculation of the uplink PhCH allocations for different TrCHs, shown in Equation (4.1), the number of PhCH bits per frame allocated to TrCHs 1 to i, for all TFCs, is given by Z_i where:

$$Z_0 = 0$$

$$Z_i = \left\lfloor \frac{\left(\sum_{m=1}^{i} RM_m \times N_{m,max}^{TTI}/F_m\right) \times N_{PhCH}}{\sum_{m=1}^{I} RM_m \times N_{m,max}^{TTI}/F_m} \right\rfloor \quad \text{for } i = 1 \text{ to } I \qquad (4.3)$$

The number of PhCH bits per frame allocated to TrCH i is then given by $Z_i - Z_{i-1}$. The number of bits that are punctured or repeated per TTI, for the largest TF on TrCH i, is given by:

$$\Delta N_{i,\max}^{\text{TTI}} = F_i \times (Z_i - Z_{i-1}) - N_{i,\max}^{\text{TTI}} \tag{4.4}$$

For lower rate TFs the rate matching parameters e_{plus} and e_{ini} are the same as the values for this maximum rate TF, but the rate matching process is run over fewer input bits. This results in a lower number of punctured or repeated bits, but approximately the same puncturing or repetition ratio.

Note that the PhCH allocations amongst the TrCHs at this stage are modified for TTIs containing one or more gaps by compressed mode by puncturing. This form of compressed mode is only allowed on the downlink and only when using fixed TrCH positions. For these TTIs the values of $\Delta N_{i,\max}^{\text{TTI}}$ are reduced to make room for the notional 'p bits' mentioned earlier. This means that more puncturing or less repetition is carried out to make room for the gap(s). See Section 4.8 for more details.

4.4.3.2 Downlink PhCH Allocation for Flexible TrCH Positions

For flexible TrCH positions, the PhCH capacity allocated to a given TrCH is not fixed and depends upon the TF l. However it is the same for all TFCs containing TF l. The formula for calculating the PhCH allocation for TrCH i and TF l is given below, using the following nomenclature:

$\Delta N_{i,l}^{\text{TTI}}$ number of punctured or repeated bits per TTI from TrCH i, for TF l;

$N_{\text{PhCH},i,l}$ number of PhCH bits per frame allocated to TrCH i for TF l;

$T F_i(j)$ TF that is used on TrCH i in TFC j.

The calculation is split into two stages:

1. First, tentative values of $N_{\text{PhCH},i,l}$ are calculated for each TF that is contained in a TFC.

$$N_{\text{PhCH},i,l} = \left\lceil \frac{\left(RM_i \times N_{i,l}^{\text{TTI}}/F_i\right) \times N_{\text{PhCH}}}{\max_{j \in \text{TFCS}} \sum_{m=1}^{I} \left(RM_m \times N_{m,T F_i(j)}^{\text{TTI}}/F_m\right)} \right\rceil \tag{4.5}$$

for all TFs l used in the TFCS. This splits the PhCH capacity between the TrCHs depending on the product of the number of bits per frame $N_{i,l}^{\text{TTI}}/F_i$ and the rate matching attribute RM_i for a given TrCH, relative to the total sum of this product across all TrCHs for the maximum rate TFC.

2. The second phase corrects these allocations in the cases when the total tentative allocation for a TFC exceeds the available PhCH capacity. This excess may occur because of the rounding up operation in Equation (4.5).

Details of the second phase correction procedure are given in [50]. In essence, every TFC is checked to ensure that the $N_{\text{PhCH},i,l}$ values for each TrCH given from stage one do not exceed the PhCH capacity N_{PhCH}. If they do then the value of $N_{\text{PhCH},i,l}$ is recalculated for all TrCHs i and each TF l used in this TFC. This recalculation ensures that the PhCH capacity is not exceeded by using the more conservative approach of cumulative TrCH allocations and rounding down, similar to that shown in Equations (4.3) and (4.1). If the allocation for a given TF is corrected, the updated value is used for all TFCs that contain this TF. Note that the PhCH capacity required by each TFC in the TFCS is checked in order of ascending TFCI; checking the TFCs in a different order may result in updating different TFs.

The number of bits to be punctured or repeated per TTI is then calculated from the final value of $N_{\text{PhCH},i,l}$

$$\Delta N_{i,l}^{\text{TTI}} = F_i \times N_{\text{PhCH},i,l} - N_{i,l}^{\text{TTI}}. \tag{4.6}$$

4.4.3.3 Downlink Rate Matching Parameter Calculation
In this section we detail how the number of punctured or repeated bits $\Delta N_{i,l}^{\text{TTI}}$ or $\Delta N_{i,\text{max}}^{\text{TTI}}$ calculated above are converted into values of X_i, ΔN_i, e_{ini}, e_{plus}, and e_{minus} for the rate matching pattern generation.

For TrCHs using convolutional coding or TrCHs where repetition is used, the parameters are given as follows. A notional total number of bits N_i is defined as follows:

$$N_i = N_{i,\text{max}}^{\text{TTI}} \text{ for fixed TrCH positions}$$
$$= N_{i,l}^{\text{TTI}} \text{ for flexible TrCH positions}$$

The rate matching parameters are then given by:

$$\Delta N_i = \Delta N_{i,\text{max}}^{\text{TTI}} \text{ for fixed TrCH positions}$$
$$= \Delta N_{i,l}^{\text{TTI}} \text{ for flexible TrCH positions}$$

$$X_i = N_{i,l}^{\text{TTI}}$$
$$e_{\text{ini}} = 1$$
$$e_{\text{plus}} = 2 \times N_i$$
$$e_{\text{minus}} = 2 \times |\Delta N_i|$$

It can be seen that the parameters e_{plus} and e_{minus}, which determine the repetition or puncturing ratio, are given based on the notional total number of bits N_i and number of bits to be punctured or repeated ΔN_i. For flexible TrCH positions, where the rate matching pattern varies with the TF used, these are given by the actual numbers for this TF, whereas for fixed TrCH positions where the rate matching pattern is fixed, they are given by the values for the maximum rate TF. Note that when using fixed TrCH positions, the value of ΔN_i above will give the actual number of bits punctured or repeated only for the maximum rate TF.

A similar approach is used for TrCHs using turbo coding where puncturing is used, where the two parity streams are punctured separately and the systematic bits are not punctured. For these TrCHs the notional total number of bits for each parity stream is given by:

$$N_i = N_{i,\text{max}}^{\text{TTI}}/3 \text{ for fixed TrCH positions}$$
$$= N_{i,l}^{\text{TTI}}/3 \text{ for flexible TrCH positions}$$

A notional total number of bits to be punctured across both parity streams is also defined as follows:

$$\Delta N_i^{\text{total}} = \Delta N_{i,\text{max}}^{\text{TTI}} \text{ for fixed TrCH positions}$$
$$= \Delta N_{i,l}^{\text{TTI}} \text{ for flexible TrCH positions}$$

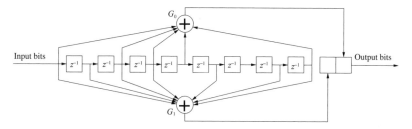

Figure 4.8. Half rate convolutional encoder.

The rate matching parameters are then given by:

$$\Delta N_i = \left\lfloor \Delta N_i^{\text{total}}/2 \right\rfloor \quad \text{for the first parity stream}$$
$$= \left\lceil \Delta N_i^{\text{total}}/2 \right\rceil \quad \text{for the second parity stream}$$

$$X_i = N_{i,l}^{\text{TTI}}/3$$
$$e_{\text{ini}} = N_i$$
$$e_{\text{plus}} = a \times N_i$$
$$e_{\text{minus}} = a \times |\Delta N_i|$$

where $a = 2$ for the first parity sequence and $a = 1$ for the second parity sequence. Note that as turbo encoding is one third rate, the number of coded bits per TTI will always be a multiple of three, so the divisions by three above will always give a whole number.

Having detailed the rate matching calculations and operation for both the uplink and downlink, in the next two sections we describe convolutional and turbo encoding schemes detailed in [50], and methods to decode these codes.

4.5 CONVOLUTIONAL ENCODING AND DECODING

Convolutional coding is the most frequently used channel coding scheme in UMTS. It is specified in Section 4.2.3.1 in [50], and was summarized in Table 4.4 in Section 4.3.4. Both half and one third rate binary codes are specified, with a constraint length K of nine and generator polynomials as shown in Table 4.4. In this section we explain what these terms mean, and how the codes are typically decoded.

4.5.1 Convolutional Encoding

In the context of convolutional codes, a binary code simply means that the encoder deals with one input bit at a time. For the half rate code, two output bits are produced for each input bit. Similarly, for the one third rate code, three output bits are given for each input bit. Constraint length nine means that each output bit is determined by a total of nine input bits, the current and the eight previous input bits.

This means that the encoders can be represented as a delay line with eight delays, as shown in Figures 4.8 and 4.9. The two or three output bits for each input bit are given by

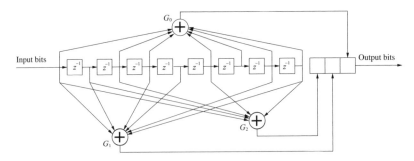

Figure 4.9. One third rate convolutional encoder.

an exclusive OR (XOR) operation on various bits in the delay line and on the input bit. To encode a block of data, the data is simply shifted through the delay line one bit at a time, and for each input bit two or three output bits are produced.

Which bits in the delay line are connected to each of the two or three XORs is denoted by a generator polynomial for each XOR operator. Each bit in the generator polynomial represents a possible connection to the XOR, a one denotes that the connection is made and a zero means the connection is not made. So, for example, a generator polynomial of nine ones, or 777 in octal notation, would mean that all nine connections to the XOR operator were made.

For the half rate code defined in UMTS, two polynomials are defined $G_0 = 561$ and $G_1 = 753$ in octal notation. This results in the structure depicted in Figure 4.8.

The one third rate code has three polynomials defined, $G_0 = 557, G_1 = 663$ and $G_2 = 711$, again in octal notation, whose structure is depicted in Figure 4.9.

These particular polynomials were chosen as, amongst all the possible polynomials for a half or one third rate binary $K = 9$ encoder, these choices give the maximum possible value for the minimum Hamming distance d_{free} [47] between different output code words from the encoders. This value d_{free} simply gives the number of bits which are different between the two 'closest' output sequences from the encoder. Maximizing this value gives the best performance in Gaussian channels. However, it is interesting to note that in other channels other polynomials can give an improved performance [69].

Note the similarity in the terminology here to the generator polynomials described in Section 4.3.3 for the CRC codes. However, following [50], we have used a different notation to represent the polynomials for the convolutional encoders. For the CRCs the generator polynomials were shown explicitly in polynomial form (e.g. $D^8 + D^7 + D^4 + D^3 + D + 1$ for the eight bit CRC), whereas here they are shown in octal form. The two forms are exactly equivalent to each other; a certain power of D being present in the polynomial form means the equivalent bit is one in the octal form. So, for example, the octal equivalent for the eight bit CRC generator polynomial above is 633.

As described above, for each input bit the outputs are determined by the current and the previous eight input bits. The previous eight input bits in the delay line memory of the encoder can be considered as the 'state' of the encoder. Each input bit can then be thought of as generating a set of two or three output bits depending on the input bit itself and on the state of the encoder. Each input bit also moves the encoder to a new state. This way of considering the encoding process as moving a state machine through a series of states, generating output bits on each transition, is useful when we consider decoding convolutional

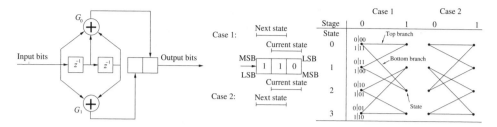

Figure 4.10. Half rate $K = 3$ convolutional encoder and the corresponding trellis diagram.

codes below. For the constraint length nine codes used in UMTS, the encoder state machine has $2^8 = 256$ states.

Before each code block is encoded, the initial state of the encoder delay line is set to all zeroes. Also eight tail bits with values zero are added to the end of each code block in order to drive the encoder state machine back to this all zero state. These eight tail bits are encoded in the same way as the data bits, giving two or three output bits for each tail bit. This is the reason for the '+8' terms in the number of output bits Y_i for each code block shown in Table 4.4.

We now expand this explanation of convolutional codes, and their representation, using a simpler four state encoder. This encoder is used later in a worked example of the operation of the algorithm used to decode convolutional codes. Also, the ideas on the representation of convolutional encoders are used in the explanation of this decoding algorithm.

4.5.1.1 Example Convolutional Encoder
Due to space limitations, we use a four state code to illustrate our discussions, rather than the 256 state codes introduced above. Our example code is half rate, with generator polynomials $G_0 = 7$ and $G_1 = 5$, and is shown on the left hand side of Figure 4.10.

The centre diagram in Figure 4.10 shows a shift register representation of the same encoder, with the three registers filled with bits 1 1 0. For clarity, the output XOR operators are not shown. This representation of the encoder is useful for deriving a trellis diagram for the encoder, as shown on the right hand side of Figure 4.10. This representation shows for each state of the encoder what the next state will be for an input bit of either zero or one. Also shown is what the corresponding output bits will be. So, for example, 1|01 at state two in the Case 1 trellis indicates that in this state if an input bit of one is received, the corresponding output bits will be zero then one. In this case the bottom branch from state two will be taken, and the new state of the encoder will be state three.

The delay line and shift register representations of the encoder shown on the left hand side and the centre of Figure 4.10 are of course exactly equivalent. Which is used depends on what is being illustrated and is often a matter of preference by the author. So, for example, the delay line representation is used in [70–73] and the shift register representation in [47, 74]. In this book we usually follow the specification [50] and use the delay line representation.

Another matter of notation is the location of the *Most Significant Bit* (MSB) within the shift register when generating the trellis diagram. Treating the most recent bit as the MSB in the state index yields the trellis shown as 'Case 1' in Figure 4.10 and used for example in [47, 73], whereas treating this as the *Least Significant Bit* (LSB) [70–72, 74] yields

'Case 2'. Again, which definition is chosen is a matter of preference by the author, and has no effect on the performance of the codes or on the decoding process. Here we have chosen to treat the most recent input bit as the MSB of the state, i.e. Case 1 in Figure 4.10. Either trellis, Case 1 or Case 2, is constructed by loading the shift register in sequence with all possible combinations. Then the content in the shift register sections denoted *current state* and *next state* are used to construct the lines between the different states in the trellis. In our example, the shift register is loaded with bits 1 1 0, therefore there exists a branch between current state two and next state three. By going through all eight possible shift register bit combinations, the trellis can be generated.

4.5.2 Terminology

Before describing how convolutional codes can be decoded, we first introduce some terminology that is used in this description and in our discussion on turbo codes and decoding in Section 4.6. To aid readability, we neglect the TrCH and code block indices i and r used in the notation o_{irk} and y_{irk} in [50] to represent the input and output bits for the encoder. Instead we use the notation u_k to represent the kth input bit to an encoder for any particular code block and TrCH. The bit index k runs from 1 to K_i, where K_i are the number of bits per code block for this TrCH. Similarly, y_{kl} are the encoded bits corresponding to the input bit u_k. The index l runs from 1 to n, where $n = 2$ for the half rate code and $n = 3$ for the one third rate code. Finally, we use the notation $\hat{y}_{kl}$ to represent the received version of each encoded bit y_{kl}, and $\hat{u}_k$ to represent the decoded version of each encoder input bit u_k.

The logical values of the encoded bits y_{kl} produced by the encoders shown earlier are either 0 or 1. When we discuss 'hard decision' decoders, the received versions of these bits are also either 0 or 1. For such decoders the demodulator makes a definite, 'hard', decision about whether each received bit $\hat{y}_{kl}$ is more likely to be a 0 or a 1 and passes just this decision to the decoder. So, for hard decision decoders, both the encoded bits y_{kl} and the received version of these bits $\hat{y}_{kl}$ are either 0 or 1.

As described later, better performance is achieved using 'soft decision' decoders. For these decoders the demodulator passes 'soft' information to the decoder about the value of the received bits. For such soft decision decoders, in a QPSK system such as UMTS, the encoded bits y_{kl} can be considered to be $+1$ or -1 and the received version $\hat{y}_{kl}$ of these bits can be thought of as a real number. A large positive number for $\hat{y}_{kl}$ would mean the demodulator thought the encoded bit y_{kl} very likely to be $+1$, while a small negative number for $\hat{y}_{kl}$ would mean the demodulator thought the encoded bit y_{kl} slightly more likely to be a -1 than a $+1$. This representation of y_{kl} and $\hat{y}_{kl}$ eases the implementation, and description, of the decoders and so we use it when considering soft decision Viterbi decoders and turbo decoders. For the purpose of a discussion on Viterbi and turbo decoders, the mapping between the logical values 0 and 1 and the real values -1 and $+1$ for the bits y_{kl} is arbitrary. We have taken logical 0 to map to -1 and logical 1 to map to $+1$.

4.5.3 Decoding of Convolutionally Encoded Data: the Viterbi Algorithm

The *Viterbi Algorithm* (VA) [66] is the usual method to decode convolutional codes. It chooses the most likely *sequence* of bits for each code block, and so is often referred to

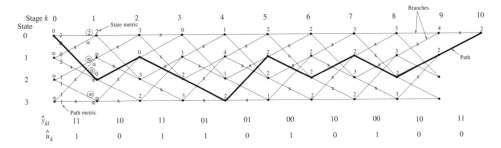

Figure 4.11. The decoding procedure based on the Viterbi algorithm.

as a *Maximum Likelihood Sequence Estimator* (MLSE). Note the emphasis on the word *sequence* above; as will be discussed later, this is not the same as choosing the most likely value for every given bit. However, in practice the difference is negligible, and so the VA is the detector or decoding algorithm of choice for convolutionally encoded data.

Note that the VA is also widely used in many applications beyond decoding convolutional codes. A discussion of some of these applications can be found in [75, 76] and [77], here we concentrate on the algorithm's use in decoding convolutional codes.

4.5.3.1 *Operation of the Viterbi Algorithm*   We now describe the operation of the Viterbi algorithm. We use the example encoder described above to illustrate our discussion.

The trellis in Figure 4.10 shows the transitions possible for our example encoder for one input bit. This trellis can be extended to show the transitions taken by the encoder whilst encoding a sequence of input bits. Such a trellis is shown in Figure 4.11. This is merely the trellis from our example encoder, extended to cover ten input bits and so ten transitions of the encoder state machine. Various additional features are shown in Figure 4.11 to represent the calculations the VA makes – these will be described later.

The VA works by using the received version of the encoded bit sequence to find the most likely path through this trellis representing the state machine of the encoder. Once this most likely path through the trellis is known, the data bits which would have caused the encoder to follow this path can be implied and these bits are the output from the VA.

For each path through the trellis (i.e. each sequence of input bits at the encoder) we can calculate the Hamming distance between the received bit sequence and the sequence that would have been transmitted if this path was taken at the encoder. It can be shown [47], given hard decisions input to the decoder, that the path through the trellis that minimizes this Hamming distance is the most likely, or *Maximum Likelihood* (ML), path.

The 'brute force' way of finding this ML path would be to calculate the Hamming distance between the received sequence and every possible transmitted sequence. However, for an input sequence of K_i bits there are 2^{K_i} possible transmitted sequences. Calculating the Hamming distance for each path independently will require of the order of K_i operations per path, as each path is of the order of K_i encoded bits long. Hence, the number of operations required for this brute force approach will be of the order of $K_i \times 2^{K_i}$. This approach very quickly becomes impractical as the number of bits per code block K_i increases. For example, for a code block of length $K_i = 100$ bits, we would have of the order of 10^{32} or 100 million, million, million, million, million operations.

The VA dramatically simplifies this search for the ML path by working through the trellis one stage at a time. Note the terminology here; we use the term 'stage' to mean the set of states that the trellis may move to as a result of a given input bit. For any binary convolutional code, including our example code, each state will have two transitions into it from states at the previous stage of the trellis. The key point in the VA is that once two paths merge at a given state and stage in the trellis, the path with the higher Hamming distance can be discarded. There is no need to consider this path any further, as we can be sure it will never become the minimum Hamming distance (i.e. the ML) path. This is because once two paths merge at a given state and stage in the trellis, they both have exactly the same possible transitions through the trellis ahead of them. Hence the path with higher Hamming distance at this state/stage will never be able to become a better path than the selected one, and can be discarded.

Also note that the Hamming distance for each of the two paths considered at each state in the trellis can be calculated very simply from the Hamming distance for the preceding state in the path, plus the incremental Hamming distance. In our example code this incremental Hamming distance is just the distance between the two bits that were received and those that would have been transmitted if the given branch was taken at the encoder. Hence, the calculation of the two Hamming distances at each state for each stage in the trellis is just a simple addition of a previous distance plus an incremental distance.

So the VA works by, for each stage in the trellis, finding for each state the Hamming distance of the best path through this state. These calculations are based on the best Hamming distances at the previous stage in the trellis, and the incremental Hamming distance for the various transitions possible from the previous stage in the trellis to this stage. For each state there are only two paths from the previous stage to this stage, and so only these two states need to be considered. The winning path is selected, and its Hamming distance is stored to be used at the next stage in the trellis. Also an indicator of which of the two paths was selected is stored. The algorithm proceeds like this through the trellis until it reaches the end of the trellis. At this point we can trace back along the winning path using the stored information about which path was selected for each state at every stage in the trellis. This trace back allows us to identify the ML path and hence the input bits that caused the encoder to take this path.

As the VA needs to consider only two paths for each state at each stage in the trellis, and the Hamming distance for each can be found based on a single sum for each path, the complexity of the VA is of the order of the number of states at each stage in the trellis multiplied by the number of stages in the trellis. Mathematically this is $K_i \times 2^{K-1}$, where K_i is the length of the code block and K is the constraint length of the codes (nine for the codes used in UMTS). For any reasonable K_i and K values this means that the VA is much less complex than the 'brute force' search described above, whose complexity was of the order $K_i \times 2^{K_i}$. For example with a code block of 100 bits and the $K = 9$ UMTS codes, we will have of the order of 25 thousand operations, as opposed to the 10^{32} operations for the brute force search. Yet the VA will find the same ML path as the brute force search.

Although for simplicity in the description above we detailed the VA minimizing the Hamming distance between the received bit sequence and possible encoded sequences, other measures tend to be used in practice. These other measures are detailed in Section 4.5.5. Therefore for generality, and also for brevity, instead of referring to Hamming distances below we refer to 'metrics'. This is the term that tends to be used to describe Viterbi decoders in the literature. For our particular example using hard decisions, the 'state metric' referred to below for a given state is just the Hamming distance between the received sequence and the best path to that state. Similarly, the 'incremental metrics' are just the incremental Hamming distances described above. Finally, the 'path metric' is the metric associated with a given path at a certain stage in the trellis. At each state two path metrics are calculated from old state metrics, and one is chosen as the new state metric.

We now summarize these ideas in a step by step procedure describing the decoding process.

4.5.3.2 *Step By Step Operation of the Viterbi Decoder*   The step by step operation of the VA can be stated as follows. We know the encoder starts with initial state zero, so we set the metric for this state to 0 and for the other states to ∞. This ensures that no path from one of these other states will be picked as the winning path. Start at state 0 at stage $k = 0$ and carry out step 1 below for the received bits for this stage. Then go through steps 2 to 4 for each state at this stage. Then move to the next stage and repeat these steps for the states at this stage. Continue like this until the end of the trellis has been reached. Then the trace back comes into play. As we know the trellis was terminated, start at state 0 at the last stage and follow the branches selected at each stage back to state 0 at stage 0. During this process the decoded bits (denoted $\hat{u}_k$) can be generated as we have stored the bits associated with each branch beforehand, i.e. step 4.

In summary, the VA comprises two processes, state metric computation and trace back, and uses the following steps:

1. Work out the incremental metric between the expected bits and the received bits for each branch at this stage.

2. Add each incremental metric to the previous state metric, this yields two competing path metrics at each state.

3. Compare the two path metrics and select the smaller one. Store this value as the new state metric for this state.

4. Store an indication of the source state of this branch, e.g. the LSB of the previous state, and the bit that caused this transition.

5. Repeat steps 2–4 for each state at this stage in the trellis. Then move to the next stage in the trellis and return to step 1.

6. Commencing at state 0 at the last stage of the trellis, use the stored trace back information from step 4 to follow the ML path backwards through the trellis.

7. At each stage of the trace back output a 0 or 1 as the decoded bit, depending on the bit stored in step 4 for the selected transition.

Note that at step 4 above, in practice it is not necessary to store both the previous state indicator and the bit that caused the selected transition. Only the previous state indicator is needed, as given a certain transition the bit that caused this transition can be implied from the properties of the encoder. In fact, because of the two bit memory of the encoder, if we store the LSB of the previous state as the indicator, this bit is actually the decoded bit from two stages earlier. As discussed in Section 4.5.6 the trace back memory is a significant percentage of the memory required by a Viterbi decoder, so this is a useful trick for real implementations.

4.5.4 Example Viterbi Decoding

We now describe the operation of the VA for our example code, shown in Figure 4.10, with a given input bit sequence. For this code we can see that the input bit sequence 1011010100 results in the encoded bit sequence 11, 10, 00, 01, 01, 00, 10, 00, 10, 11. The encoder state machine, starting at state zero, runs through the state sequence 2,1,2,3,1,2,1,2,1,0 with this set of input bits. Notice that this sequence of input bits includes two tail bits (0,0) to drive the encoder state machine back to its all zero state.

Now let us assume that the receiver observes the sequence 11, 10, 11, 01, 01, 00, 10, 00, 10, 11 as the received version of these encoded bits. This sequence has two encoded bits in error due to channel and noise artefacts.

The example in Figure 4.11 shows the path metric (Hamming distance in our example) on the left and the updated state metric on the right for the transitions between stage $k = 0$ and $k = 1$ in the trellis. We have encircled the path metric with the smaller value at each state, and discarded the branch with the larger value (marked X). The encircled value becomes the new state metric for stage $k = 1$ in the trellis, and is shown on top of each state. These updated state metrics, together with the paths that are discarded at each state, are shown throughout the trellis. It is a useful exercise for the reader to verify the selected path chosen and the updated state metric at each state/stage.

Starting at stage $k = 10$ and state 0, the winning path is traced back through the trellis. For this transition we observe that the winning path was the one from state 1 in the previous stage of the trellis, and the associated bit that would have caused this transition is a zero. Thus we follow this path and output a decoded bit of zero. Similarly at stage $k = 9$ we note that the selected branch back from state 1 is from state 2, so we follow this branch back. At each state, there is only one branch available, therefore we have a unique path back through the trellis. In our example this ML path is shown in bold. We observe that for our example the Viterbi algorithm has correctly identified the path through the trellis that the encoder took, and so correctly estimated the transmitted bits sequence. This is despite the fact that we corrupted bits 5 and 6 of the encoded sequence, and is an example of the error correction capability inherent in convolutional codes.

4.5.5 Use of Other Metrics in the Viterbi Algorithm

As described above, the VA determines the most likely path through the encoder's trellis by calculating for each state, at every stage of the trellis, a 'path metric' for the most likely path through this state. In our example above this path metric was merely the Hamming distance between the received bit sequence $\hat{y}_{kl}$ and the sequence y_{kl} that would have been transmitted if the encoder had followed this path to this point in the trellis. The VA finds the path through the trellis that minimizes this Hamming distance.

More generally, the path metric can be thought of as a representation of how likely it is that the encoder took the given path through the trellis, given the version of the encoded bits that was received. By working through the trellis choosing which of the two possible paths has the best metric for each state at each stage in the trellis, the Viterbi algorithm identifies the maximum likelihood path.

There are various measures that can be used to calculate these metrics. Although for simplicity in our example above we used hard decisions and the Hamming distance metric, better performance can be obtained using soft decisions. In fact as shown in [47] using soft decisions gives an additional gain of about 2 dB over hard decisions. Hence in practice soft decisions are used in most applications. The operation of a Viterbi decoder using such soft decisions is very similar to the example described above; again two path metrics are computed for each state at each stage in the trellis, and the best is selected as the survivor and the new state metric.

A suitable 'soft' path metric for the selected path at state $\grave{s}$ at stage $k - 1$ in the trellis, to state s at stage k in the trellis, is $M_k(s, \grave{s})$, where

$$M_k(s, \grave{s}) = M_{k-1}(\grave{s}) + \sum_{l=1}^{n} \hat{y}_{kl} y_{kl} \tag{4.7}$$

Here, y_{kl} are the bits (values -1 or $+1$) that would have been transmitted for the input bit u_k if the encoder took the transition from state $\grave{s}$ at stage $k - 1$ in the trellis to state s at stage k, and $\hat{y}_{kl}$ are the received versions of these bits. The summation is taken over the $n = 2$ or $n = 3$ encoded bits for each input bit for the half, or one third, rate code. $M_{k-1}(\grave{s})$ is the metric at state $\grave{s}$ at the previous stage in the trellis. Notice that, as for the Hamming distance metric, this soft metric is calculated recursively based on the metrics at the previous stage in the trellis and so it too can be calculated relatively simply.

For an *Additive White Gaussian Noise* (AWGN) channel and the QPSK modulation used in UMTS, this path metric is proportional[5] to the log of the probability that the encoder took the selected path to state s at stage k in the trellis. Hence, maximizing this metric means the VA will choose the most likely path through the trellis. Notice the contrast with the hard decision decoder described earlier, where the path with the *lowest* metric is the more likely and so is selected in the comparison.

Next we describe techniques to ease the implementation of the VA using the soft metric detailed above.

4.5.6 Implementation of the Viterbi Algorithm

The implementation of a Viterbi decoder is limited by the number of operations which are needed, and the amount of memory needed to store working variables. In this section we discuss how these operations and memory can be minimized. First we detail the operations needed in the metric calculations, and how these can be considered as sets of 'butterfly' operations. We then discuss the bit width needed in calculating and storing the metrics. Finally, we detail the memory needed for the trace back information, and how to reduce this memory.

4.5.6.1 *Metric Calculation and Butterflies*   The bulk of the operations needed in the VA are to calculate the path metrics $M_k(s, \grave{s})$ given in Equation (4.7). For each state s at each stage k in the trellis, this path metric is calculated for each of the two paths leading to this state. The path with the largest (i.e. the best) metric is selected as the survivor, and the other path is discarded. The $\sum_l \hat{y}_{kl} y_{kl}$ term in Equation (4.7) is the 'incremental metric'; it

---

[5] Other than a fixed additive term that is the constant for all paths through the trellis to a given stage k, and so does not affect which path is chosen.

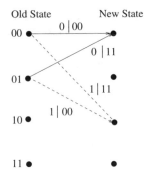

Figure 4.12. Viterbi butterfly in four state trellis.

is the increase in the metric due to a transition from state $\grave{s}$ at stage $k-1$ in the trellis to state s at stage k in the trellis. Let us denote this incremental metric as $\gamma_k(s, \grave{s})$, with

$$\gamma_k(s, \grave{s}) = \sum_{l=1}^{n} \hat{y}_{kl} \, y_{kl} \tag{4.8}$$

where $\hat{y}_{kl}$ are the n received symbols at the given stage in the trellis, and y_{kl} are the n transmitted symbols for the transition from state $\grave{s}$ to state s. In a QPSK system such as UMTS, the transmitted bits can be considered to be $+1$ or -1, so the calculation of an incremental metric reduces to $n-1$ additions/subtractions. So, for our example code shown in Figure 4.10, $\gamma_k(0, 0)$, giving the incremental metric for the transition from state 0 to state 0 is given by

$$\gamma_k(0, 0) = -\hat{y}_{k1} - \hat{y}_{k2}$$

Note that as there are $n=2$ or $n=3$ output bits per transition, there are only $2^2 = 4$ or $2^3 = 8$ different sets of bits y_{kl} that can be generated by the encoder. Also, half of these are the inverse of the other half. Hence there are only two or four different γ values that need to be calculated from the two or three received soft decisions for each stage of the trellis.

Using these incremental metrics, the winning state metric at stage k in the trellis for state s is given by

$$M_k(s) = \max\left((M_{k-1}(\grave{s}) + \gamma_k(s, \grave{s})), (M_{k-1}(\tilde{s}) + \gamma_k(s, \tilde{s}))\right) \tag{4.9}$$

where $\grave{s}$ and $\tilde{s}$ are the two states with paths leading to the state s.

As noted in Section 4.5.1 the convolutional codes used in UMTS have 256 states. Thus at each stage we have 256 states for which to calculate new state metrics $M_k(s)$. The transitions from the states at one stage in the trellis to those at the next stage can be considered as 128 'butterfly' operations. One such butterfly is shown in Figure 4.12, for the simple case of our four state example code.

In each butterfly two paths each from two old states are considered. As can be seen from Figure 4.12, one path from each of these two old states leads to one new state, the other two paths from the two old states lead to another new state. From Equation (4.9), the new metrics for states 0 and 2 in Figure 4.12 are given by:

$$M_k(0) = \max\left((M_{k-1}(0) + \gamma_k(0, 0)), (M_{k-1}(1) + \gamma_k(0, 1))\right)$$
$$M_k(2) = \max\left((M_{k-1}(0) + \gamma_k(2, 0)), (M_{k-1}(1) + \gamma_k(2, 1))\right) \tag{4.10}$$

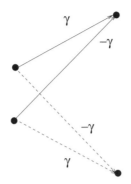

Figure 4.13. Incremental metrics in Viterbi butterfly.

Thus for each butterfly operation, we need to do four additions, and two compare/select operations. These operations are often referred to as 'Add, Compare, Select', and are the bulk of the operations required in a Viterbi decoder.

Careful examination of the encoders shows that, as the MSB and LSB of all the n generator polynomials is one for both our example $K = 3$ encoder and the $K = 9$ encoders used in UMTS, the four transitions in a certain butterfly give only two different sets of transmitted bits y_{kl}. Also, one of these sets is the inverse of the other. This can be seen from Figure 4.12 for our example code; the transmitted bits for all four transitions are either 00 or 11. Hence, the incremental metrics used for all four transitions will be given by either $+\gamma$ or $-\gamma$, where γ is the incremental metric for any one of the four transitions in the butterfly. So the butterfly from Figure 4.12 can be redrawn as shown in Figure 4.13.

Considering the 256 states of the trellis for the UMTS convolutional code as 128 butterflies, and calculating 'new' metrics from 'old' using operations like those shown in Equation (4.10), is useful as it allows two new state metrics to be calculated in one set of operations from two old state metrics. Only one incremental metric γ is used, and this is calculated using $n - 1$ additions or subtractions of the received symbols $\hat{y}_{kl}$.

4.5.6.2 *Soft Decision and Metric Bit Widths*   Quantization of the input soft decisions to a Viterbi decoder is an important factor in both the performance of the decoder and its complexity. At one extreme are hard decisions, which can be considered to be soft decisions quantized with one bit per decision. At the other extreme the soft decisions can be quantized as very high precision floating bit numbers, using perhaps 32 or 64 bits per input. Such high precision soft decisions increase the size and complexity of the adders needed in the *Add Compare Select* (ACS) units, and the memory needed to store the 256 'old' and 'new' metrics for each stage of the trellis. Real Viterbi decoders typically use either three or four bit input soft decisions as in practice it is found this gives performance very close to that with no quantization.

We now consider the size of the path metrics given quantized input soft decisions. With three bit soft decisions and a half rate code, as the metrics at the first stage of the trellis are given by the addition of two three bit numbers, they may be up to four bits long. At the second stage in the trellis the metrics will have had another two three bit numbers added, and so may be up to five bits long. More generally, for a trellis L stages long, with three bit input soft decisions, the metrics may be up to $\lceil \log_2(L) + 3 + \log_2(n) \rceil$ bits long. The trellis size in UMTS is limited to $L = 512$ stages (as the maximum code block size is 504 and we have eight termination bits), and so the maximum metric size for the one third rate code is 14 bits. Hence the adders in the ACS units need to be able to cope with 14 bit inputs, and the 512 storage units for 256 old and new metrics should be at least 14 bits wide.

Limiting the input soft decisions to three bits each, and the fact that the trellis length in UMTS is limited to 512 stages, has allowed us to limit the maximum size of the metrics to 14 bits. The bit width used to store and compute these metrics can be further reduced by noting that it is only the difference between metrics, and not their absolute values, that is used in the VA. Hence if we subtract a constant from all metrics at a given stage in the trellis, we will not affect which path is chosen as the ML path. Typically, this fact is used to reduce the width of the metrics by periodically subtracting a constant from each metric to keep the maximum metric within a lower number of bits. This is known as metric rescaling. Another approach is to allow the metrics to overflow, but in a controlled manner. It can be shown [78] that if the maximum difference between two metrics is M then, if the metrics are represented with at least $\lceil \log_2(2 \times M) \rceil$ bits, the metrics can be allowed to overflow whilst still allowing the VA to determine which of two merging paths has the largest metric. Hence this overflow does not affect which path is chosen as the ML path by the VA.

Using either of these techniques with the convolutional codes used in UMTS allows the metrics to be represented using eight bit numbers. Thus eight bit adders can be used in the ACS units, and $2 \times 256 \times 8 = 4096$ bits (or 512 bytes) of storage is needed to store the 256 'old' and 'new' metrics for each stage of the trellis. Typically, the 'controlled overflow' [78] technique is used in dedicated hardware implementations of the Viterbi decoder, whilst the metric rescaling approach is used in software implementations on embedded DSPs.

4.5.6.3 Trace Back Memory

The other major memory requirement for the VA comes from the trace back information stored for each state at each stage in the trellis to record which of the two merging paths was selected as the survivor. This information is used to trace back and release the decoded bits once the ML path has been found. Assuming one bit of trace back information is stored for each state at each stage of the trellis, as the trellis used in UMTS is up to 512 stages long, storing all this trace back information requires $512 \times 256 = 131\,072$ bits, or 16 kbytes, of trace back information. Because the trellis size has been limited to 512 stages by code block segmentation, this memory requirement is not excessive and so the trace back information can be stored like this. Alternatively it may be reduced, at the cost of extra complexity, using what is known as segmented trace back.

Observations of the Viterbi algorithm show that all the surviving paths at a stage l in the trellis will normally have come from the same path (which will be the ML path) within at most δ transitions before l. The value of δ is usually set to be around five times the constraint length of the convolutional code. This means that the value of the decoded bit $\hat{u}_k$ on the ML path can be calculated by tracing back from any state at stage $k + \delta$ in the trellis. Whichever state we trace back from, the surviving path at this state will at some point have diverged from the ML path and we will reach the same transition at stage k in the trellis. This means the value of the decoded bit $\hat{u}_k$ will be the same regardless of which state we trace back from.

This principle can be seen from our example trellis shown in Figure 4.11. This trellis is for a $K = 3$ encoder, and so taking δ to be five times the constraint length we have $\delta = 15$. At stage $k = 9$ in the trellis the ML path passes through state 1. Careful examination of the trellis shows that the surviving paths for the three other states at this stage diverge from the ML path at stage 2 in the trellis. Hence we would correctly decode u_0 and u_1 by tracing back from any of the four states at stage 9 in the trellis.

In our example the trellis is only ten stages long, so there is no saving in trace back memory to be made from this observation. However, for the 256 state code used in UMTS storing trace back information for only δ transitions for each state, where δ is set to five times the constraint length of the code, reduces the memory requirement for trace back information from 16 kbytes to $45 \times 256 = 11\,520$ bits or less than 1.5 kbytes.

4.5.6.4 *Overall Memory and Calculation Requirements for VA*   With the techniques described above we can see that a Viterbi decoder for the convolutional codes defined for UMTS can be implemented using 128 butterfly ACS operations for each stage in the trellis. With four additions or subtractions per butterfly, this gives $128 \times 4 = 512$ addition/subtraction operations and 128 compare/select operations per trellis stage. Using metric rescaling 256 subtractions may also be required. Also, at each stage in the trellis two or four incremental metrics must be calculated using one or two addition/subtraction operations. This gives an additional $4 \times 2 = 8$ addition/subtraction operations per trellis stage for the one third rate code. Using the techniques described above to limit the size of the metrics, and segmented trace back, about 2 kbytes of memory is needed. This makes implementation of the Viterbi decoder perfectly achievable in either dedicated hardware or on an embedded DSP in UMTS terminals. In the next section we discuss the performance of convolutional codes and the Viterbi decoder.

4.5.7 Performance Results

Here we present a set of Monte Carlo simulation curves showing the performance of convolutional codes in an additive white Gaussian noise (AWGN) channel. These curves illustrate the performance gains given by convolutional codes over uncoded transmission and the additional gain given by increasing the constraint length or decreasing the coding rate. They should also enable the reader to verify his own implementation. All simulation results were derived using models in C, use soft decisions and floating point implementations.

Our results show the *Bit Error Ratio* (BER) against the bit energy to noise power spectral density ratio E_b/N_0. The transmitter amplitude is constant and set to unity, and noise power σ^2 is varied to satisfy a given E_b/N_0 ratio according to

$$\sigma^2 = \frac{1}{2 \times 10^{\frac{E_b/N_0 \text{(dB)}}{10}}} \tag{4.11}$$

for a half rate code. Care must be taken when comparing results like the ones presented below with other results. The term *Signal to Noise Ratio* (SNR) may be used to denote the energy for all bits per block, while E_b may be used to denote only the information bits. This means that the curves are shifted (left or right) but have the same shape.

The first graph, Figure 4.14, shows the BER in an AWGN channel for a 1/2 rate code with constraint length $K = 3, 7$, and 9, and a 1/3 rate code with $K = 9$ against the BER for an uncoded QPSK system. The generator polynomials used in all cases are the maximum d_free choices given in [47], which as noted earlier for $K = 9$ are also the codes defined for UMTS.

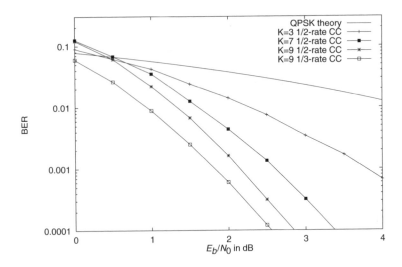

Figure 4.14. The BER versus E_b/N_0 for 1/2 and 1/3 rate CC for different constraint lengths in AWGN.

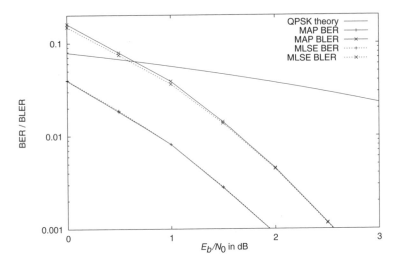

Figure 4.15. The BER and BLER versus E_b/N_0 for an MLSE and MAP detector in AWGN.

As mentioned earlier, the Viterbi decoder is a maximum likelihood *sequence* estimator (MLSE), and so does not minimize the number of bit errors. The optimum decoder, with respect to the bit error ratio, is the *Maximum A Posteriori* (MAP) detector [79]. This decoder will be described in more detail when we discuss decoding turbo codes in Section 4.6, but is much more complex in terms of computational operations than the VA. Theoretically we would expect the MLSE (Viterbi) to be the optimum decoder in terms of the *BLock Error Rate* (BLER – i.e. the number of code blocks with one or more errors), and the MAP decoder to be optimal in terms of the BER. Figure 4.15 shows the BER and BLER for our half rate $K = 9$ code decoded with both the MAP and MLSE decoders. As expected, the MLSE decoder gives the best performance in terms of BLER, and the MAP decoder

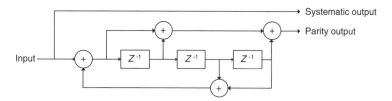

Figure 4.16. Recursive systematic convolutional (RSC) codes used in the UMTS turbo
         encoder.

gives the best performance in terms of BER. However, note from Figure 4.15 that these
differences in performance are marginal even at low E_b/N_0, and at high E_b/N_0 values the
curves converge.

4.6 TURBO ENCODING AND DECODING

Parts of this section are based on [80] © IEEE.

Turbo codes were introduced in 1993 by Berrou, Glavieux and Thitimajshima [67],
who reported results close to the Shannon limit [81]. In their paper Berrou *et al.* [67] used
a parallel concatenation of two recursive systematic convolutional (RSC) codes, with an
interleaver between the two encoders. An iterative structure with a modified version of the
MAP algorithm by Bahl *et al.* [79] was used to decode the codes. Since then much work
has been done in the area, reducing the decoder complexity [82, 83], understanding the
excellent performance of the codes [84–86], and extending the concept to use concatenated
block codes [87, 88]. Turbo codes have been used as part of several standardized systems,
including digital video broadcasting [89], INMARSAT's M4 digital satellite system and
both the 3GPP [50] and 3GPP2 [90] next generation terrestial mobile communications
systems.

In this section we begin by describing the encoder standardized by 3GPP for use in
UMTS. We then describe the iterative decoder usually used to decode turbo codes, and
give a detailed description of the component, soft in and soft out, decoders used within this
iterative decoder. Next, we give an example of the decoding procedure. We then describe
the choice of interleaver used within the encoder and decoder before describing how turbo
decoders can be implemented in real systems.

Other introductory descriptions of turbo codes can be found in [80, 91, 92]

4.6.1 Encoder

During the 3GPP standardization process there was much discussion and debate over the
channel coding scheme to be used for high rate services. After considering traditional concatenated codes and various turbo encoders, a turbo encoder using two eight state recursive
systematic convolutional (RSC) codes was agreed. In this section we first describe the RSC
codes used to build the turbo encoder, before describing the encoder itself.

The RSC codes used, shown in Figure 4.16, are a form of convolutional code similar to
those described in Section 4.5.1. However, unlike standard convolutional codes these codes

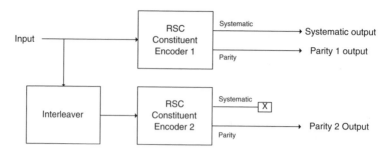

Figure 4.17. Turbo encoder used in UMTS.

are, as their name suggests, systematic and recursive. We explain what these terms mean below.

Using the notation described in Section 4.5.1, the RSC codes are constraint length $K = 4$ half rate codes with generator polynomials 1 and G_1/G_0, where $G_0 = 13$ and $G_1 = 15$ in octal. Hence for each input bit, one of the outputs (from the first generator polynomial) is the input bit itself. This is called the systematic output, a systematic code is one where the original data bits appear in the encoded data sequence. The second output, called the parity output, is formed from an exclusive OR operation on the bits in a delay line. However, unlike the standard form of convolutional codes, the inputs to this delay line are given by the input data bits and a feedback term determined by previous entries in the delay line determined by G_0. Hence the code is recursive.

The turbo encoder uses two of these RSC codes as shown in Figure 4.17. The two RSC codes take the same input bits, but the order of these bits is interleaved between the two encoders. The choice of interleaver has a vital effect on the performance of the code; the interleaver used in UMTS is detailed in Section 4.6.6. The systematic bits, the parity bits from both encoders and the termination bits are all transmitted, giving an approximately one third rate coder. Notice that the systematic bits are only transmitted from the first RSC encoder; those from the second encoder can be discarded as they are identical to the first systematic stream.

Figure 4.17 shows the encoder configuration for encoding the data bits. After these bits have been encoded, the two RSC encoders are separately terminated by driving their delay lines back to the all zero state. This requires three input bits for each encoder. Unlike traditional convolutional codes, due to the recursive nature of RSC codes, the input bits needed to terminate the codes will not necessarily be zeroes. Instead the termination bits needed are the output from the lower XOR gate shown in Figure 4.16 after encoding the data sequence. Each of the three termination bits for each encoder gives a systematic and parity output bit, giving a total of 12 output bits for termination. This is the reason for the $+12$ term in the equation for the number of coded bits in Table 4.4.

Like traditional convolutional codes, RSC codes can be considered as a state machine with a trellis and so can be decoded using the Viterbi algorithm. Also they are found [93] to have the same minimum Hamming distance as the equivalent nonrecursive code, and so in concatenated codes give similar performance to nonrecursive codes. As discussed in [84,93] it is their combination in a concatenated code with an interleaver that gives the remarkable performance seen with turbo codes.

We now describe decoding these codes. For this description, the terminology introduced in Section 4.5.2 is used again. Also we supplement this terminology as follows. From each

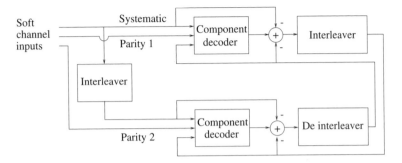

Figure 4.18. Turbo decoder schematic.

component encoder there are always two transmitted bits y_{kl}, with $l = 1$ or 2, for a particular input bit u_k. One of these will be the systematic bit, which we denote as y_{ks}, and the other the parity bit which we denote as y_{kp}. Similarly $\hat{y}_{ks}$ and $\hat{y}_{kp}$ represent the received versions of these two bits. We use only soft decisions in this section, so the transmitted bits y_{ks} and y_{kp} are always considered to be -1 (logical 0) or $+1$ (logical 1). Finally we use the notation $\underline{y}$ to represent the entire encoded sequence, and $\underline{\hat{y}}$ to represent the received version of this encoded sequence.

4.6.2 Turbo Decoder Structure

The technique generally used to decode turbo codes is illustrated in Figure 4.18. Two 'soft in soft out' component decoders are linked by interleavers in a structure similar to that of the encoder.

By 'soft in soft out' decoders we mean decoders that take soft decision inputs and give soft decision outputs. These soft decisions give not just an estimate for each bit, but also a measure of the probability that this estimate is correct. Various algorithms can be used in turbo decoders to implement these component decoders. Suitable decoders are the Maximum A Posteriori (MAP) [79] algorithm, the Max Log–MAP algorithm derived from it [94, 95], and the *Soft Output Viterbi Algorithm* (SOVA) [96]. The SOVA decoder is described in Section 4.6.3 whilst the Max Log–MAP algorithm is described in Section 4.6.5. In this section we give an overview of the the operation of the decoder shown in Figure 4.18, more details of the soft outputs from the component decoders and finally a more detailed description of the operation of the decoder.

4.6.2.1 *Overview of the Turbo Decoder*   The decoder in Figure 4.18 operates iteratively, with each component decoder feeding values to the other, and both decoders typically being run several times. In the first iteration the first component decoder takes the received channel values only, and produces a soft output as its estimate of the data bits. This soft output from the first decoder is then used as additional information for the second decoder, which uses it along with the channel outputs to calculate its estimate of the data bits. Now the second iteration can begin, and the first decoder decodes the received channel values again, but now with additional information about the value of the input bits provided by the output of the second decoder in the first iteration. This additional, or a-priori, information allows the first decoder to obtain a more accurate set of soft outputs than it did in the first iteration.

These more accurate soft outputs are then used by the second decoder as improved a-priori information. This cycle is repeated, and with every iteration the BER of the decoded bits tends to fall. However, the improvement in performance obtained with increasing numbers of iterations decreases as the number of iterations increases. Hence, for complexity reasons, usually only about four to eight iterations are used.

Due to the interleaving used at the encoder, care must be taken to properly interleave and deinterleave the soft decisions for the bits, as seen in Figure 4.18. Furthermore, because of the iterative nature of the decoding, care must be taken not to reuse the same information more than once at each decoding step. For this reason the concept of so-called extrinsic and intrinsic information was used in the original paper describing the iterative decoding of turbo codes [67]. These concepts and the reason for the subtraction circles shown in Figure 4.18 are described below. First we detail the soft outputs given by the component decoders.

4.6.2.2 Soft Outputs and Log Likelihood Ratios

The soft outputs from the component decoders are typically represented in terms of *Log Likelihood Ratios* (LLRs). These LLRs are simply, as their name implies, the logarithm of the ratio of two probabilities. The component decoders attempt to give the LLR for each bit u_k given the received version $\hat{\underline{y}}$ of the encoded sequence $\underline{y}$. This output LLR from the component decoders is represented as $L(u_k|\hat{\underline{y}})$, where

$$L(u_k|\hat{\underline{y}}) = \ln \left(\frac{P(u_k = +1|\hat{\underline{y}})}{P(u_k = -1|\hat{\underline{y}})} \right) \tag{4.12}$$

Here $P(u_k = +1|\hat{\underline{y}})$ is the probability that the bit u_k was a $+1$ given the received sequence $\hat{\underline{y}}$, and similarly for $P(u_k = -1|\hat{\underline{y}})$. Careful examination of Equation (4.12) shows that the sign of the LLR gives a hard decision about whether u_k is more likely to be $+1$ or -1, and the amplitude of the LLR gives a representation of how likely it is that this hard decision is correct. So, for example, if a component decoder calculated that a given bit u_k was very likely to be a $+1$, then the LLR $L(u_k|\hat{\underline{y}})$ would be large and positive. Similarly if u_k was calculated to be only slightly more likely to be a -1 than a $+1$, $L(u_k|\hat{\underline{y}})$ would be small and negative.

4.6.2.3 Detailed Description of Iterative Turbo Decoder Operation

We now give more detail on the decoder structure shown in Figure 4.18. Conceptually each component decoder shown takes three inputs:

- the received versions of the systematically encoded bits, denoted $L_c\hat{y}_{ks}$ for a given input bit u_k;

- the received versions of parity bits from the associated component encoder, denoted $L_c\hat{y}_{kp}$;

- information from the other component decoder about the likely values of the bits concerned. This is known as a-priori information, and is denoted as $L(u_k)$.

The term L_c in the received versions of the systematic and parity bits is a measure of the channel reliability, for an AWGN channel it is a constant proportional to the SNR. Each component decoder exploits these inputs to determine metrics for the various transitions in the trellis representing the RSC encoder. These metrics and the possible transitions through the trellis are then used to produce the output LLRs $L(u_k|\hat{\underline{y}})$.

It can be shown [67] that, for a systematic code such as an RSC code, the output from the MAP decoder can be written as

$$L(u_k|\hat{\underline{y}}) = L(u_k) + L_c\hat{y}_{ks} + L_e(u_k) \qquad (4.13)$$

Hence, the output from the MAP decoder, giving the LLR for each input bit u_k given the received channel sequence $\hat{\underline{y}}$, is equal to the input a-priori signal $L(u_k)$ plus the input received systematic information $L_c\hat{y}_{ks}$ plus a third term $L_e(u_k)$. This third term is derived by the component decoder, using the constraints imposed by the RSC encoder, from the a-priori information sequence $L(\underline{u})$ and the received channel information sequence $\hat{\underline{y}}$, *excluding* the received systematic bit y_{ks} and the a-priori information $L(u_k)$ for the bit u_k. Hence it is called the *extrinsic* LLR for the bit u_k. It is this extrinsic LLR that is passed from one component decoder as a-priori information to the other.

We now describe in more detail how the iterative decoding of turbo codes is carried out. Consider initially the first component decoder in the first iteration. This decoder receives the channel sequence containing the received versions of the transmitted systematic bits, $L_c\hat{y}_{ks}$, and the parity bits, $L_c\hat{y}_{kp}$, from the first encoder. The first component decoder can then process the soft channel inputs and produce its estimate $L_{11}(u_k|\hat{\underline{y}})$ of the conditional LLRs of the data bits u_k, $k = 1, 2 \ldots K_i - 1$. In this notation we take $L_{ab}(u_k|\hat{\underline{y}})$ to mean the output from the ath decoder in the bth iteration. Note that in this first iteration the first component decoder will have no a-priori information about the bits, and hence will have $L(u_k) = 0$.

Next the second component decoder comes into operation. It receives the channel sequence containing the *interleaved* version of the received systematic bits, and the parity bits from the second encoder. However, now the decoder can also use the conditional LLR $L_{11}(u_k|\hat{\underline{y}})$ provided by the first component decoder to generate a-priori LLRs $L(u_k)$ to be used by the second component decoder. Equation (4.13) shows that the extrinsic information from this decoder is given by the output of the decoder with the a-priori information $L(u_k)$ and the received systematic bits $L_c\hat{y}_{ks}$ removed. Hence the subtraction circles shown in Figure 4.18. The extrinsic information is interleaved to give the a-priori information for the second decoder, which uses it along with the received channel sequence to produce its output LLRs $L_{21}(u_k|\hat{\underline{y}})$. This is then the end of the first iteration.

For the second iteration, the first component decoder again processes its received channel sequence, but now it also has a-priori LLRs $L(u_k)$ provided by the extrinsic portion $L_e(u_k)$ of the output LLRs $L_{21}(u_k|\hat{\underline{y}})$ from the second component decoder, and hence it can produce an improved output LLR $L_{12}(u_k|\hat{\underline{y}})$. The second iteration then continues with the second component decoder using these improved LLRs from the first decoder to derive, through Equation (4.13), improved a-priori LLRs $L(u_k)$ which it uses in conjunction with its received channel sequence to calculate $L_{22}(u_k|\hat{\underline{y}})$.

This iterative process continues, and with each iteration on average the BER of the decoded bits will fall. However, the improvement in performance for each additional iteration carried out falls as the number of iterations increases. Hence for complexity reasons usually only around four to eight iterations are carried out, as no significant improvement in performance is obtained with a higher number of iterations.

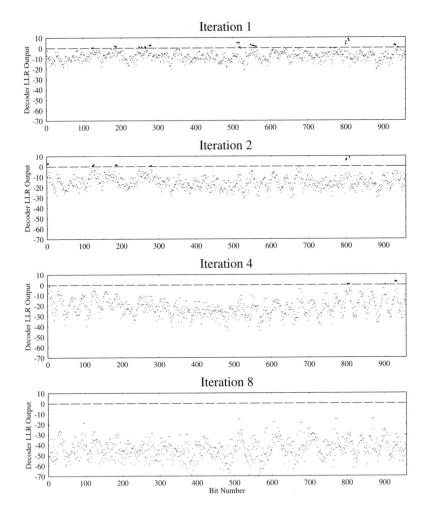

Figure 4.19. Soft outputs from component decoder in an iterative turbo decoder for a transmitted stream of all -1s.

Figure 4.19 shows how the output LLRs $L(u_k|\hat{y})$ from the component decoders in an iterative decoder vary with the number of iterations used. The output from the second component decoder is shown after one, two, four and eight iterations. The input sequence consisted entirely of -1 values, hence negative LLR $L(u_k|\hat{y})$ values correspond to a correct hard decision, and positive values to an incorrect hard decision. The encoded bits were transmitted over an AWGN channel at a channel SNR of -1 dB, and then decoded using an iterative turbo decoder. It can be seen that as the number of iterations used increases, the number of positive LLR $L(u_k|\hat{y})$ values, and hence the BER, decreases until after eight iterations there are no incorrectly decoded values. Furthermore, as the number of iterations increases, the decoders become more certain about the value of the bits and hence the magnitudes of the LLRs gradually become larger. Also note that the erroneous decisions in the figure for one, two and four iterations appear in bursts, since deviating from the error free trellis path typically inflicts several bit errors.

When the series of iterations finishes the output from the turbo decoder is given by the deinterleaved LLRs of the second component decoder, $L_{2I}(u_k|\underline{y})$, where I is the number of iterations used. The sign of these LLRs gives the hard decision output from the turbo decoder.

We now describe the *Soft Output Viterbi Algorithm* (SOVA), that can be used as one of the soft in soft out component decoders shown in Figure 4.18.

4.6.3 The SOVA Decoder

The SOVA decoder is the simplest of the component decoders that are typically used in turbo decoders. It is modified from the Viterbi algorithm described in Section 4.5.3 to take account of any a-priori information known, and to produce soft outputs. These two modifications are needed in a turbo decoder to allow the SOVA decoder to be used in the iterative decoding described above. In this section we describe this algorithm, before detailing an example of a turbo decoder using it.

When two paths merge at a given state, the Viterbi algorithm selects the path with the best metric as the survivor. This add compare select (ACS) operation is also used in the SOVA decoder. However, the metric calculated for each state at each stage in the trellis is modified to include the a-priori information from the other decoder. It can be shown [80] that a suitable path metric for the path to state s at stage k in the trellis from state $\grave{s}$ at the previous stage is $M_k(s, \grave{s})$ as shown in Equation (4.14):

$$M_k(s, \grave{s}) = M_{k-1}(\grave{s}) + \frac{1}{2}\hat{u}_k L(u_k) + \frac{L_c}{2}\sum_l \hat{y}_{kl} y_{kl} \qquad (4.14)$$

This is similar to Equation (4.7) for the metric in the standard VA. The additional term of $\frac{1}{2}\hat{u}_k L(u_k)$ gives the contribution of the a-priori information; $\hat{u}_k$ is the decoded bit associated with the transition from state $\grave{s}$ to s and $L(u_k)$ is the a-priori information for this transition.

In addition to the traditional ACS cycle with this modified metric, the SOVA decoder also stores information about the reliability of the decision to select the winning path and discard the other. This reliability can be calculated [96] from the difference between the two metrics. So this metric difference is calculated and stored for each state at each stage in the trellis. Once the maximum likelihood (ML) path has been selected, hard decisions about the decoded bits are given by the bits along this path, as for the traditional VA. Soft decisions for the bits are given by considering the metric differences for the paths along the ML path that merged with it but were discarded.

Consider Figure 4.20 which, for simplicity, shows a four state trellis. The ML path is the all -1 path shown in bold. Various paths are shown merging with this ML path but being discarded. Transitions which correspond to an input bit of -1 are shown as solid lines, those which correspond to an input bit of $+1$ are shown as dashed lines. The metric difference for the path that was discarded at state s and stage k of the trellis is labeled Δ_k^s in Figure 4.20. The soft decision $L(u_k|\underline{y})$ for the input bit u_k associated with the transition to stage k in the trellis is given by the hard decision for that bit on the ML path, scaled by a function of the metric differences for paths merging with the ML path that would have given a different value for u_k.

For example in Figure 4.20 the ML path gives a value of -1 for u_{k+2} whereas the paths merging with the ML path at stages $k + 2$ and $k + 3$ would have given a value of $+1$ for u_{k+2}

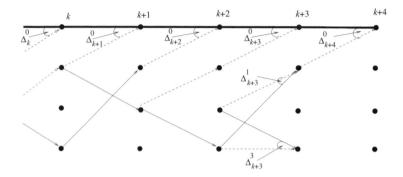

Figure 4.20. Simplified section of the trellis for SOVA decoding of a four state code.

if they had been selected rather than discarded. Therefore if either of the metric differences Δ_{k+2}^0 or Δ_{k+3}^0 is small, the decision of -1 for u_{k+2} will be unreliable. On the other hand, the value of Δ_{k+4}^0 will not affect the soft decision for u_{k+2} because the path merging with the ML path at stage $k+4$ in the trellis would have given the same hard decision for u_{k+2} as the ML path. It was shown by Hagenauer and Hoeher [96] that the soft output $L(u_k|\hat{y})$ for the input bit u_k can be approximated by the hard decision for u_k scaled by the minimum metric difference for paths merging with the ML path that would have given a different value for u_k had they been selected rather than discarded.

As noted in Section 4.5.6, the hard decision for a decoded bit u_k is typically the same for all the surviving paths at a later stage $k+\delta$ in the trellis, as long as δ is set to be sufficiently long. Typically a value of five times the constraint length of the encoder is used. Any paths merging with the ML path at a later stage in the trellis will not affect the decoded value of u_k. Thus, when calculating the LLR of the bit u_k, the soft output Viterbi algorithm must take account of the probability that the paths merging with the ML path from stage k to stage $k+\delta$ in the trellis were incorrectly discarded. This is done by considering the values of the metric difference $\Delta_i^{s_i}$ for all states s_i along the ML path from trellis stage $i=k$ to $i=k+\delta$.

The SOVA decoder can be implemented as follows. For each state at each stage in the trellis the metric is calculated for both paths merging at that state, as in the VA. Again as in the VA, one path is selected as the survivor, and the metric associated with this path is stored. Also, a pointer to the previous state along the winning path is stored in the trace back memory for that state. Additionally in the SOVA decoder, in order to allow the reliability of the decoded bits to be calculated, we also need to store the metric difference for that state and a binary vector containing δ bits which indicate whether or not the discarded path would have given different hard decisions to the selected path for the previous δ bits. This series of bits is called the 'update sequence' in [97], and can be calculated by a modulo 2 addition between the trace back memory for the winning path and the discarded path (as these two sets of trace back memory give the previous δ bits along the two paths).

As the SOVA progresses through the received bits, at stage $k+\delta$ in the trellis the soft decision for the bit u_k can be calculated and released by the decoder. The state with the largest metric at stage $k+\delta$ in the trellis is selected, and the algorithm traces back along the path leading to this state. At each stage in the trellis, from stage $k+\delta$ back to stage k, the algorithm examines the bit in the 'update sequence' for this state that indicates whether the discarded path at that stage would have given a different hard decision for u_k. If this bit indicates that a different hard decision would have been given, then the algorithm compares the metric difference stored at this state to the smallest metric difference found so far, and selects the minimum of the two. The soft decision for the bit u_k is then given by the hard decision for this bit on the selected path, scaled by the minimum metric difference found.

We now illustrate these ideas by considering a specific example.

4.6.4 Turbo Decoding Example

In this section we discuss an example of turbo decoding using the SOVA algorithm detailed in the previous section. This example serves to illustrate the details of the SOVA component decoder and the iterative decoding of turbo codes.

We consider a one third rate turbo code, similar to that used in UMTS. However, for simplicity rather than use the eight state component codes and the interleaver specified in UMTS, we use a simpler four state code with four input bits and a 2×2 rectangular interleaver. Like the UMTS turbo encoder, both component encoders are terminated separately. This gives two extra input bits for each encoder. These, and the two resulting parity bits from each encoder, give a total of eight extra bits for termination. Hence we have a total of $(4 \times 3) + 8 = 20$ encoded bits for our four input data bits.

The four state component code used in our example is shown in Figure 4.21. Notice that this is the recursive form of the code shown in Figure 4.10 and used in our Viterbi decoding example. Instead of the generator polynomial $G_0 = 7$ being used to give the first of each pair of output bits, here it is used to give the feedback term.

The 2×2 rectangular interleaver is shown in Figure 4.22. The four input bits are written into the interleaver along the rows of the square, and read out along the columns. This means that an input sequence of 0, 1, 2, 3 is reordered into 0, 2, 1, 3 after interleaving.

Notice the example encoder described above would not be used in practice. It gives 20 output bits for only four input bits, and decoding it using a turbo decoder as described below is relatively complex for such a short code. However, its parameters are similar to

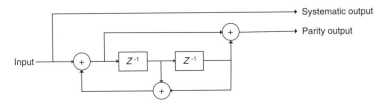

Figure 4.21. Component RSC codes used in our example turbo code.

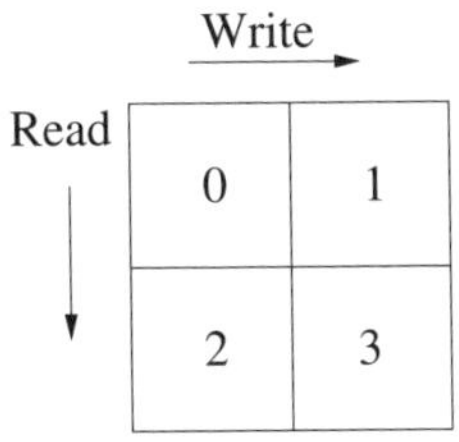

Figure 4.22. 2×2 rectangular interleaver used in our example turbo code.

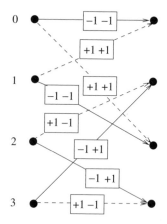

Figure 4.23. State transition diagram for our (2, 1, 3) RSC component codes.

those specified for UMTS, in that two half rate RSC codes are concatenated in parallel to give a one third rate code, and these two codes are separately terminated. Hence it allows a relatively simple description of the decoding process which illustrates the procedures (with larger trellises and a different interleaver) that could be used to decode the turbo encoder specified for UMTS.

The state transition diagram for the component RSC codes is shown in Figure 4.23. As in all our diagrams in this section, a solid line denotes a transition resulting from a −1 input bit, and a dashed line represents an input bit of +1. The figures within the boxes along the transition lines give the output bits associated with that transition – the first bit is the systematic bit, which is the same as the input bit, and the second is the parity bit.

For the sake of simplicity we assume that an all −1 input sequence is used. Thus there will be four input bits which are −1, and the encoder trellis will remain in state 0 and generate four parity bits all of −1. The two bits necessary to terminate the trellis will be −1, −1 in this case and the resulting parity bits will also be −1. Similarly for the second encoder, after interleaving the four input bits will again be −1,−1,−1,−1 leaving the encoder trellis in state 0, and the two bits needed to terminate the trellis will again be −1,−1. Thus all 20 of the transmitted bits will be −1 for this all −1 input sequence. The received channel output sequence for our example, together with the input and the parity bits detailed above, are shown in Table 4.8. It can be seen that from the 20 coded bits which were transmitted, all of which were −1, six would be decoded as +1 if hard decision demodulation was used. Also note that of the transmitted systematic bits corresponding to the four input data bits, two would be incorrectly decoded as +1 if hard decision demodulation was used.

The path metric for the SOVA decoder is given by Equation (4.14), which is repeated here for convenience:

$$M_k(s, \grave{s}) = M_{k-1}(\grave{s}) + \frac{1}{2}\hat{u}_k L(u_k) + \frac{L_c}{2}\sum_l \hat{y}_{kl} y_{kl}$$

Figure 4.24 shows the trellis for the first component decoder in the first iteration. The metrics calculated using Equation (4.14) are shown for each state at each stage in the trellis. Where two paths merge and one is discarded, the metric for the path that is discarded is shown crossed out. Notice how, as the decoder knows that the trellis is terminated at the

Table 4.8 Input and transmitted bits and received soft
decisions for turbo decoding example.

Input	Parity 1	Parity 2	$\hat{y}_{ks}$	Received $\hat{y}_{kp1}$	$\hat{y}_{kp2}$
−1	−1	−1	+0.9	−0.3	−0.9
−1	−1	−1	−0.2	−1.3	−1.4
−1	−1	−1	+0.4	−0.5	−0.3
−1	−1	−1	−0.9	−0.6	+0.5
−1	−1	—	+0.4	−1.7	—
−1	−1	—	−1.1	−0.4	—
−1	—	−1	+1.2	—	−0.2
−1	—	−1	+1.0	—	−1.1

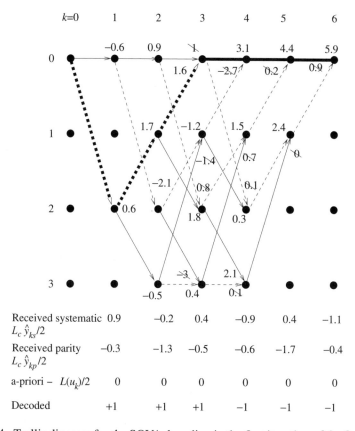

Figure 4.24. Trellis diagram for the SOVA decoding in the first iteration of the first decoder.

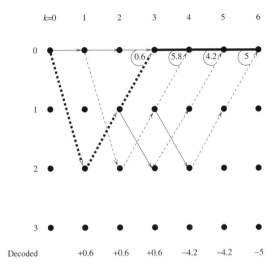

Figure 4.25. Simplified trellis diagram for the SOVA decoding in the first iteration of the first decoder.

encoder, only paths leading to state 0 at stage 6 of the trellis are considered. The a-priori and channel values shown are given as $L(u_k)/2$ and $L_c y_{kl}/2$ so that the metric values are calculated by simple addition and subtraction of the values shown. For simplicity we assume that the channel quality is such that the channel reliability measure L_c is equal to two, so that the multiplicative factor $L_c/2$ used to scale the received channel values in Equation (4.14) is one. Also, as initially we are considering the operation of the first decoder in the first iteration, there is no a-priori information and so we have $L(u_k) = 0$ for all k.

The maximum likelihood (ML) path chosen by the SOVA is shown in bold in Figure 4.24. It can be seen that an incorrect decision is made at stage 3 in the trellis where the path from state 1 is chosen rather than the all zero path. This results in three bits being decoded as $+1$ rather than -1. Notice how this process of finding the metrics for various paths through the trellis, and hence finding the ML path, is similar to that described in Section 4.5.3 for the VA.

We now discuss how, having determined the ML path, the SOVA algorithm finds the LLRs for the decoded bits. Figure 4.25 is a simplified version of the trellis from Figure 4.24, which shows only the ML path and the paths that merge with this ML path and are discarded. Also shown are the metric differences between the ML and the discarded paths. These metric differences, together with the update sequences that indicate for which of the bits the survivor and discarded paths would have given different values, are stored by the SOVA algorithm for each state at each stage in the trellis. When the ML path has been identified, the algorithm uses these stored values along the ML path to find the LLR for each decoded bit. Table 4.9 shows these stored values for our example trellis shown in Figures 4.24 and 4.25. The calculation of the decoded LLRs shown in this table is detailed below.

Notice in Table 4.9 that at trellis stages $k = 1$ and $k = 2$ there is no metric difference or update sequence stored because, as can be seen from Figures 4.24 and 4.25, there are no paths merging with the ML path at these stages. For all subsequent stages there is a merging path, and values of the metric differences and update sequences are stored. For the update sequence a 1 indicates that the ML and the discarded merging path would have given different values for a particular bit. At stage k in the trellis we have taken the MSB, on

Table 4.9   SOVA output for the first iteration of the first decoder.

Trellis stage k	Decoded bit u_k	Metric difference	Update sequence	Decoded LLR
1	+1	–	—	+0.6
2	+1	–	—	+0.6
3	+1	0.6	111	+0.6
4	−1	5.8	1001	−4.2
5	−1	4.2	11100	−4.2
6	−1	5.0	110001	−5

the left hand side, to represent u_k, the next bit to represent u_{k-1}, etc. until the LSB, which represents u_1. For our RSC code any two paths merging at trellis stage k give different values for the bit u_k, and so the MSB in the update sequences in Table 4.9 is always 1. Notice furthermore, that although in our example the update sequences are all of different lengths, this is only because of the very short frame length we have used. More generally, as explained in Section 4.6.3, all the stored update sequences will be $\delta + 1$ bits long, where δ is usually set to be around five times the constraint length of the convolutional code.

We now explain how the SOVA algorithm can use the stored update sequences and metric differences along the ML path to calculate the LLRs for the decoded bits. As explained in Section 4.6.3, the decoded a-posteriori LLR $L(u_k|\underline{y})$ for a bit u_k is given by the hard decision for this bit scaled by the minimum metric difference of paths merging with the ML path that give a different hard decision for this bit. Whether or not the merging paths give a different value for u_k is determined using the stored update sequences. Denoting the update sequence stored at stage l along the ML path as $\underline{e}_l$, for each bit u_k the SOVA algorithm examines the MSB of $\underline{e}_k$, the second MSB of $\underline{e}_{k+1}$, etc. up to the $(\delta + 1)$th bit (which will be the LSB) of $\underline{e}_{k+\delta}$. For our example this examination of the update sequences is limited because of our short frame length, but the same principles are used. Taking the fourth bit u_4 as an example, to determine the decoded LLR $L(u_4|\hat{y})$ for this bit the algorithm examines the MSB of $\underline{e}_4$ in row four of Table 4.9, the second MSB of $\underline{e}_5$ in row five and the third MSB of $\underline{e}_6$ in the final row. It can be seen that only the paths merging at stages $k = 4$ and $k = 5$ of the trellis give values different from the ML path for the bit u_4. Hence the decoded LLR $L(u_4|\hat{y})$ from the SOVA algorithm is given by the hard decision on the ML path (-1) times the minimum of the metric differences stored at stages 4 and 5 of the trellis (5.8 and 4.2), yielding $L(u_4|\hat{y}) = -4.2$.

The remaining decoded LLR values in Table 4.9 are computed following a similar procedure. However, it is worth noting explicitly that the low value (0.6) of the metric difference for the merging path at stage 3 in the trellis, which is where the incorrect path is chosen as the survivor, gives the LLR for the bits where this path and the ML path give different values. Hence, the LLRs for the three incorrectly decoded bits, i.e. u_1, u_2 and u_3, have the lowest magnitudes of any of the decoded bits.

We now move on to describing the operation of the second component decoder in the first iteration. This decoder uses the extrinsic information from the first decoder as a-priori information to assist its operation, and therefore should be able to provide a better estimate of the encoded sequence than the first decoder was. Equation (4.13) gives the extrinsic information $L_e(u_k)$ from a component decoder as the soft output $L(u_k|\underline{y})$ from the decoder

Table 4.10   Calculation of the extrinsic information from the first decoder in the first iteration.

Bit index k	Output $L_{11}(u_k|\hat{y})$	a-priori $L(u_k)$	Systematic information $L_c y_{ks}$	Extrinsic information
1	+0.6	0	+1.8	−1.2
2	+0.6	0	−0.4	+1.0
3	+0.6	0	+0.8	−0.2
4	−4.2	0	−1.8	−2.4

with the a-priori information $L(u_k)$ (if any was available) and the received systematic channel information $L_c \hat{y}_{ks}$ subtracted. Table 4.10 shows the extrinsic information calculated from this equation from the first decoder.

Notice how the channel values shown in Table 4.10 are twice those shown in Figure 4.24. This is because, as mentioned earlier, the values in Figure 4.14 include the factor of $\frac{1}{2}$ used in the metric calculation shown in Equation (4.14). Also notice that we show the extrinsic information calculation only for the first four decoded bits u_1 to u_4. This is because, as the two component codes are terminated separately, the second component decoder does not use extrinsic information on the decoded bits used to terminate the first trellis.

Figure 4.26 shows the trellis for the SOVA decoding of the second decoder in the first iteration. The extrinsic information values from Table 4.10 are shown as $L(u_k)/2$ after being interleaved, by our 2×2 rectangular interleaver, and divided by two. Also shown is the channel information $L_c y_{ks}/2$ used by this decoder. For the four information bits these are the values used by the first component decoder after being interleaved. Finally, we have the received systematic bits for the two termination bits used to terminate the second encoder, and the parity bits for this encoder ($L_c \hat{y}_{kp}/2$).

The ML path chosen by the second component decoder is shown by a bold line in Figure 4.26, together with the LLR values output by the decoder. These are calculated, using update sequences and minimum metric differences, in the same way as was explained for the first decoder using Figure 4.25 and Table 4.9. It can be seen that the decoder makes an incorrect decision at stage $k = 5$ in the trellis and selects a path other than the all zero path as the survivor. However, the incorrectly chosen path gives decoded bits of +1 for only two transitions, and hence only two, rather than three, decoding errors are made. Furthermore, the difference in the metrics between the correct and the chosen path at trellis stage $k = 5$ is only 0.2, and so the magnitude of the decoded LLRs $L(u_k|\hat{y})$ for the two incorrectly decoded bits, u_2 and u_5, is only 0.2. This is significantly lower than the magnitudes of the LLRs for the other bits, and indicates that the algorithm is less certain about these two bits being +1 than it is about the other bits being −1.

Having calculated the LLRs from the second component decoder, the turbo decoder has now completed one iteration. The first four soft output LLR values from the second component decoder, which correspond to the four data bits input to the encoder, could now be deinterleaved and used as the output from the turbo decoder. After deinterleaving this would result in an output sequence which gave negative LLRs for all the decoded bits except u_2, which would be incorrectly decoded as a +1 as its LLR is +0.2. Thus, even after only one iteration, the turbo decoder has corrected one of the two errors in the

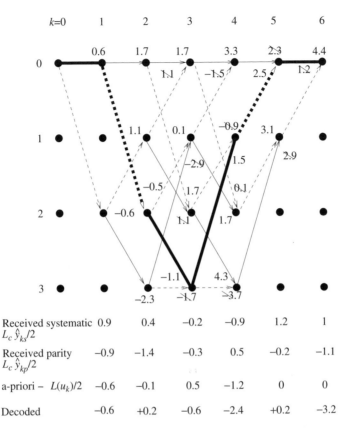

Figure 4.26. Trellis diagram for the SOVA decoding in the first iteration of the second decoder.

received systematic bits. However, generally better results are achieved with more iterations, and so we now progress to describe the operation of the turbo decoder in the second iteration.

In the second, and all subsequent, iterations the first component decoder is able to use the extrinsic information from the second decoder in the previous iteration as a-priori information. Table 4.11 shows the calculation of this extrinsic information using Equation (4.13) from the outputs $L_{21}(u_k|\hat{y})$ of the second decoder in the first iteration. It can be seen that it gives negative extrinsic LLRs for all the bits except u_4. This extrinsic information is then deinterleaved and used as the a-priori information for the first decoder in the next (second) iteration.

The trellis for this decoder is shown in Figure 4.27. It can be seen that this decoder uses the same channel information as it did in the first iteration. However now, in contrast to Figure 4.24, it also has a-priori information to assist it in finding the correct path through the trellis. The selected ML path is again shown by a bold line, and it can be seen that now the correct all zero path is chosen. The second iteration is then completed by finding the extrinsic information from the first decoder, interleaving it and using it as a-priori information for the second decoder. It can be shown that this decoder will also now select the all zero path

Table 4.11   Calculation of the extrinsic information from the second decoder in the first iteration.

Bit index k	Output $L_{21}(u_k|\underline{\hat{y}})$	a-priori $L(u_k)$	Systematic information $L_c y_{ks}$	Extrinsic information
1	−0.6	−1.2	+1.8	−1.2
2	+0.2	−0.2	+0.8	−0.4
3	−0.6	+1.0	−0.4	−1.2
4	−2.4	−2.4	−1.8	+1.8

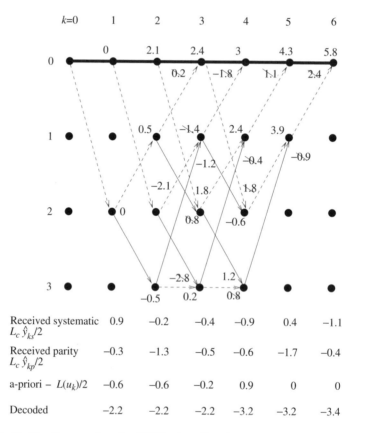

Figure 4.27. Trellis diagram for the SOVA decoding in the second iteration of the first decoder.

as the ML path, and hence the output from the turbo decoder after the second iteration will be the correct all −1 sequence.

This concludes our example of the operation of an iterative turbo decoder using the SOVA algorithm. As mentioned earlier, a real decoder for the turbo code defined in UMTS would be more complex due to the eight state component codes used and the much longer trellis lengths (between 40 and 5114, rather than four, input bits). However, principles similar to those described above could be used for a decoder using the SOVA algorithm.

We now describe an alternative soft input soft output decoder that is often used as the component decoder in a turbo decoder.

4.6.5 The Max Log–MAP Algorithm

The SOVA decoder described in Section 4.6.3 operates with a set of standard Viterbi ACS operations running forward through the trellis calculating metrics and deciding on surviving paths. Trace back and soft output calculations are then carried out. In its conceptually simplest form, the Max Log–MAP algorithm [94, 95] can be thought of as two sets of Viterbi like ACS algorithms and a 'dual maxima computation' [98]. One of the two sets of ACS recursions progresses through the trellis in the forward direction as usual, calculating metrics for merging paths in the trellis and for each state at each stage in the trellis choosing a survivor. However, instead of keeping a record of only two sets of metrics (the new and old metrics), the metrics for all the states in the trellis are stored for every stage in the trellis. As for the VA and the SOVA we denote these state metrics, for a state s in the trellis at stage k, as $M_k(s)$. The storage of these metrics imposes a heavy memory demand that in practice can be largely avoided using a technique [98] that will be described later. However, this complicates the description of the algorithm, and so is ignored initially. Also note that a trace back is not required, and so the trace back memory needed in the traditional Viterbi algorithm and in the SOVA decoder is not needed.

The other ACS recursion also works through the trellis calculating metrics and selecting survivors, but progresses backwards through the trellis. These 'backward' metrics are denoted as $\beta_k(s)$. At each stage in the trellis the backward metrics just calculated, and the forward metrics calculated previously and stored, are used to calculate the soft output for the bit associated with that stage in the trellis. Effectively, for every stage in the trellis the algorithm calculates the probability of the most likely path through the trellis which gave a $+1$ at that stage, and the most likely path that gave a -1. The LLR for this bit is then calculated from these probabilities.

This calculation is relatively simple and is done as follows. It can be shown that the probability (in the LLR domain) of the most likely path through the trellis, passing through state $\acute{s}$ at stage $k-1$ in the trellis and state s at stage k in the trellis, is given by $M_{k-1}(\acute{s}) + \gamma_k(\acute{s}, s) + \beta_k(s)$. Here $\gamma_k(\acute{s}, s)$ is the incremental metric for the transition from state $\acute{s}$ at stage $k-1$ in the trellis to state s at stage k in the trellis, calculated as usual using addition and subtraction of the soft inputs to the decoder. The soft output $L(u_k|\hat{y})$ from the Max Log–MAP algorithm is given by the difference between the maximum value of $M_{k-1}(\acute{s}) + \gamma_k(\acute{s}, s) + \beta_k(s)$ for transitions which give a decoded bit of $+1$, and the maximum for those that give a decoded bit of -1. This is the 'dual-maxima' computation. It can be combined with the backward metric calculation to give a total complexity for the Max Log–MAP algorithm of no greater than three times that of the Viterbi algorithm [98].

Note the performance of the Max Log–MAP algorithm described above can be improved by adding a correction term [82] to compensate for discarding one of the merging paths at each ACS stage. A similar correction term is also added in the 'dual maxima' calculation used to give the soft outputs. This is known as the Log–MAP algorithm, and as described in [82] gives performance identical to that of the optimal MAP algorithm [79].

A more detailed description of the MAP, Log–MAP and Max Log–MAP algorithms, and a comparison between them, can be found in [80].

The problem with the Max Log–MAP algorithm as described above is the memory needed to store the forward metrics $M_k(s)$ for each state at each stage in the trellis. This requirement, and methods to reduce it, are discussed in Section 4.6.7. We first describe the turbo interleaver used in UMTS.

4.6.6   Turbo Interleaver Choice

It is well known that the choice of interleaver plays a vital part in the performance of turbo codes. In their original paper [67], Berrou *et al.* used a pseudo-random interleaver between the two component encoders to show the remarkable performance of turbo codes.

Various other methods for designing good interleavers have been proposed, for example [99, 100]. These interleavers work well for a fixed code block size where a good interleaver can be found for this particular size, and a copy of the interleaver pattern stored at both the encoder and decoder. However, for the UMTS turbo encoder an interleaver pattern is required for every possible code block size from 40 bits up to 5114 bits. The simple block interleavers with inter-column permutations, used for the TrCH and PhCH interleavers described above, are not suitable as they give a significant performance degradation over other interleaver patterns. On the other hand, storing individual interleaver patterns optimized for each of the required code block sizes would require 5075 different interleaver tables, with sizes from 40 indices to 5114 indices, to be stored. This would require over 13 million interleaver indices to be stored, which is not a feasible option. Hence, an algorithm for generating good interleavers 'on demand' for this range of interleaver sizes is required.

After much debate, the interleaver algorithm described below was selected. A brief description only is given, for more details see Section 4.2.3.2.3 of [50]. This choice was a combination of several previous candidates from different contributing companies, and was found to give good performance for various code block sizes over the required range. It is based on a block interleaver with a number of columns derived from a prime number p. Inter-row permutations are used, similar to the inter-column permutations used in the TrCH and PhCH interleavers earlier. But intra-row permutations are also used, which means the elements within each row are individually interleaved using different patterns for each row. These intra-row permutations are based on Galois Field arithmetic using the primitive root of the prime number p. The interleaver operates as follows:

1. Select a number of rows R for the block interleaver, based on the code block size K_i. Either 5, 10 or 20 rows are used.

2. Choose a prime number p, and a number of columns C, to use. The prime number p is chosen to be the smallest, from an allowed set of primes, which satisfies $R \times (p+1) \geq K_i$. The number of columns C is then chosen to be the smallest of either $p+1$, p or $p-1$ such that the K_i bits fit into the block (i.e. such that $R \times C \geq K_i$).

3. Write the data bits to be interleaved into a block of R rows and C columns, row by row as for the TrCH and PhCH interleavers described earlier. Fill any unused positions with dummy bits (which are pruned later).

4. Perform an intra-row permutation on each row. A different pattern is used for each row, with the patterns calculated using the primitive root v of the prime p, with mod p and mod $p - 1$ residual number arithmetic.

5. Depending on the block size perform one of four fixed inter-row permutations.

6. Read out the interleaved data bits, column by column, omitting any dummy bits inserted in stage 3.

The performance of turbo decoders using interleavers generated with this algorithm will be shown in Section 4.6.8. Good results are obtained with various code block sizes K_i. Also, this algorithm can be implemented in the UE with relatively little memory or computational demands. Typically, an interleaver table is generated once per decode, and used multiple times in the iterative decoding.

We now move on to discuss other aspects relevant in the implementation of turbo decoders.

4.6.7 Implementation of Turbo Decoders

Like the implementation of a Viterbi decoder, described in Section 4.5.6, real implementations of turbo decoders are limited by the number of operations used and the amount of memory needed. In this section we discuss both these issues for the iterative turbo decoders described above.

In terms of the number of operations required the UMTS turbo decoder is, perhaps surprisingly, similar to the Viterbi decoder used for the convolutional code. This is because the extra complexity of the soft output decoders and multiple iterations is offset by the fact that the component codes have only eight, rather than 256, states. For the Max Log–MAP component decoder, the complexity of a turbo decoder relative to a Viterbi decoder for the convolutional code is approximately:

$$\frac{8}{256} \times 2 \times I \times 3$$

Here the factor of 2 comes because we have two component decodes per iteration, the factor of 3 comes because, as noted earlier, the complexity of the Max Log–MAP component decoder is about three times that of a Viterbi decoder, and the factor 8/256 comes from the different number of states. Finally, I is the number of iterations used. Hence, a turbo decoder, using the Max Log–MAP algorithm with six iterations, has about the same number of operations per decoded bit as the Viterbi decoder for the convolutional codes.

As for the Viterbi decoder, quantization of the input soft decisions is an important consideration in both the memory used by a turbo decoder and the complexity of the ACS type operations used. Similarly important in the turbo decoder is the quantization of the extrinsic LLRs passed from one component decoder to another, which, due to the iterative feedback used, tend to have a higher dynamic range than the input soft decisions. Various studies [101–103] show that with between four and six bits for the input soft decisions, and one or two additional bits for the extrinsic information, the performance of the decoder is very close to that without any quantization.

As described in Section 4.5.6 for the Viterbi decoder, the turbo decoder can break metric calculations down into ACS butterflies. Also, similar techniques (metric rescaling or 'controlled overflow' [78, 104]) can be used to limit the bit width of the $M_k(s)$ and $\beta_k(s)$ metrics and so the bit width of the adders needed in the ACS units. The number of bits needed for these metrics depends on whether metric rescaling or controlled overflow is used, and on the size of the extrinsic and input soft decisions. Typically around eight to ten bits are used.

We now discuss the memory needed for a turbo decoder. As well as the input soft decisions, the extrinsic information passed from one decoder to another needs to be stored. Assuming seven bits for the extrinsic information, this extrinsic information needs around 35 kbits of memory for the maximum code block size. In addition to this, the component decoders will need memory for their internal calculations. As noted in Section 4.6.5, the Max Log–MAP algorithm needs to store the forward metrics $M_k(s)$ for each state at each stage in the trellis until the corresponding backward metrics $\beta_k(s)$ are calculated and the two sets of metrics used to find soft outputs. With the maximum size code block of 5114 bits, we have a trellis 5117 stages long. With the eight state code and ten bit metrics this would mean storing up to $5117 \times 8 \times 10 = 409\,360$ bits of metric information if all the forward metrics were stored before being used. Fortunately, as described in [98], the algorithm can be modified to reduce drastically this memory requirement, at the cost of increasing the number of metric calculations needed.

This memory reduction for the Max Log–MAP decoder is achieved by decoding the bits in blocks of L_d bits, rather than decoding all K_i bits at once. Instead of calculating the forward metrics for the whole trellis and storing them, a smaller section of forward metrics of length L_d is calculated and stored. The backward metrics are then calculated. But rather than start at the end of the trellis, the backward recursion starts at some delay L_b after the end of the section of the trellis for which the forward metrics have been calculated and stored. The backward metrics at the stage in the trellis where the backward recursion starts are all set to be equal (as no information is available for the true values of the backward metrics at this stage). Initially, the backward metrics calculated are practically useless. However, as the backward recursion progresses the metrics generated become more reliable, and if L_b is chosen to be sufficiently large, then after L_b stages the metrics will be as reliable as if the backward recursion had started at the end of the trellis. Typically, L_b is chosen to be around five times the constraint length of the code. Note this is very similar to the idea of segmented trace back and the parameter σ described earlier and used in Viterbi decoding of convolutional codes. After L_b stages of the backward recursion, useful values of $\beta_k(s)$ are being generated, and the recursion starts to calculate values of $\beta_k(s)$ for the L_d stages in the trellis for which the $M_k(s)$ values have been calculated and stored. These backward metrics are used, together with the forward metrics $M_k(s)$ which have been calculated and stored, to generate soft decisions for the block L_d of data using the 'dual maxima' calculation. The process is then repeated for the next L_d stages of the trellis, and so on until the whole trellis has been covered.

In terms of memory storage we would like to keep the value of L_d as small as possible. However for every block of L_d soft decisions generated, L_b unreliable backward metrics must be calculated and discarded before the L_d reliable backward metrics can be calculated. L_b must be sufficiently large or the backward metrics calculated for the L_d stages in the trellis would not be reliable enough to give good soft decisions for the associated bits. Thus a compromise must be reached between the memory needed to store the L_d forward metrics, and the computation wasted calculating L_b discarded backward metrics for each L_d generated soft decisions. It is suggested in [98] that we use $L_d = L_b$, which means that in effect two backward Viterbi recursions are needed to run in parallel. This increases the complexity of the algorithm by about a third, but with $L_d = L_b = 5K$ the number of bits needed to store the forward metrics is reduced to $5K \times 8 \times 10 = 1600$ bits, rather than 409 360 bits.

If the SOVA component decoder is used, metric differences and update sequences are stored for each state at each stage in the trellis, as described in Section 4.6.3. This is effectively 'soft' trace back information that is used to give the decoded soft decisions. Similarly to the standard Viterbi decoder this memory can be reduced using segmented trace back as described earlier.

Using these techniques it is possible to implement a turbo decoder for UMTS without using an excessive number of operations or amount of memory. The number of operations used will be similar to the Viterbi decoder used for the convolutional codes. The amount of memory needed will be larger, but as described above can be reduced to a reasonable level.

Finally, note that often a correct decode is achieved after fewer iterations than the maximum number used. Therefore, if we could tell when a correct decode has been achieved we could often stop the decoder early without affecting the performance of the decoder. This early stopping can be done for example using CRC checks [105] or the SNR of the extrinsic information [106]. This can significantly reduce the average number of iterations, and so the average number of operations and amount of power used by the turbo decoder.

4.6.8 Performance Results for Turbo Coding

In this section we show performance results for the turbo codes described above. As for the results shown for convolutional codes in Section 4.5.7, an AWGN channel is used with QPSK modulation.

Figure 4.28 shows the performance of a turbo decoder, for a code block size of 1000 bits, with a varying number of iterations. It can be seen that the performance

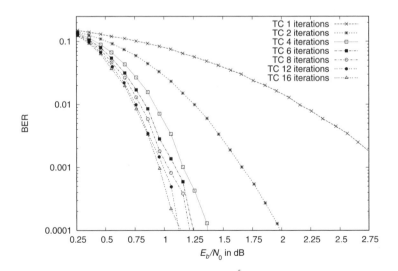

Figure 4.28. The BER versus E_b/N_0 for turbo decoders with varying numbers of iterations I.

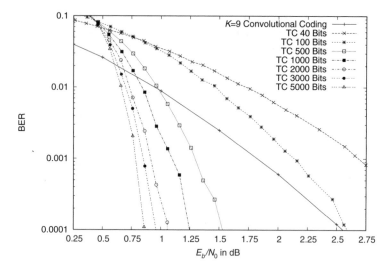

Figure 4.29. The BER versus E_b/N_0 for turbo decoders with varying code block sizes K_i.

increases dramatically with increasing numbers of iterations. However, also note that even four iterations are enough to give most of the performance gain available, and the increase in performance beyond eight iterations is negligible.

Figure 4.29 shows how the performance of the code improves with increasing code block size. Also shown in this figure for comparison is the performance of the one third rate convolutional code described earlier. It can be seen that for very short block sizes the convolutional code outperforms the turbo code. However, as the block size increases so does the performance of the turbo code. For a block size of 100 bits the turbo code performs approximately equivalently to the convolutional code at a BER of 10^{-4}. For a block size of 500 bits it performs about 1 dB better, and for larger code block sizes the performance benefit over convolutional codes increases. These results are all generated with six iterations of a turbo decoder using the Max Log–MAP component decoder

This concludes our discussion of turbo codes and their decoders. We now move on to describe TFC detection, through either TFCI decoding or BTFD.

4.7 TFC DETECTION

As discussed earlier, the TF used by each TrCH can change on a TTI by TTI basis. In order to decode the received data the UE must know which TF was used for each TrCH it decodes. The Node B may send control information each frame, called the *Transport Format Combination Indicator* (TFCI), to signal which TFC was used for the frame. In this case the TFCI information is decoded at the UE to determine the TFC that was used at the Node B, as described in Section 4.7.1 below.

If TFCI bits are transmitted, then they will always occupy at least 30 bits per frame. For low rate services this takes up a significant amount of the available physical channel capacity. For example with an $SF = 256$ service, sending TFCI information uses 3 kbits/s or 10 % of the PhCH capacity. With the pilot and power control signalling that must also be sent for a DCH, using TFCI could reduce the PhCH capacity available to coded data by as much as 20 %. So it is advantageous to be able to detect which TFs were used without this control information. This is called *Blind Transport Format Detection* (BTFD) and is described in Section 4.7.2.

4.7.1 TFCI Transmission and Reception

A given CCTrCH can have up to 1024 TFCs, and so when TFCI bits are sent to indicate which TFC was used, up to ten bits are needed. Usually a CCTrCH will have far fewer TFCs, and so fewer bits are needed to represent the TFCI. When TFCIs are sent, it is vital that they are received correctly, as incorrect TFCI reception will usually result in errors on all TrCHs. Therefore the TFCI is heavily coded by the physical layer. In this section we describe this coding and possible decoding techniques.

For channels where TFCI bits are sent, space for either 30 or 120 coded bits per frame is allocated on the PhCH. On the downlink with high spreading factors (SF = 128 or higher), and on the uplink, 30 coded bits per frame are sent. On the downlink with lower spreading factors, the number of TFCI bits per frame is increased to compensate for the lower spreading gain, and 120 coded TFCI bits are sent per frame. In compressed frames, the slot format is changed so no TFCI information is lost, and on the downlink, space is allocated for at least 32 or 128 bits. On the uplink, space is allocated for at least 30 bits.

As correct reception of the TFCI bits is so important, great care was taken in choosing a code to use that gives good performance with these various numbers of input and output bits, but was still relatively simple. A coding scheme based on Reed–Muller codes using a 32 × 32 Hadamard matrix was chosen. Ten 'basis sequences' of 32 bits each are defined, and each input bit is used to mask a basis sequence. The 32 output bits are then given by the modulo two addition of the masked basis sequence. Mathematically, if $M_{i,n}$ are the basis sequences, with $n = 0, 1 \ldots 9$ indicating the sequence number and $i = 0, 1 \ldots 31$ giving the bit number within a sequence, and a_n are the input TFCI bits, then the output codeword is given by:

$$b_i = \left(\sum_{n=0}^{N-1} a_n \times M_{i,n} \right) \bmod 2 \tag{4.15}$$

where $N = 1, 2 \ldots 10$ is the number of TFCI bits to be encoded. The resulting 32 bits are then either punctured to give 30 bits to be transmitted, or repeated to give 120 bits to be transmitted.

When puncturing is used the final two coded bits are punctured, and for this reason the Hadamard sequences were rearranged so that the final two bits are less important to the performance of the code. The order of the sequences was also carefully selected to give better distance properties of the codes, and therefore better decoding performance, when fewer TFCI bits are sent.

The received TFCI bits can be decoded using a variety of techniques. The number of possible code words is at most 1024, and for low capability UEs will be much lower. For example, for the 32 kbits/s capability class defined in [68] only 32 different TFCs can be used. Therefore, a 'brute force' decode, comparing the received sequence to each possible transmitted sequence and choosing the transmitted sequence which is closest to the received sequence, may be appropriate. Alternatively, one or several *Fast Hadamard Transforms*

(FHTs) may be performed to find the transmitted sequence with the best correlation to the received sequence, as described in [107]. Both these techniques will choose the maximum likelihood sequence and so will give the best TFCI decoding performance. Finally, a Reed–Muller decoder may be used, which for large numbers of possible TFCIs will be more efficient but is not a maximum likelihood decoder.

As well as TFC information, when a DSCH is received, the TFCI on the DCH is also used to carry information about the spreading factor and OVSF code to be used by the DSCH. This is called split mode as the TFCI is split between control information for the DCH and the DSCH. Two forms of split mode are defined:

Hard Split. In this case the DCH and DSCH are allocated five TFCI bits each, which are separately encoded using the masked basis sequence technique described above, but with five basis sequences of 16 bits each for each code.

Soft Split. In this case the DCH is allocated N TFCI bits, and the DSCH the remaining $10 - N$ bits. The input bits from both TFCIs are combined and the resulting ten bits are encoded as shown in Equation (4.15).

Soft split mode gives better coding performance and allows a more flexible split of TFCI capacity between the DCH and DSCH. However, the hard split mode is necessary because the control information for the DCH and DSCH can come from different entities in the network and so may not be able to be encoded together when soft handover is used. Later versions of [50] extend hard split mode to give a more flexible split of capacity between the DCH and DSCH.

4.7.2   Blind Transport Format Detection (BTFD)

Blind transport format detection (BTFD) is used to determine the TFs used on a CCTrCH that is transmitted without the overhead of the TFCI information described above. In this section we describe how BTFD can be carried out at the UE.

In order to reduce the complexity of BTFD at the UE, guidelines are given in [50] on different types of detection that can be performed on certain types of TrCH. These types of detection are:

Guided Detection. Because only certain combinations of TFs, as given by the TFC set, are allowed, the TF of some TrCHs may be able to be implied from the TF of others. So, for example, in the TFC set shown in Table 4.7 in Section 4.4.3, the TF of $TrCH_2$ can be implied from the TF of $TrCH_1$, but not vice versa. This is called guided TF detection – the TF of the guided TrCH is implied from the TF of another TrCH, called the guiding TrCH. For this to be possible the guided and guiding TrCHs will also have the same TTI length.

Single Detection. This form of detection can be used on TrCHs having only one TF with any transport blocks. These TrCHs have either one TF only, or have a second TF with no transport blocks. It may be possible to treat these TrCHs as guided, in which case the UE should decide their TF based on the guiding TrCH. However, if this is not possible [50] states that the UE can decode these TrCHs by assuming the TF with transport blocks was sent every TTI. In the case where the TF with no transport blocks was sent, the data received will not be valid, but this will usually be spotted with a CRC fail and the invalid data can be discarded by higher layers.

Explicit Detection. For TrCHs whose TF cannot be implied or treated as fixed, the TF must be determined explicitly. These TrCHs will be sent with at least one set of CRC bits on all TFs, so the UE can determine the TF using trial and error decodes until a TF which gives a CRC pass is found. See below for more details. The guiding TrCH mentioned above will be an explicit TrCH. Note that as CRC bits are sent for all TFs on explicitly detected TrCHs, if there is a TF with no data bits it must have at least one transport block of zero size. For this TF only CRC bits are sent, as described in Section 4.3.3.

Additionally, [50] places restrictions on the circumstances in which the UE must be able to perform BTFD. These restrictions aim to limit the complexity of performing BTFD at the UE, and include the following:

- fixed TrCH positions must be used;

- all TrCHs must be able to be detected using one of the techniques above. If there are any guided TrCHs, then there must be one TrCH that can be used as the guiding TrCH for all these TrCHs;

- explicitly detected TrCHs must use convolutional coding with one code block only;

- at most three TrCHs should need to be explicitly detected, with at most 16 possible TFs between them.

A UE can implement BTFD as follows. As discussed above, any single detection TrCHs can be decoded by assuming the TF with transport blocks has been transmitted. Explicit detection TrCHs have their TFs determined based on speculative decoding and CRC checks, as described below. Once the TF of the explicit TrCHs has been determined, one of these (the guiding TrCH) can be used to imply the TF of any guided TrCHs.

The key to BTFD is therefore detection of the TFs of explicitly detected TrCHs. From the first restriction above, a UE can implement its BTFD algorithm assuming that fixed TrCH positions are used. As explained in Section 4.4.3, this means that each TrCH has a fixed allocation of PhCH capacity, regardless of the TF used. This PhCH capacity is entirely used by the TrCH at its maximum rate TF, and partially used for lower rate TFs. For these lower rate TFs the unused capacity is filled with DTX indicators, inserted at the end of the rate matched bits just before they are interleaved by the TrCH interleaver. Also, the rate matching pattern applied to each TrCH at the Node B will be the same regardless of the TF used, for higher rate TFs this pattern is just run over more bits. So for lower rate TFs the coded bits are rate matched and interleaved in exactly the same way as the corresponding bits are for the maximum rate TF. This allows the UE to carry out the deinterleaving, demultiplexing and derate matching processes by assuming that the maximum rate TF has been transmitted. At the end of this process the UE will have a set of deinterleaved and derate matched soft decisions the size of the encoder output sequence for the maximum rate TF. If a lower rate TF was in fact transmitted, there will be some soft decisions at the end of this set that correspond to DTX indicators in the transmitter. These will be noise. But the received soft decisions for the coded bits that were sent will have been deinterleaved and derate matched exactly as if the UE had known the actual TF used at the transmitter.

At this stage, the UE can try decoding the received soft decisions for each possible TF and run a CRC check on each set of decoded bits to see which gives a CRC pass. As convolutional coding is used and we have only one code block on each TrCH, we do not need to redo the entire Viterbi decoding for each possible TF. Instead it is sufficient to just redo the trace back for each of the TFs. As described in Section 4.5 this trace back is relatively simple; most of the complexity of the Viterbi decoding lies in calculating metrics using the ACS operations.

Unfortunately, there is a finite chance that we will get a CRC pass even with an incorrect TF. For example with a 12 bit CRC there is a one in $2^{12} = 4096$ chance that an incorrect TF will give a CRC pass. Hence, if the UE tries trace backs and CRC checks on all possible TFs it will occasionally find two TFs in a given TTI that give a CRC pass, the correct TF and another that gives a CRC pass by chance. When this happens the UE needs some form of tie-break to decide which is the correct TF, choosing one arbitrarily means it will not meet the stringent false TF detection rate requirements in [9]. The metrics in the Viterbi trellis can be used for this purpose. As the convolutional encoder is terminated at the transmitter, at the correct end position in the Viterbi trellis the metric of the all zero state should be the highest of the 256 metrics. So the UE can use information regarding how 'good' this all zero metric is compared to other metrics to help choose the correct TF to decode. This information can also be used to avoid doing trace backs and CRC checks at all for some TFs, and so reduce the complexity of BTFD.

In summary, using the restrictions in [50], the UE can perform BTFD with the same set of operations as for other convolutionally encoded TrCHs. The only additional requirement is for speculative trace backs and CRC checks for several possible TFs (up to 16) on up to three explicitly detected TrCHs. The TFs of other TrCHs can either be considered to be fixed, or can be implied from the TF of one of the explicitly detected TrCHs.

4.8 COMPRESSED MODE AND THE BRP

As discussed in Section 3.11, compressed mode is used to create gaps in the DCH transmissions to allow the UE to make measurements on other frequencies. The gaps may be in the downlink only, the uplink only, or both. The following techniques can be used by the physical layer to create these gaps:

Higher Layer Scheduling (HLS). The MAC knows where gaps are, and so sends less data (i.e. selects the TFC to use from a subset of the total TFC set) during frames containing a gap. This allows a gap to be created without losing data. This form of compressed mode can be used on the uplink or on the downlink when flexible TrCH positions are used.

Spreading Factor Reduction (SFR). The spreading factor is halved so that twice as much data can be transmitted in what is left of the frame. This means that up to half the frame can be a gap without losing data. Note this form of compressed mode can be used on both the uplink and downlink, with either fixed or flexible TrCH positions, but cannot be used when the spreading factor in normal mode is four.

Puncturing. Layer 1 performs extra puncturing (or less repetition) so that there is less coded and rate matched data to be transmitted on the physical channel so a gap can be created. This form of compressed mode can only be used on the downlink and only with fixed TrCH positions.

Table 4.12   Allowed forms of compressed mode on the uplink and downlink.

	Higher layer scheduling (HLS)	Spreading factor reduction (SFR)	Puncturing
Uplink	Yes	Yes	No
DL fixed positions	No	Yes	Yes
DL flexible positions	Yes	Yes	No

Table 4.12 summarizes these different forms of compressed mode and when they can be used.

In this section we discuss the effect of these gaps on the BRP in both the uplink and the downlink. We use the terms 'compressed frame' and 'normal frame' below to mean a frame with and without a gap.

4.8.1   Compressed Mode Effects on Downlink BRP

Typically, gaps in the downlink are needed for UEs that do not have dual receivers. For such receivers to make measurements on a different frequency, the UE must temporarily stop reception on the original frequency. So at the start of the gap the UE will stop receiving on the original frequency and switch to the frequency on which measurements are to be made. It will then make the measurements and switch back to the original frequency in time for the end of the gap. Depending on whether fixed or flexible TrCH positions are used, compressed mode by either SFR or HLS (for flexible TrCH positions), or by puncturing or SFR (for fixed TrCH positions) can be used. In this section we describe the modifications needed to the BRP chain described in Section 4.3.2 for each of these modes.

4.8.1.1   Compressed Mode by Spreading Factor Reduction (SFR)   For the downlink BRP, SFR is the simplest form of compressed mode. It can only be used if the spreading factor in normal frames is higher than four. In a compressed frame the spreading factor is halved, giving twice as many data bits per slot, but the data and DTX symbols are mapped to only 7.5 slots per frame. This means we have exactly the same amount of data and 'normal' DTX as in normal frames, and so the channel encoding, rate matching, DTX insertion and interleaving all operate exactly as in normal mode. Extra DTX is added in the physical channel mapping block (after the PhCH interleaver) to fill up the rest of the transmitted frame. See Figure 4.4 in Section 4.3.2 for a reminder of this BRP chain in the downlink. Between 8 and 14 slots will be transmitted, so between 0.5 and 6.5 slots of extra DTX are added in the PhCH mapping block.

This 'extra' DTX is positioned so that as little data and 'normal' DTX as possible is positioned after the gap. Any data that must be positioned after the gap (because there is not enough room before the gap) is positioned at the end of the frame. This is because the quality of the received soft decisions just after the gap will be lower than usual, due to the gap in the CRP. The extra DTX is added in the most vulnerable positions (just after the gap). This is illustrated in Figure 4.30, which shows how the data and normal DTX, and the extra DTX, is mapped around a gap falling in slots 10–12 or slots 4–6.

Figure 4.30. PhCH mapping for compressed mode by SFR in the downlink.

4.8.1.2 Compressed Mode by Higher Layer Scheduling (HLS)

In the downlink, compressed mode by HLS is only allowed for flexible transport channel positions. In TTIs with a gap, higher layers select a low rate TFC so that the data to be transmitted will fit into what is left of the TTI. The channel encoding, rate matching and TrCH interleaving operate as for normal TTIs. However the second DTX insertion is modified to insert less DTX than it would usually, to leave room for the gap. If we denote the PhCH capacity available in normal mode as N_{PhCH}, and the number of bits produced at the output of the TrCH multiplexing as S, then in normal frames we would add $(N_{\text{PhCH}} - S)$ DTX indicators. In compressed frames the slot format will be modified if TFCI bits are carried, so as not to lose any TFCI capacity. This means the capacity for data bits across all 15 slots of the frame is reduced to $\tilde{N}_{\text{PhCH}}$. In a frame where only N_{tr} slots are transmitted due to an HLS gap we would add $(N_{\text{tr}}/15 \times \tilde{N}_{\text{PhCH}} - S)$ DTX indicators. The fact that we add less DTX to leave room for the gap means that the number of bits and DTX indicators input to the PhCH (second) interleaver will not be a multiple of 30. This means, as described in Section 4.3.7, the final row in the interleaver will be partially empty which slightly complicates the interleaver (and corresponding deinterleaver) functions.

4.8.1.3 Compressed Mode by Puncturing

Compressed mode by puncturing is used in the downlink only, and is only allowed when fixed TrCH positions are used. Due to limitations on the other two forms of compressed mode described above, it is the only way to create gaps for CCTrCHs using fixed TrCH positions with spreading factor four. It is also the most complex form of compressed mode for the downlink BRP to handle. As the name suggests, a gap is created by effectively puncturing some of the coded and rate matched information bits to be transmitted.

In reality, rather than repeat or puncture coded bits in the rate matching stage and then puncture some of these rate matched bits to make room for the gap, the rate matching is modified to repeat fewer bits, or puncture more bits, to make room for the gap. In [50], the space for the gap is considered to be marked with 'p bits'. These p bits are added in the TrCH interleaver, and removed prior to the PhCH interleaver. As discussed earlier, they form no part of the signal transmitted by the Node B or received by the UE. In effect they merely distort the pattern of the TrCH deinterleaver. Note that although they are useful conceptually, and so we discuss them here, they are merely conceptual and need form no part of a real implementation.

The downlink BRP is modified as follows for compressed mode by puncturing. At the start of each TTI for each TrCH the rate matching block checks to see if there are any gaps by puncturing in any of the frames of the TTI. For every frame with a gap, it calculates how many p bits are needed to make room for the gap. These p bits are split between the different TrCHs using the same technique as is used to split the PhCH capacity between the different TrCHs (using a formula similar to Equation (4.3)). Once the number of p bits required in each frame of the TTI is known, the total number of p bits required over the whole TTI is calculated. Then the number of bits to be repeated or punctured across this TTI, for the maximum rate TF of the TrCH, $\Delta N_{i,\max}^{\mathrm{TTI}}$, is reduced by the total number of p bits allocated to this TrCH. For the maximum rate TF, the amount of repetition or puncturing will be adjusted by the number of p bits. For lower rate TFs, the amount of repetition or puncturing will be adjusted relative to the rate of the TF. The first DTX insertion stage then adds DTX to fill the proportion of the PhCH allocated to this TrCH, minus the number of p bits that are needed. This means that for each TrCH the amount of repetition or puncturing, and for lower rate TFs the amount of DTX added, is adjusted to make room for any gaps. The effects of any extra puncturing (or less repetition) that is needed is spread across the TTI of the TrCH.

In the TrCH (first) interleaver, each frame of the TTI for a TrCH corresponds to a column in the matrix interleaver. Hence, the p bits are conceptually added at this stage; for each column corresponding to a frame with a gap the number of p bits needed in that frame for this TrCH is added at the start of the column. Effectively, this just changes the pattern of the TrCH interleaver. The radio frame segmentation, TrCH multiplexing and second insertion of DTX indication blocks operate exactly as normal. The p bits are conceptually removed at the PhCH segmentation stage, before the remaining bits are interleaved and mapped onto the physical channel. This means that, as for compressed mode by HLS, the number of bits passing through the PhCH interleaver will no longer be a multiple of 30, and so the structure of this interleaver changes.

In summary, compressed mode by puncturing modifies the number of bits added or removed by rate matching, the amount of DTX added and the patterns of the TrCH and PhCH interleavers. It therefore has a significant impact on the downlink BRP chain and is an important factor to consider in the design of the UE BRP receiver.

4.8.1.4 *TFCI Mapping*

For all forms of compressed mode, the slot formats are modified for frames with gaps so that no TFCI information is lost. For any frame with a gap, the number of TFCI bits per slot is doubled. This means that even with the longest possible gap in one frame (seven slots), there is space for more TFCI bits in compressed frames than in normal frames.

As described in Section 4.7.1, the TFCI encoder produces 32 encoded bits and in normal frames sends either 30 or 120 bits, by puncturing two bits or repeating some bits three times and other bits four times. In compressed frames there is always space for at least 32 or 128 TFCI bits. So, either no TFCI bits are punctured, to give 32 bits, or all TFCI bits are repeated four times, to give 128 bits. If more than eight slots are transmitted in the frame there will be space for more bits. Rather than use repetition, as on the uplink, these positions are filled with DTX mapped so as to keep the TFCI bits away from the gap, similar to the mapping of data and 'normal DTX' for compressed mode by SFR shown in Figure 4.30.

4.8.2 Compressed Mode Effects on the Uplink BRP

In the case where there is a downlink gap, a gap may also be needed on the uplink because the UE has a fixed duplex RF so that the receive frequency cannot be moved without moving the uplink frequency. Also, it may be that the frequency on which measurements are to be made is close to the uplink frequency on which the UE is transmitting, so that the UE needs to stop transmitting to allow the measurements to be made. These gaps may or may not be accompanied by downlink gaps, depending on whether the UE has dual receive capabilities. In either case, the changes to the uplink BRP are relatively minor. The calculations by the rate matching control described in Section 4.4.2 are modified to take account of the gap and change the $\Delta N_{i,j}$ values so that the modified PhCH capacity is filled.

In the case of compressed mode by HLS, the number of bits available on each PhCH combination is reduced to take account of the gap when each combination is checked against the number of bits needed for the current TFC. A PhCH combination is chosen as described earlier using the modified capacities. The choice made may be the same as it would be for normal frames, in which case the $\Delta N_{i,j}$ values are reduced so more puncturing or less repetition is done. Alternatively, the next higher PhCH combination may be chosen. In this case, depending on the gap size and the number of PhCHs used, there will usually be more capacity on compressed frames, even allowing for the gap, and so the $\Delta N_{i,j}$ values will be increased.

For compressed mode by SFR, initially the PhCH combination is chosen without taking the gap, or the change in spreading factor, into account. This form of compressed mode can only be used for TFCs where the PhCH chosen in normal frames is higher than SF $= 4$. The chosen SF is then halved, and the $\Delta N_{i,j}$ values are calculated using this lower SF offset by the size of the gap. For example, if the PhCH capacity for the SF originally chosen is $N_{\text{PhCH},j}$, and we have a frame where only N_{tr} rather than 15 slots are transmitted, the PhCH capacity becomes $2 \times N_{\text{PhCH}} \times N_{\text{tr}}/15$. In this case the PhCH capacity in the compressed frame will always be higher than in normal frames, as the minimum number of transmitted slots per frame N_{tr} is eight. Therefore the $\Delta N_{i,j}$ values will be increased.

In either case, the rate matching fills the PhCH capacity allowing for the gap, and the rest of the BRP chain described in Section 4.3.1 behaves as for normal frames. The only difference is in the PhCH interleaver which, unlike normal frames, will have an incomplete last row and will use padding bits as described in Section 4.3.7.

For the uplink BRP the only other modification is to the TFCI coding and mapping. Similar to the downlink, for uplink compressed frames the DPCCH slot format is changed so that no TFCI coded bits are lost. Unlike the downlink, the change in slot format depends on how many slots are transmitted. For frames where between 10 and 14 slots are transmitted, the number of TFCI bits per slot increases from two to three. For frames where only eight or nine slots are transmitted, four TFCI bits are sent per slot. In normal frames, 30 TFCI bits are transmitted, whereas in compressed frames, between 30 and 42 TFCI bits will be transmitted (depending on how many slots are transmitted). As described in Section 4.7.1, the TFCI encoder produces 32 coded bits, which are usually punctured down to 30 bits. For compressed frames this puncturing is usually not needed, and any extra TFCI bits are filled by repeating some of the 32 encoded TFCI bits, as described in Section 4.3.5.2.1 of [50].

4.9 BRP LIMITATIONS FOR DIFFERENT TrCHs AND CCTrCHs

As described earlier, the BRP in UMTS is very flexible, offering various forms of channel coding, with different QoS on different TrCHs, and is able to match a wide variety of input data sizes to the PhCH sizes offered by CRP. The complexity of the BRP in a UE is limited by the capabilities it signals to the network, see [68] for details. These capabilities include parameters such as whether turbo encoding is supported, the maximum number of data bits or PhCH, bits that can be decoded or received, the maximum number of TrCHs, etc. These capabilities are an extremely important consideration in the design of the BRP implementation.

Sections 4.2.13 and 4.2.14 in [50] also give some restrictions that are true for all UEs, depending on the type of the TrCH being carried. These are summarized below:

BCH The BCH TrCH for each cell must be carried on its own CCTrCH, with one TrCH only. The BCH has fixed parameters to allow the UE to receive it without any prior knowledge from the network – it has a 20 ms TTI, one transport block with 246 data bits and a 16 bit CRC, and half rate convolutional encoding. These parameters give an exact match onto an SF = 256 PhCH so that rate matching is not required.

RACH The RACH TrCH is also carried on its own CCTrCH, with one TrCH only. It can have a 10 or 20 ms TTI, and a minimum spreading factor of 32.

CPCH Like the RACH, the CPCH has one TrCH only per CCTrCH. Its minimum SF is either limited by the UE's capabilities or is four, but it can use only one PhCH.

PCH The PCH may be on its own CCTrCH, or may share a CCTrCH with one or several FACH TrCHs. It will always have a 10 ms TTI, and in the case it shares a CCTrCH with FACHs, will always be the first TrCH multiplexed into the CCTrCH. Only one PhCH will be used, with any SF other than SF = 512.

FACH One or several FACHs can share a CCTrCH with a single PCH. This is the only case in which different types of TrCH can share a CCTrCH. As for the PCH, only one PhCH will be used, with any SF other than SF = 512.

DCH DCH TrCHs cannot share a CCTrCH with any other type of TrCH, but several DCH TrCHs can share one CCTrCH. Depending on the UE capabilities [68], several downlink PhCHs of any SF can be used, and on the uplink up to six DPDCHs with SF = 4 can be used.

DSCH Like the DCH, one or more DSCH TrCHs will be sent on their own CCTrCH, but otherwise the DSCH is limited only by the capabilities of the UE.

Finally, note that on the uplink one UE can send only one CCTrCH at any time. On the downlink, depending on the UE capabilities [68], several CCTrCHs may be received simultaneously. In fact, in order to meet measurement requirements, all UEs need to be able to receive a DCH CCTrCH and a BCH CCTrCH simultaneously. Higher capability UEs may also be able to receive a DSCH and/or a FACH CCTrCH simultaneously.

4.10 CONCLUSIONS

In this chapter we have given a detailed description of the bit rate processing defined in [50] for the FDD mode of UMTS. Most other mobile communications standards, such as GSM/GPRS [108] and 3GPP2 [90] define several fixed channels with a given number of input bits and defined coding, interleaving, etc. In contrast, the BRP for UMTS allows a much more flexible set of configurations. The number of input bits for a channel can take any value subject only to restrictions on the total number of bits across all TrCHs being received or transmitted. Several different TrCHs can be transmitted or received simultaneously, and the BRP will multiplex them together and match their different error rate and delay requirements. A range of different error detection (CRC) and error correction (convolutional or turbo coding) options are provided, and each TrCH can have these configured independently depending on its quality of service requirements.

We have described in detail this BRP chain for both the uplink and downlink. We have also discussed the theory behind decoding techniques for the channel coding options, and how these decoders could be implemented.

5

Type Approval Testing: A Case Study

Rudolf Tanner

5.1 INTRODUCTION

The previous chapters have discussed the physical layer and the issues associated with implementation. Assuming that at this stage a prototype is available, then we intend now to test the prototype. *User Equipment* (UE), for example a handset, goes through many stages of testing. There are, for instance, tests that prove the functionality of the design, tests that verify the performance, tests that prove safety, radiation limits and regulatory compliance.

In this chapter, we address a subset of the 3GPP user equipment *Type Approval* (TA) test specification, namely for the *Dedicated Physical Channel* (DPCH) *Block Error Rate* (BLER), which is carried out in a lab environment.

It goes without saying that testing is a fundamental part of every product development and requires a significant amount of resources and time.

Serial production testing needs differ from the R&D test needs because the purpose of testing is here to identify faulty components, while it can be assumed that the product will meet the technical specifications provided that the hardware functions.

The manufacturer of handsets, for example, requires a complete chip set comprising RF circuitry, baseband chips and support circuitry for low cost and low power reasons. A baseband chip may be developed with a *System On Chip* (SOC) approach [109–111], which could have the following steps:

WCDMA – Requirements and Practical Design. Edited by R. Tanner and J. Woodard.
© 2004 John Wiley & Sons, Ltd. ISBN: 0-470-86177-0.

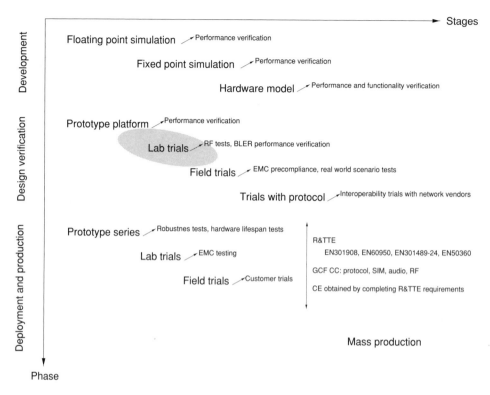

Figure 5.1. UE development phases with examples of tests carried out at each stage.

1. System design within a floating point environment ([112]);

2. Fixed point simulator ([113]);

3. Verify the design with a reference model;

4. Build prototype [114];

5. Test the end-to-end performance, field trials under realistic conditions [115, 116];

6. Repeat tests with final chip.

In all cases, the development engineers need to know how well their design needs to perform, hence clear performance requirements are needed.

Testing is carried out during the whole product development life cycle and different test strategies are employed at each stage. Figure 5.1 presents a series of product stages during the development phases and lists some tests which may be carried out.

During the development stages, the receiver performance will be analysed in isolation in order to gain confidence that the receiver can exceed the BLER[1] performance, under different propagation conditions. A reference model may then be produced in software which matches the hardware blocks. Now simulation results, e.g. test vectors, can be compared and so problems can be rectified at an early stage. Once the first piece of working hardware, a prototype for example, is available, it is common to start with functional tests in

[1]We use the terms ratio and rate interchangeably.

order to verify that the implemented functions work correctly. This may already involve protocol software to exercise handover [117]. For example, the cell search procedure is executed and the *Broadcast Channel* (BCH) information is decoded. Later, the test envelope is expanded to also cover the BLER performance. At this stage, the product design, e.g. baseband processing, is well characterized and verified. The purpose of development and design verification testing is to confirm that the implemented design can meet the product specification requirements.

The European *Radio & Telecom Terminal Equipment* (R&TTE) Directive 1995/5/EC tests are also concerned with safety aspects. Another body, which is concerned with the verification of GSM and WCDMA handsets for example, is the *General Certification Forum* (GCF) which defines the *Certification Criteria* (CC). The CE label *CE* can be obtained by meeting harmonized standards that cover the essential requirements of the R&TTE Directive, or submitting a *Technical Construction File* (TCF) to a notified body or using the full quality assurance procedure.

3GPP has produced several documents which detail the different aspects of performance testing and type approval testing, such as physical layer, protocol and safety aspects. A set of relevant specifications is listed below:

- TS25.101 [9] defines the minimum RF characteristics for user equipment, which are divided into transmitter and receiver characteristics and receiver performance requirements.

- TS34.108 [118] defines default values for test cases such as transport and physical channel parameters.

- TS34.109 [119] specifies the functions required for UE conformance testing.

- TS34.121 [10] specifies the measurement procedures for the UE (type approval) conformance tests, i.e. TS25.101.

- TS34.123-1 [120] specifies protocol and contains conformance testing, i.e. descriptions of test cases for the UE and Uu interface.

- TS34.123-2 [121] 3GPP Release 99 document is not completed yet, but Release 5 document specifies a recommended applicability of the test cases in TS34.123-1.

- TS34.123-3 [122] specifies the protocol conformance testing for user equipment at the Uu interface.

- TS34.124 [123] contains the essential *Electro Magnetic Compatibility* (EMC) requirements.

- TR34.901 [124] explains the theory applied to UE conformance tests in order to reduce the test duration.

The basic UE transmission and reception requirements needed by the design engineers are outlined in TS25.101 [9]. The more comprehensive test descriptions needed by the test engineer are given in TS34.121 [10], which comprises test configuration, test environment, test equipment accuracy and much more. In the remainder of this chapter, we address the DPCH BLER performance characteristics according to TS34.121.

This chapter is organized as follows. First, we turn the wheel of time backwards and explain briefly how 3GPP arrived at the DPCH BLER performance requirements. Then we provide an overview of the relevant 3GPP specifications. Section 5.3 explains what

Table 5.1   Required I_{or}/I_{oc} in dB for 12.2 kbps at BLER $= 0.01$ for four different propagation channel conditions and results of four different contributing organizations [126].

	Static	Case 1	Case 2	Case 3
Company A	-19.6	-21.0	-13.7	-16.0
Company B	-19.5	-21.3	-13.5	-15.7
Company C	-19.5	-21.4	-14.0	-16.0
Company D	-19.6	-21.3	-14.2	-15.9

Table 5.2   The agreed I_{or}/I_{oc} requirement for 12.2 kbps at BLER $= 0.01$.

	Static	Case 1	Case 2	Case 3
TS25.101 [9]	-16.6 dB	-15 dB	-7.7 dB	-11.8 dB
Implementation margin	3 dB	6 dB	6 dB	4 dB

equipment is needed in order to carry out the tests in an R&D lab environment. Finally, Section 5.4 shows a few results that confirm that the 3GPP requirements can be met. Note that this chapter will not discuss protocol testing, field trials, network capacity testing and interoperability testing.

5.2   HISTORY: THE MAKING OF THE 3GPP DPCH BLER REQUIREMENTS

Back in 1999, the 3GPP participating groups produced floating point simulation results and tabulated them in [125,126] based on the simulation parameters agreed to in TS25.942 [127], upon which 3GPP then based the final requirements in TS25.101 [9] by allowing for more than the typical 2 dB implementation margin. Table 5.1 lists a subset of the results and shows that the results obtained by the different contributing parties match each other closely. Based on the results listed in Table 5.1, 3GPP agreed to the requirements listed in Table 5.2.

5.3   LAB TESTING

The characterization of the DPCH BLER performance will be carried out in a lab environment. Here we do not attempt to detail every single test and choose to describe by example the process which can then be modified and applied to suit all the other test cases defined in [10].

5.3.1   Typical Test Configuration for Downlink Fading Tests

Annex A in [10] proposes a test equipment setup with the aid of connection diagrams from which a test engineer can build a test rig. Here we discuss two test configurations used to obtain the BLER performance of the UE under test, or *Device Under Test* (DUT).

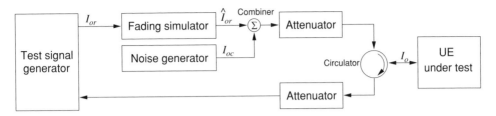

Figure 5.2. Demodulation test of DCH in multipath fading propagation.

Measuring the BLER performance of a UE when multipath is present in a lab environment requires a special purpose-built fading simulator and a noise generator. For ease of operation, it is desirable to have a combined fading simulator and noise generator, as depicted in Figure 5.2. From the left to the right in Figure 5.2, we have a test signal generator, a fading and noise generator, an attenuator, a circulator, the UE under test and an uplink path with an attenuator.

I_{or} denotes the total transmitted power spectral density as measured at the test signal generator antenna connector, i.e. Node B. $\hat{I}_{or}$ is the total received power spectral density of the downlink signal as measured at the UE antenna connector, and I_{oc} denotes the power spectral density of a band limited white noise source as measured at the UE antenna connector.

Strictly speaking, there is no real need for an uplink connection. Testing the receiver's BLER performance does not require a comparison between transmitted and decoded bits because the error is obtained from the *Cyclic Redundancy Check* (CRC), see Section 4.3.3. Further, most of the specified demodulation tests have downlink power control disabled. However, loop back tests are described where the received, demodulated and decoded signal is transmitted back to the signal source in order to compute the *Bit Error Ratio* (BER) through a bit comparison. The uplink signal is transmitted at high power and is not impaired by channel and noise artefacts. Thus it can be assumed that no bit errors occur during the retransmission back to the signal source. The BER is then obtained by comparing the bits sent in a frame, recall from the previous chapters, that each frame has a number which eases synchronization, with the received bits in error. An alternative method is to make use of a PN9 generator[2]. The transmitter sends a repetitive bit sequence to the UE via the DPCH, which, once synchronization has been established, allows the UE to compute the BER itself, without the need for an uplink, i.e. loop back.

When a second transmit antenna is present, e.g. in the case of *Space Time Transmit Diversity* (STTD) (see Chapter 3), then the hardware setup becomes more complicated. Figure 5.3 shows an example when two Node B antennas are required.

5.3.2 Selecting Test Equipment

Different pieces of equipment are required to build a test rig in order to perform TA testing [128] as shown in Figures 5.2 and 5.3. In addition, we also need measurement equipment to verify various parameters such as power. Note that the accepted uncertainty of test

---

[2] A *Pseudo Noise* (PN) sequence generator produces a bit sequence which is (pseudo) random for a short period of time, and repeats itself after a certain time, dependent on the length of the polynomial, see also Chapter 4.

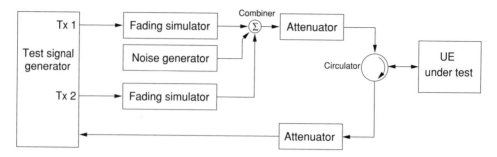

Figure 5.3. Demodulation test of DCH in open loop transmit diversity mode.

requirements can be found in Section F.2 of [10]. The range of equipment required can span simple devices and sophisticated kit:

- power meter;
- Node B (base station) signal generator;
- arbitrary signal waveform generator;
- noise and interference generator;
- propagation channel simulator;
- spectrum analyser;
- vector signal analyser.

Power meters are a rather simple but useful device for measuring RF signal power [129]. Common mistakes are that an unsuitable probe (sensor) is used or that loading effects cause a measurement error due to impedance mismatch. The test engineer should be aware that a small error may have a big impact. For example, some *Dedicated Channel* (DCH) demodulation tests measure the receiver performance at two *Signal to Interference Ratio* (SIR) points 0.1 dB apart. Modern power meters use sensors that use thermocouples for accuracy, or diodes for fast measurements, e.g. peak envelope power, and have sophisticated correction tables built in to correct various thermal related errors. Thermocouple based probes have a good zero stability and range from about 1 μW onwards but one should avoid heating up the probe by either holding it in the hand or connecting it to a warm antenna connector. Diode-based probes are available for powers greater than 100 pW.

Perhaps the greatest challenge is the selection of the test signal generation methodology. Different options are illustrated in Figure 5.4. The dashed boxes highlight which features can be found in a single piece of equipment. For example, the second column shows that there are combined fading channel simulators with an internal noise source.

The left hand side in Figure 5.4 depicts a signal chain with separate test apparatus, while the outermost right in Figure 5.4 shows a single device that can do it all in one. The flexibility is limited by using a baseband signal generator which only produces the reference channels as addressed in subsection 5.3.3, used for DCH demodulation tests. Thus R&D departments tend to use an *Arbitrary Waveform Generator* (ARB) which allows them to generate any desired WCDMA baseband signal for supporting the development. However,

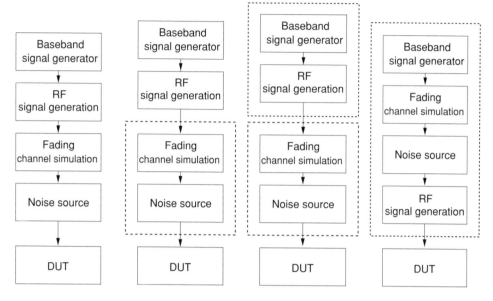

Figure 5.4. Different equipment options are available in order to generate the desired RF signal and optionally the channel impairment.

this requires a special software tool which allows the generation of test vectors which in turn can be loaded into a baseband signal generator or ARB. Selecting the most suitable piece of equipment is not trivial, and is dependent on the technical specification, the chosen test strategy and the available budget. Important equipment specifications are the power range of the RF generator and the associated accuracy, the *Error Vector Magnitude* (EVM), the ability to automate tests, frequency stability, the ability to synchronize the signal generators for transmit diversity tests and the ability to embed markers in the data in order to create time delays between different signals, i.e. cells [130, 131].

There are two basic types of fading channel simulator. In the first type, the fading is introduced at the baseband signal level prior to the RF up converter. Alternatively, the RF signal is first down converted to baseband followed by the fading stage, and then up converted back to RF. Both schemes are illustrated in Figure 5.5. The former approach takes advantage of the fact that the baseband signal is already available and thus is more economic. The latter method either employs an external PC to control the fading, or employs a real time fading generator inside the equipment. The up and down conversion approach tends to be more expensive but provides a much greater degree of flexibility in its usage in a lab environment where equipment is shared amongst the engineers and moved around. It is important that a good filter strategy is employed to ensure that the up and down signal conversion does not corrupt the signal spectrum in any way.

RF cables and connectors are often dismissed as trivial commodities, but it is important that high quality cables and connectors are used in the lab, such as RG213/U or RG223/U, while working in the 2 GHz band. An attenuator is required if the signal generator cannot output a signal at the desired low power level with a sufficiently small absolute accuracy, e.g. ± 0.7 dB. We may choose, for example, to transmit at a higher level, say -40 dBm, and use a 20 dB attenuator to obtain a -60 dBm power level at the receiver. Important

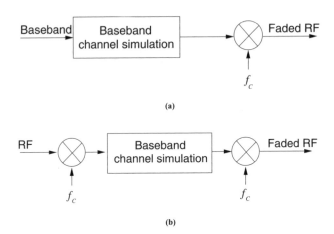

Figure 5.5. Two different approaches to introducing channel fading. (a) Baseband channel simulator; (b) RF channel simulator.

selection criteria are the flatness over the frequency range, the maximum RF power and the impedance. Step attenuators are generally less accurate than fixed attenuators, but it is always possible to get hold of a calibration sheet from the supplier or manufacturer.

Passive RF signal combiners are sometimes found in catalogues under the heading 'splitters' [132]. A combiner is required for merging two RF signals, e.g. from two transmit antennas (the signals are added vectorially), while a splitter may be used to generate a tap for monitoring the signal power. Proper termination of the output of a combiner is important because it yields greatest isolation between the two input ports and improves the measurement accuracy because the reflected power is insignificant. If the carrier frequencies of two input signals differ due to drift, then the output signal oscillates (known as 'beating') and the typical insertion loss applies, e.g. 3 dB for a 2–1 port combiner [132].

The traditional spectrum analyser is a general purpose piece of RF test equipment that is used for analysing RF signals in the frequency domain and can be used for making measurements on the UE transmit signal. Many spectrum analysers on the market today are customized to make specific measurements for different applications. For instance, considering WCDMA, a high quality spectrum analyser will be able to make transmit power measurements in a 3.84 MHz bandwidth. It will also be able to measure adjacent channel leakage rejection and the transmit spectrum mask. The spectrum analyser may go further and include a demodulator that will allow measurements of the transmit signal constellation, including error vector magnitude and frequency error. Also, code domain power measurements are possible where the individual channels are decoded and their characteristics measured.

A *Vector Signal Analyser* (VSA) is similar in many ways to a spectrum analyser in that it can analyse the RF signal in the frequency domain. Generally it will have a much narrower bandwidth but will include a demodulator as standard. It will have various options for decoding signals of differing standards, one of which will be for WCDMA. It can perform code domain analysis, measuring the individual channels on either the uplink or the downlink. It can also display the baseband constellation with error vector magnitude, frequency error and more [133]. VSAs tend to be generally less expensive than spectrum analysers since their RF (input stage) capabilities are less sophisticated.

The interested reader is referred to the web pages of the leading manufacturers of RF test equipment, where additional information can be found, such as datasheets, application notes and even tutorials.

5.3.3 Downlink Reference Measurement Channel Test Signal

The reference measurement channels, a set of well defined physical channels, are employed to obtain a consistent set of measurement results for type approval tests. Annexes C and E in [10] specify radio access bearers for different data rates and the relative power setting for each *Physical Channel* (PhCH) respectively. The 12.2 kbps measurement channel shall be supported by any UE and is thus discussed herein. All reference measurement channels are in fact DCH *Transport Channels* (TrCHs), with a *Dedicated Traffic Channel* (DTCH) and a *Dedicated Control Channel* (DCCH) component, see Table 5.3. The specification for the 12.2 kbps reference physical channel is reproduced in Table 5.4. Note that uplink DPCCH slot format 2 is used for closed loop tests.

The set of downlink physical channels requires further specification. The required PhCHs for a connection setup are both synchronization channels P-SCH, S-SCH, *Primary Common Pilot Channel* (P-CPICH), *Primary Common Control Channel* (P-CCPCH), *Secondary Common Control Channel* (S-CCPCH), *Paging Indicator channel* (PICH), *Acquisition Indicator Channel* (AICH) and a DPCH. Some of the measurements, like uplink open loop power control and transmitter ON/OFF time mask, do not require a DPCH, for which the relative powers listed in Table 5.5 apply. Similar tables exist for RF type receive and transmit tests. The relative power for each PhCH in connection with a DPCH is tabulated in Table 5.6.

The S-CPICH is only transmitted when the test requires the UE to use it as the phase reference. Multipath and interference (I_{oc}) may be disabled during the call setup, i.e. cell search procedure, in order to ease synchronization. The DPCH E_c/I_{or} requirement is stated for each test case, together with the other PhCHs but not the *Orthogonal Channel Noise Simulator* (OCNS). OCNS represents the intracell interference, interference from other users in the same cell where all users employ the same scrambling code but a different *Orthogonal Variable Spreading Factor* (OVSF) code, see Table 5.7. Note that the 16 OVSF

Table 5.3   12.2 kbps reference measurement transport channel specification [9].

Parameter	DTCH	DCCH
TrCH number	1	2
Transport block size	244	100
Transport block set size	244	100
TTI	20 ms	40 ms
Coding type	Convolutional	Convolutional
Coding rate	1/3	1/3
Rate matching attribute	256	256
Length of CRC block	16	12
DL position of TrCH in radio frame	fixed	fixed

Table 5.4   12.2 kbps reference measurement physical channel specification [9].

Parameter	Value
Bit rate	12.2 kbps
UL DPDCH	60 kbps
UL DPCCH	15 kbps
UL DPCCH slot format	0
UL DPCCH/DPDCH power ratio	-5.46 dB
UL TFCI	On
UL repetition factor	23 %
DL DPCH	30 ksps
DL slot format	11
DL TFCI	On
DL power offset	0 dB
DL puncturing	14.7 %

Table 5.5   PhCH power ratios for connections without a DPCH.

Ratio	Relative power
P-CPICH / I_{or}	$-$ 3.3 dB
P-CCPCH / I_{or}	$-$ 5.3 dB
SCH / I_{or}	$-$ 5.3 dB
PICH / I_{or}	$-$ 8.3 dB
C-CCPCH / I_{or}	-10.3 dB
$\hat{I}_{or}$	Test dependent

Table 5.6   PhCH power ratios for connections with a DPCH.

Ratio	Relative power
P-CPICH_E_c/I_{or}	-10 dB
S-CPICH_E_c/I_{or}	-10 dB
P-CCPCH_E_c/I_{or}	-12 dB
P-SCH_E_c/I_{or}	-15 dB
S-SCH_E_c/I_{or}	-15 dB
PICH_E_c/I_{or}	-15 dB
DPCH_E_c/I_{or}	Test dependent
OCNS	Remaining power

Table 5.7   The relative power for each OCNS
subchannel [9].

Channelization code number	Relative power
2	-1 dB
11	-3 dB
17	-3 dB
23	-5 dB
31	-2 dB
38	-4 dB
47	-8 dB
55	-7 dB
62	-4 dB
69	-6 dB
78	-5 dB
85	-9 dB
94	-10 dB
125	-8 dB
113	-6 dB
119	0 dB

codes specified in 3GPP R99 were chosen in order to provide a typical (average) Crest factor. The operator can calculate the OCNS (linear) power from the relative power ratios in dB, according to:

$$
\begin{aligned}
P_{\text{OCNS}} = {} & 1 - 10^{(\text{P}-\text{CPICH}\_E_c/I_{or})/10} - \\
& 0.9 \cdot 10^{(\text{P}-\text{CCPCH}\_E_c/I_{or})/10} - \\
& 0.1 \cdot 10^{(\text{SCH}\_E_c/I_{or})/10} - \\
& 10^{(\text{PICH}\_E_c/I_{or})/10} - \\
& 10^{(\text{DPCH}\_E_c/I_{or})/10}
\end{aligned} \tag{5.1}
$$

since all powers shall add up to unity (see Table E.3.3 in [10]). The OCNS signal has been defined such that it simulates realistic intracell interference with respect to an average or realistic *Peak to Average Power Ratio*[3] (PAPR) level. It comprises 16 DPCH channels with spreading factor 128. It is important that the data remains uncorrelated for the duration of a test measurement. There is no need to define a time offset for each OCNS channel any more since there are no specific pilot bits which can yield a high peak power. The total (linear) OCNS power is obtained from Equation (5.1) and is then distributed to each OCNS subchannel according to the relative power listed in Table 5.7.

With the help of Equation (5.1), we can now calculate the signal power for each physical channel of a 12.2 kbps downlink test signal and generate a WCDMA RF signal in order to measure the performance of a UE. A typical example is given in Table 5.8.

---

[3] A WCDMA signal is a superposition of many physical channels. This can yield to large peaks and the PAPR is a measure to quantify this, like the Crest factor. A high PAPR value means that expensive power amplifiers are needed (if linear modulation is required) or that signal clipping can occur which then affects the performance.

Table 5.8   Practical example for power allocation.

Ratio	Relative power	Linear power
P-CPICH_E_c / I_{or}	-10 dB	0.1
S-CPICH_E_c / I_{or}	-10 dB	0.1
P-CCPCH_E_c / I_{or}	-10 dB	0.1
P-SCH_E_c / I_{or}	-15 dB	0.032
S-SCH_E_c / I_{or}	-15 dB	0.032
PICH_E_c / I_{or}	-10 dB	0.1
DPCH_E_c / I_{or}	-10 dB	0.1
OCNS	Remaining power	0.436

Table 5.9   Test 1 in Section 7.3 [10].

Case	Speed	Taps	Delay profile	Power profile
1	3 km/h	2	0, 976 ns	0 dB, -10 dB
2	3 km/h	3	0, 976 ns	all 0 dB
3	120 km/h	4	0, 260 ns, 521 ns, 781 ns	0 dB, -3 dB, -6 dB, -9 dB
4	3 km/h	2	0, 976 ns	both 0 dB
5	50 km/h	2	0, 976 ns	0dB, -10 dB
6	250 km/h	4	0, 260 ns, 521 ns, 781 ns	0 dB, -3 dB, -6 dB, -9 dB

5.3.4 Propagation Channels

Annex D in [10] lists all propagation channels used for testing the UE's DPCH performance.
The aim of each condition is to mimic a certain aspect of a wireless channel in order to
verify the baseband processing implementation. There are four basic conditions which we
subsequently describe, namely:

- static propagation condition;

- multipath fading propagation condition;

- moving propagation condition;

- birth–death propagation condition.

In addition to the aforelisted four conditions, reference [134] defines a set of more realistic
propagation channels, the deployment scenarios. However, to date, neither TS25.101 nor
TS34.121 state performance requirements for such deployment scenarios.

The static propagation condition is the most simple case and represents the well known
Additive White Gaussian Noise (AWGN) channel [47]. It is well suited to comparing different receiver designs against theoretical results.

The multipath fading propagation condition is subdivided into the six cases listed in
Table 5.9. Figure 5.6 shows a snapshot of a Case 3 channel. All taps, or multipath rays,
are required to comply with the classical Doppler spectrum (see page 181 in [7]). The

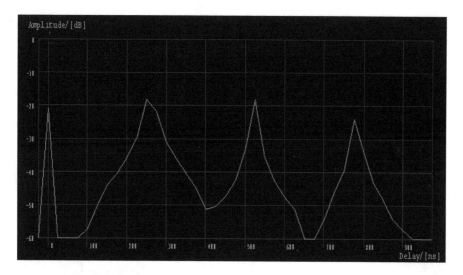

Figure 5.6. The Case 3 multipath channel with four taps.

maximum Doppler frequency f_D is calculated according to $f_D = (v/\lambda)\cos(\alpha)$, where the speed v is given in m/s, λ is the wavelength in m, and α is the spatial angle which is zero for determining the maximum Doppler frequency (see page 18 in [135]). The purpose of the test is to determine the RAKE receiver's capability to detect, track and assign RAKE fingers reliably and to cope with interchip interference, see also Chapter 3.

The moving propagation condition has two nonfading paths of equal strength and phase. The first ray is the reference ray and is static, i.e. it has a fixed position. The second ray is oscillating, according to the relationship:

$$\Delta\tau = a_1 + \frac{a_2}{2}(1 + \sin(\Delta\omega t)) \tag{5.2}$$

with the constants $a_1 = 5$ μs, $a_2 = 1$ μs, the running time index t and the frequency term $\Delta\omega = 0.04$ s^{-1}. Term $\Delta\tau$ denotes the delay, i.e. the position of the oscillating ray. The purpose is to determine the RAKE receiver's capability to track quickly and reliably the moving ray with a RAKE finger, otherwise the diversity loss will prevent the receiver from meeting the BLER target. Figure 5.7 depicts a snapshot of the moving propagation channel [136].

The birth–death propagation condition simulates a channel which has two nonfading paths. Both paths are allowed to appear on time positions taken from the set $\{-5, -4, -3, -2, -1, 0, 1, 2, 3, 4, 5\}$ μs but at any time there is only one path at any given time position. In alternating order, one path at a time disappears and reappears every 191 ms. The new time position is chosen at random. The test gives insight into the RAKE receiver's capability to detect and assign a RAKE finger to an emerging ray, otherwise the diversity loss will prevent the receiver from meeting the BLER target. Figure 5.8 shows a snapshot of a birth–death channel [136].

A representative of the set of deployment scenarios [134], the hilly terrain propagation channel, is depicted in Figure 5.9. Note the typical large delay spread of up to 20 μs. The

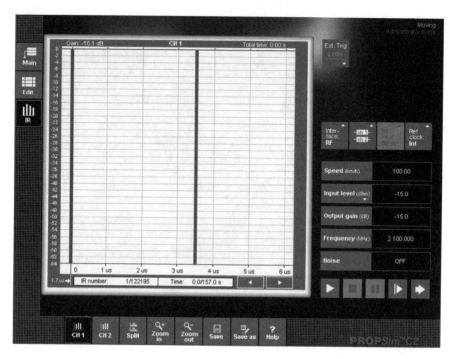

Figure 5.7. A snapshot of the moving propagation channel.

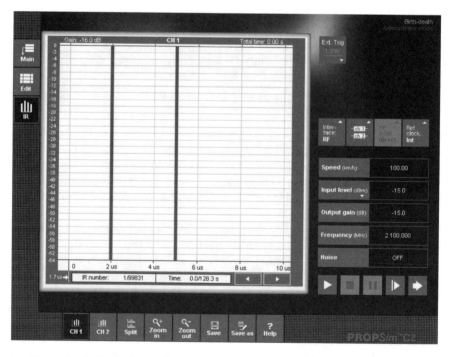

Figure 5.8. A birth–death channel snapshot, the location of one ray changes every 191 ms.

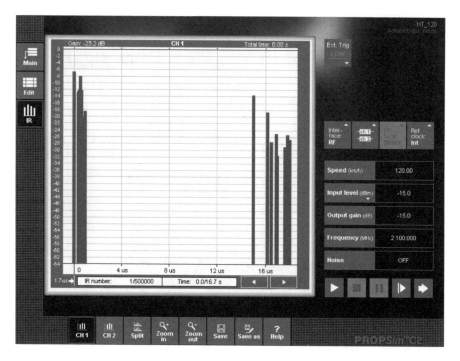

Figure 5.9. A typical hilly terrain channel as specified in the 3GPP deployment scenarios.

Table 5.10   Specification and accuracy of the test environment [10].

Type	Nominal value	Accuracy
Pressure	NA	$\pm$ 5 kPa
Normal temperature	+15 °C to +35 °C	$\pm$ 2 degrees
Relative humidity	25 % to 75 %	$\pm$ 5 %
DC voltage	NA	$\pm$ 1 %
AC voltage	nominal e.g. 240 V	$\pm$ 1.5 %
Vibration	0.96 m^2/s^3	10 %
Vibration frequency	5 Hz – 20 Hz and 20 Hz – 500 Hz	$\pm$ 0.1 Hz

early rays are from reflections in the vicinity followed by another group of rays which can stem from large reflectors such as mountains [136].

5.3.5 Test Environment and Test Uncertainty

Annexes F and G in [10] define the relevant requirements regarding the test condition and the environment respectively. A 95 % value is commonly used as the confidence level or acceptable uncertainty measure of the system [137].

The measurement accuracy of the test environment defined in Annex G of [10] is summarized in Table 5.10.

Table 5.11   Maximum uncertainty of the test system used for some DCH demodulation tests.

Test	$\hat{I}_{or}/I_{oc}$	I_{oc}	DPCH_E_c/I_{or}
Static condition	± 0.3 dB	± 1 dB	± 0.1 dB
Multipath fading	± 0.56 dB	± 1 dB	± 0.1 dB
Moving propagation	± 0.6 dB	± 1 dB	± 0.1 dB
Birth–death propagation	± 0.6 dB	± 1 dB	± 0.1 dB

Table 5.12   Typical absolute uncertainty figures of standard test rig components.

Type	Quoted accuracy
RF signal generator	± 0.7 dB
Fading generator	± 0.5 dB
Noise generator	± 0.5 dB
Fixed attenuator	± 0.3 dB
Step attenuator	± 1.8 dB

Annex F [10] contains informative requirements on the general test environment, such as the test tolerances, the derivation of the test requirements, the acceptable uncertainty of test equipment and the practical use of statistical testing. Strict requirements are listed for the test system uncertainty on the transmitter, receiver, performance and *Radio Resource Management* (RRM) tests, and on the interpretation of the measurement results and the rules for statistical testing. Table 5.11 lists the system uncertainty requirements for the performance tests for the aforementioned DCH BLER measurement tests. The small uncertainty in the DPCH_E_c ratio is justified since it is possible to achieve an accurate (relative) power distribution in the transmitted signal. The accuracy of the $\hat{I}_{or}/I_{oc}$ ratio is based on the uncertainty of a typical commercial power meter used to measure the power after the combiner, refer to Figure 5.2. The accuracy for I_{oc} is in fact of limited importance since the BLER performance is generally unaffected by the absolute signal level, e.g. the BLER results are equal for tests carried out at an I_{oc} level of -60 dBm/3.84 MHz or -62 dBm/3.84 MHz since both levels are well above the required receiver sensitivity.

A test system as shown in Figure 5.2 needs to be assessed in order to compare its accuracy with the limits set out in Table 5.11. Two types of uncertainty are encountered: first the test system accuracy and secondly the accuracy through a measurement. Table 5.12 lists realistic accuracies of commercial RF equipment in order to derive the former figure. Table 5.13 states the rated accuracies for different RF measurement apparatus [128] to obtain the total measurement uncertainty. The relative accuracy is generally much better and ensures that subsequent measurements meet the 3GPP requirements.

The overall test system error for I_{oc}, the absolute signal power at the receiver's antenna connection can be obtained by [137]:

$$I_{oc}^{\text{error}} = \sqrt{0.7^2 + 0.3^2} = 0.76\,\text{dB}, \tag{5.3}$$

Table 5.13   Quoted test and measurement equipment uncertainty for signal power measurements.

Type	Typical accuracy
RF signal generator	± 0.7 dB
Power meter (RF)	± 0.2 dB
Spectrum analyser	± 0.3 dB
Vector signal analyser	± 0.7 dB
SIR	± 0.2 to ± 0.5 dB

which includes the quoted absolute transmitted signal power uncertainty and the quoted error of a fixed attenuator. The purpose of the attenuator is to achieve a signal power level as low as −60 dBm (provided that the signal generator does not provide the required dynamic range), or to achieve a higher accuracy since signal generators can have a greater error at very low signal levels. The difficulty with the latter approach is that the total error increases with every device put into the RF signal chain and makes it difficult to achieve the required test system accuracies. Alternatively, we may use an appropriate measurement apparatus to verify the absolute and relative signal powers at the receiver's antenna connector. This approach allows the test engineer to measure the different signal levels quickly and economically since less accurate RF equipment can be used.

The quality of the measurement results, e.g. BLER, is linked to the error statistics, known in Monte Carlo simulation theory [112]. Therefore the test duration is another factor which needs to be considered in addition to the accuracy of the equipment employed. The DCH demodulation tests use the BLER as the figure of merit. The BLER is the result of processing encoded frames where the error events are not independent, as opposed to the BER obtained in an uncoded communication system, hence more errors should be counted, see Section 5.6.1.5 in [112]. It has been suggested, as a rule-of-thumb, to count ten times more frame errors than bit errors [138]. The number of required transmitted bits or blocks, to represent reliably a given error statistic (BLER), can be obtained from [112]:

$$N_{\text{Blocks}} = \frac{\epsilon}{\sigma_e^2 \times \text{BLER}} \tag{5.4}$$

where ϵ denotes the number of counted errors, BLER denotes the error level at which we want to operate, e.g. 0.001. The term σ_e^2 is the desired error variance of the result for which a value of 0.1 has been suggested [138].

The method following Equation (5.4) is suited to characterizing the receiver design (performance) of a UE. Conducting such tests during mass production of handsets is obviously not economical since the test duration is prohibitively long when the order of units lies in the millions. Hence, a different approach, used in the GSM community, was adopted, namely statistical testing.

Statistical testing is described in [10]. Its purpose is to minimize the time required to test each UE while maintaining a high probability of detecting a false unit. The minimum test time is obtained based on the statistics of the propagation condition, for example, at least ten wavelengths at the given speed shall be crossed in a fading scenario. Table 5.14 summarizes the minimum length of test time for each channel condition.

Table 5.14 Required minimum test time for DCH demodulation tests during mass production.

Propagation condition	Speed	Duration
Multipath	3 km/h	1.8 s
Multipath	50 km/h	0.1 s
Multipath	120 km/h	45 ms
Multipath	250 km/h	22 ms
Birth–death	NA	1.91 s
Moving propagation	NA	157 s

For example, the 1.8 s duration stated in Table 5.14 is based on $10 \times T_\lambda$ (for ten wavelengths), where $T_\lambda = 1/f_D$ and f_D denotes the Doppler frequency for a vehicle speed of 3 km/h. The probability that a unit will fail the test equals the probability that a bad unit will pass the test and shall be 0.2 %, or every 500th unit, see Table F.6.1.8-2 [10].

5.3.6 Test Requirements

Recall that reference [10] contains a variety of tests, such as those for:

- transmitter characteristics;
- receiver characteristics;
- performance requirements;
- requirements for support of RRM.

The first two groups focus on RF related issues, such as spectrum emission mask and blocking characteristics, while the fourth group addresses higher layer interaction and measurements. Next we discuss a single representative test example of the third group since the total number of tests is too big to be treated individually here.

Each demodulation test, e.g. demodulation in multipath fading propagation conditions, specifies the following basic test requirements, see also Section 7.3 in [10].

The BLER requirements for the different channel conditions were obtained through simulations during the early times of 3GPP, see Section 5.2. The BLER results were obtained from optimum floating point simulations and then an implementation loss of 2 dB or more was added in order to arrive at the required $DPCH\_E_c/I_{or}$ for a given BLER now found in [9].

In addition to the entries shown in Table 5.15, the test engineer must also adhere to the aforementioned requirements regarding

- test equipment requirements;
- test setup;
- test signal;

Table 5.15   Test 1 in Section 7.3 [10].

Parameter	Value
$\hat{I}_{or}/I_{oc}$	9 dB
I_{oc}	$-$ 60 dBm/3.84 MHz
Data rate	12.2 kbps
BLER level	0.01
DPCH_E_c/I_{or} requirement	-15 dB

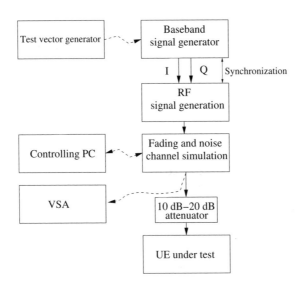

Figure 5.10. Example of an R&D lab test suite. The dashed lines are optional connections.

- test channel;

- overall measurement uncertainty;

- test environment.

Assuming that all requirements are met, the UE under test shall then show a BLER performance better than or equal to 0.01 for a given DPCH_E_c/I_{or} ratio. How this measure is obtained is the subject of the next section on lab testing.

5.3.7 Lab Testing

We assume that the R&D engineer has a signal source, a fading simulator and noise generator available as discussed in Section 5.3.2, and that all meet the test uncertainty requirements. Figure 5.10 illustrates what the test suite in the laboratory might look like [128].

The process of obtaining one BLER result from a single DCH demodulation test comprises several steps:

1. Generate the downlink signal;

2. Calibrate the signal power;

3. Calibrate the noise (interference) power;

4. Enable the fading propagation channel and add noise;

5. Perform the test over the required test duration and log the result.

The appropriate downlink test signal for the given $DPCH\_E_c/I_{or}$ ratio is generated automatically or manually (entering parameters) or with an arbitrary waveform generator, e.g. according to Section 5.3.3. In reality, however, a number of test signals will be required in order to produce a BLER curve since the performance will not be known at first. Later, only a small subset of test signals may be required with, for example, three different $DPCH\_E_c/I_{or}$ ratios centred around the required $DPCH\_E_c/I_{or}$ level [9].

The signal power is measured with a vector signal analyser before the attenuator. A stable result can be obtained by selecting the static channel in the fading simulator.

The interference power is represented by added Gaussian noise. It is measured by, for example, switching off the signal power and enabling the noise generator. It is important that the power is measured over the correct amount of (signal) bandwidth as specified in [10].

One effect of fading is that the measured signal power is varying with time, which makes it difficult to measure accurately unless we average over a long duration. Some fading simulators thus suggest performing the measurements with the static (AWGN) propagation channel first. A correction factor is applied which the manufacturer supplies for each fading (multipath) propagation channel in order to compensate the power setting due to the loss of (multipath) fading.

The test duration is conservatively estimated, based on the assumption that we want to collect at least 100 blocks in error for a required BLER value equal to 0.01. A typical test can last for 2000 seconds based on Equation (5.4). Then the BLER is derived and recorded by dividing the number of blocks that fail the CRC pass criteria by the number of transmitted blocks.

Bit error tests are a special test case because the receiver requires knowledge of the transmitted bit sequence embedded in the TrCH channel, i.e. DCH. Some test equipment manufacturers offer a loop back mode, where the demodulated and estimated data bits are fed back to the transmitter device in order to be compared against the initial bit sequence. A second method is to transmit a specified bit sequence, whose sequence can be locally reproduced in the UE and which is compared against the estimated (detected) bit sequence. For the latter purpose, two PN sequences are available in a number of pieces of test equipment, PN9 and PN15.

5.4 EXEMPLARY MEASUREMENT RESULTS

In the previous sections, we introduced the process with which 3GPP arrived at the different BLER performance and discussed a test procedure for measuring the BLER performance of a UE. In this last section, we present some results obtained in the lab with a test mobile and show that the 3GPP requirements can be met. Each graph shows two curves, *Dedicated*

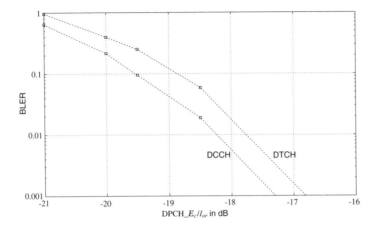

Figure 5.11. Static 12.2 kbps performance results obtained in a lab with a test mobile.

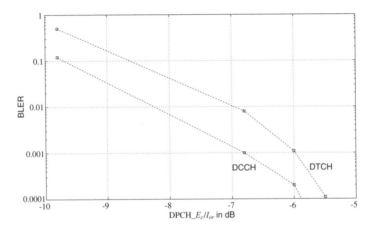

Figure 5.12. Case 6 144 kbps performance results obtained in a lab with a test mobile.

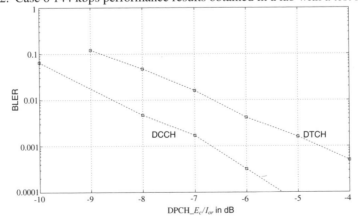

Figure 5.13. Static 2 Mbps performance results obtained in a lab with a test mobile.

Control Channel (DCCH) and *Dedicated Traffic Channel* (DTCH) [9], of which the DTCH curve yields the figure of merit.

Figure 5.11 shows the BLER curves for a typical speech service in a static propagation condition. The DPCH_E_c/I_{or} requirements must be better than -16.6 dB for the DTCH at BLER $= 0.01$. We observe that a test mobile exceeds this requirement. The DPCH_E_c/I_{or} requirement for a Case 6 channel [9] with the 144 kbps reference channel is -5.5 dB at BLER $= 0.01$ and SIR $= 3$ dB. Figure 5.12 shows that it is possible to exceed the specified 3GPP requirements. Figure 5.13 shows the BLER results for a typical 2 Mbps radio bearer service [118]. We cannot compare the performance against any 3GPP requirements here because so far no requirements have been released.

In conclusion, it can be said that it is possible to exceed the 3GPP requirements, and hence the implementation margins are reasonable.

6
Medium Access Control

Joby Allen

6.1 INTRODUCTION

This chapter discusses the structure and operation of the *Medium Access Control* (MAC) layer. The MAC sits between the physical layer and *Radio Link Control* (RLC) in Layer 2 of the *Access Stratum* (AS), see Figure 6.1.

The MAC block itself can be further divided into subblocks, as shown in Figure 6.2. The MAC can be partitioned into three functional components, MAC configuration, MAC Rx/Tx and MAC measurements. These components and the communication paths between them are illustrated in Figure 6.2. The elliptical points in Figure 6.1 are *Service Access Points* (SAPs), the interfaces to and from the MAC block. The structure of the data which passes through each SAP is stored in data frames or blocks, and will be explained later.

The following points list the responsibilities of the MAC:

- data exchange between RLC and the physical layer;
- selection of *Transport Format Combination* (TFC) for transmission;
- *Random Access Channel* (RACH) transmission control;
- identification of UE on common transport channels;
- traffic volume measurements;
- Ciphering.

3GPP has produced a set of documents which are relevant to the design engineer or to the reader who wants to know more about the details:

WCDMA – Requirements and Practical Design. Edited by R. Tanner and J. Woodard.
© 2004 John Wiley & Sons, Ltd. ISBN: 0-470-86177-0.

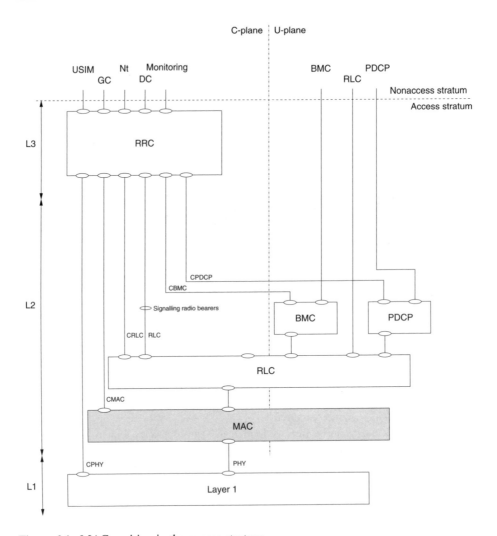

Figure 6.1. MAC position in the access stratum.

- TS25.321 [139] specifies the MAC;

- TS25.322 [140] specifies the RLC;

- TS25.331 [141] specifies the *Radio Resource Controller* (RRC).

6.1.1 Data Exchange Between RLC and the Physical Layer

Data is transferred between RLC and the physical layer via logical channels [142] in the MAC. A logical channel is a mapping between an RLC entity and a *Transport Channel* (TrCH.) To this end, the MAC functions as a switch. The current switch configuration describes the mapping between the currently active set of RLC entities, an entity is responsible for managing the data transfer across the radio link, and transport channels for the purpose of data routeing.

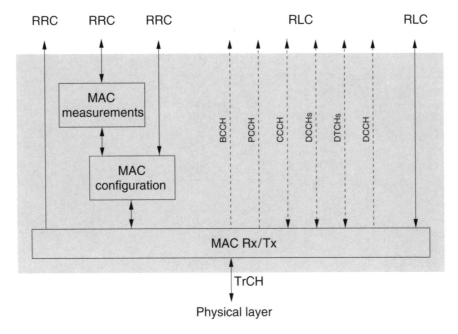

Figure 6.2. MAC functional partitioning.

There are four defined logical channels for control data, i.e. for data routed to the control plane[1]:

Broadcast Control Channel (BCCH). This downlink channel is used for the broadcast of system information. It maps the *Broadcast Channel* (BCH) or *Forward Access Channel* (FACH) to a *Transparent Mode* (TM) RLC entity.

Paging Control Channel (PCCH). This downlink channel is used for the broadcast of paging information. It maps a *Paging Channel* (PCH) to a TM RLC entity.

Common Control Channel (CCCH). This bidirectional channel is used to exchange control information between the UE and the network. In the downlink direction, it maps the FACH to an *Unacknowledged Mode* (UM) RLC entity. In the uplink direction, it maps the *Random Access Channel* (RACH) to a TM RLC entity.

Dedicated Control Channel (DCCH). This bidirectional channel is used to exchange dedicated control information between the UE and the network. In the downlink direction, it maps either a FACH, *Dedicated Channel* (DCH) or *Downlink Shared Channel* (DSCH) to any type of RLC entity. In the uplink direction it maps a RACH or DCH to any type of RLC entity.

In addition to the control channels there are a further two logical channels for user traffic, i.e. for data routed to the user plane:

Dedicated Traffic Channel (DTCH). This bidirectional channel is used to exchange dedicated user data between the UE and the network. In the downlink direction it maps

---

[1] A plane is an imaginary divide between control and information (voice, Internet packet) data.

Table 6.1   Downlink logical channel to RLC mapping.

	TM	UM	AM
BCCH	Yes	No	No
PCCH	Yes	No	No
CCCH	No	Yes	No
DCCH	No	Yes	Yes
DTCH	Yes	Yes	Yes
CTCH	No	Yes	No

Table 6.2   Uplink logical channel to RLC mapping.

	TM	UM	AM
CCCH	Yes	No	No
DCCH	No	Yes	Yes
DTCH	Yes	Yes	Yes

either a FACH, DCH or DSCH to any type of RLC entity, and in the uplink direction it maps a RACH or DCH to any type of RLC entity.

Common Traffic Channel (CTCH). This downlink channel is used to broadcast dedicated user information. It maps the FACH to a UM RLC entity.

Tables 6.1 and 6.2 summarize the above mappings.

There is only a single instance of the BCCH, PCCH, CCCH and CTCH in a UE, but there may be multiple instances of the DCCH and DTCH.

Peer-to-peer communication is achieved by the exchange of *Protocol Data Units* (PDUs). A MAC PDU, the structure of which is illustrated in Figure 6.3, consists of a MAC header and a MAC *Service Data Unit* SDU [139]. The MAC header is a portion of control information that describes the logical channel routeing for the SDU. It optionally contains the following fields:

Target Channel Type Field (TCTF). The TCTF identifies the logical channel routeing for data on FACH and RACH TrCHs.

UE ID type and UE ID. This pair of fields identify the UE on common channels. The UE ID type field indicates whether the subsequent UE ID is of type U-RNTI (*Radio Network Temporary Identity*), C-RNTI or DSCH-RNTI.

C/T Field. This field identifies the logical channel when multiple DCCHs or DTCHs are mapped on to a single TrCH.

Note that there is no explicit signalling between the peer (Node B, UE) MAC entities, all data exchanged is user data for the higher layers.

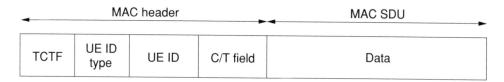

Figure 6.3. Structure of a MAC protocol data unit.

6.1.2 Selection of TFC for Transmission

Uplink logical channels are assigned a priority by the RRC. When transmitting data, it is the job of the MAC to select a suitable *Transport Format Combination* (TFC), see Chapter 4, to ensure that the maximum amount of data on the high priority logical channels is transmitted.

6.1.3 RACH Transmission Control

The RACH is a contention-based channel, because it is an asynchronous channel which is requested, and the MAC is responsible for controlling the timing of transmissions on the RACH. Part of this procedure involves the selection of the *Access Service Class* (ASC). An ASC reflects or describes classes of resources.

6.1.4 Traffic Volume Measurements

The MAC is responsible for measuring the amount of data being transmitted on the logical channels and reporting the measurements to the radio resource control. These measurements, which are reported to the network by the RRC, are used to control the configuration of the radio bearers. If the traffic volume is high, the network may configure the radio bearers in such a way that higher bandwidth is available. In contrast, if traffic volume is low, the network may reduce the amount of bandwidth available.

6.1.5 Ciphering

The MAC is responsible for enciphering and deciphering data on DCCH and DTCH logical channels that are mapped to TM RLC entities. Note that ciphering of data for UM and AM (acknowledged mode) RLC entities is done by the RLC.

The purpose of ciphering is to provide privacy over the air interface [143]. The cipher algorithms used to be kept secret and a special agreement had to be signed between 3GPP and a manufacturer in order to get hold of them. Nowadays, with the change in the general perception of the whole issue of ciphering and secrecy, the actual algorithms are now public. However, the (cipher) keys are not public and thus knowledge of the ciphering algorithms alone is of limited use for eavesdropping.

6.2 MAC FUNCTIONAL PARTITIONING

Recall that the MAC can be partitioned into the three functional components, namely MAC configuration, MAC Rx/Tx and MAC measurements.

6.2.1 MAC Configuration

The MAC configuration component is responsible for storing and maintaining the current MAC configuration. On request from the RRC, it stores new configuration data, modifies existing configuration data and deletes old configuration data. The MAC configuration component is driven by the RRC over the CMAC SAP. A MAC configuration typically contains the following information.

6.2.1.1 *CCTrCH Configuration* The *Coded Composite Transport Channel* (CC-TrCH), see also Chapter 4, part of the configuration contains information regarding the configured TrCHs and uplink TFCs.

6.2.1.2 *Logical Channel Configuration* The logical channel part of the configuration identifies the list of configured logical channels. The following information is stored for each logical channel:

Logical channel type. This may be BCCH, CCCH, DCCH, etc.

RLC entity. The RLC entity to which this logical channel is mapped.

RLC PDU type. In general a single logical channel is mapped to a single RLC entity. However, an AM RLC entity may have two logical channels mapped to it. In this case, one logical channel would be used to carry RLC control PDUs and the other to carry RLC data PDUs. This field indicates whether the logical channel is used for control or data in the case where two logical channels are mapped to a single AM RLC entity.

TrCH ID. The identity of the TrCH to which this logical channel is mapped.

C/T field. This field is used to identify a particular instance of the DCCH or DTCH logical channel and is used when several logical channels are multiplexed on a single TrCH.

Priority. The transmit priority of the logical channel. This is used for TFC selection and ASC selection.

All the information in the above list is controlled by the RRC. In addition to this, the *Buffer Occupancy* (BO) must be stored for each logical channel. The BO is a measure of the amount of data available for transmission in the RLC entity for a logical channel.

6.2.1.3 *UE Configuration* The UE part of the configuration stores the currently configured UE IDs. Three types of UE ID may be configured:

C-RNTI. This identifies the UE within a cell. The C-RNTI is used in the uplink direction on a DCCH or DTCH mapped to the RACH.

U-RNTI. This identifies the UE within UTRAN as a whole. It is only used in the downlink direction for the transfer of RRC PDUs in specific procedures such as cell update.

DSCH-RNTI. Identifies the UE on the DSCH and is used only on the DSCH.

6.2.1.4 RACH Configuration

The RACH part of the configuration contains the parameters used to control the RACH transmission procedure, as follows:

M_{max} denotes the maximum number of preamble ramping cycles.

N_{BO1min} and N_{BO1max}. represent the range used when interpolating the backoff timer value, see also RACH access procedure in Section 6.4.3.

Access service class information is the list of PRACH partitions and persistence values.

NumASC. This value contains either the ASC chosen by the RRC or, in cases where the MAC is to choose the ASC, the number of ASCs. The RRC may pass a value to indicate which it contains, or the MAC can work it out, if the UE ID U-RNTI is configured then it contains the number of ASCs, otherwise it contains the ASC value as calculated by the RRC.

The information in this list is controlled by the RRC. Other RACH information is stored by the MAC for use in the RACH transmission procedure, such as a parameter which maintains a count of the preamble cycles for the current RACH access attempt.

6.2.1.5 Ciphering Configuration

The ciphering part of the configuration contains the parameters required for enciphering and deciphering data carried by DCCHs and DTCHs mapped to TM RLC entities. The list of TM RLC entities to be enciphered and deciphered and the set of parameters required as input to the ciphering engine, such as the ciphering key, are stored. The set of parameters will vary depending on the interface to the ciphering algorithm that is used.

6.2.1.6 Communication Between MAC and RRC

The RRC updates the configuration information in the MAC by sending a special message on the CMAC SAP, and the MAC may then respond with a confirmation message [139].

6.2.2 MAC Rx/Tx

This component facilitates the exchange of data between the RLC and the physical layer. It is partitioned into three entities as illustrated in Figure 6.4.

6.2.2.1 MAC-b

The MAC-b routes PDUs received on the BCH to the BCCH.

6.2.2.2 MAC-c/sh

MAC-c/sh is responsible for the PCH, FACH, DSCH and RACH. PDUs received on the PCH are routed to the PCCH. PDUs received on the FACH are routed to the BCCH, CCCH, CTCH or MAC-d (for DCCH and DTCH). PDUs transmitted on the CCCH are routed to the RACH. MAC-c/sh implements functionality to encode and

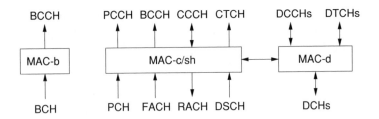

Figure 6.4. MAC Rx/Tx partitioning.

decode MAC headers for received and transmitted PDUs. It also controls the RACH access procedure.

6.2.2.3 *MAC-d*

MAC-d is responsible for dedicated channels. It routes PDUs transmitted on the DCCH and DTCH logical channels to either the DCHs or the MAC-c/sh (where the DCCH or DTCH is mapped to RACH) and routes PDUs received on the DCHs or from MAC-c/sh to the DCCH or DTCH logical channels. MAC-d implements functionality to encode and decode MAC headers for received and transmitted PDUs. It also implements the TFC selection algorithm and is responsible for enciphering and deciphering DCCHs and DTCHs mapped to TM RLC entities.

6.2.2.4 *Communication Between MAC and RLC*

Communication between the RLC and the MAC is done by exchanging different types of message over the MAC SAP:

MAC_DATA_REQ The RLC uses this message to pass data to the MAC for transmission. It contains parameters to carry the data and to identify the source RLC entity. If the source RLC entity is an AM entity, then the message must also indicate whether the control or data logical channel is to be used.

MAC_DATA_IND The MAC uses this message to pass received data to the RLC. It contains parameters to carry the data and identify the destination RLC entity.

MAC_STATUS_IND This message may be used by the MAC to indicate status information to the RLC, for example, that it is ready to transmit data. It may also be used by the MAC to instruct the RLC to provide the current BO values for the RLC entities.

MAC_STATUS_RSP This message may be sent by the RLC as a response to the MAC_STATUS_IND message, for example to transfer the current BO values.

6.2.2.5 *Communication Between MAC and Physical Layer (L1)*

Communication between the physical layer and the MAC is done by exchanging messages over the PHY SAP. The following messages are defined [144]:

PHY_DATA_IND. This message is sent by the physical layer to indicate received data. One PHY_DATA_IND is sent every TTI for each active downlink TrCH. Depending on the implementation of the physical layer, the PHY_DATA_IND may only be sent for TrCHs that actually received data on that TTI.

PHY_DATA_REQ. This message is sent to the physical layer to instruct it to transmit data. One PHY_DATA_REQ is sent each TTI for each active uplink TrCH.

PHY_ACCESS_REQ. This is sent by the MAC to indicate to the physical layer that it wishes to initiate a RACH transmission. It contains parameters to inform the physical layer of the RACH parameters such as, for example, which ASC to use.

PHY_ACCESS_CNF. This is sent by the physical layer on completion of the preamble cycle part of the RACH access procedure, to indicate the response on the *Acquisition Indicator Channel* (AICH).

6.2.2.6 Communication Between MAC and RRC Some RRC procedures, such as RRC connection establishment, need to know when a RACH transmission cycle is complete and whether the transmission was successful or not. After completion of the RACH transmission procedure, the MAC sends an indicator message to the RRC. This message contains a parameter to indicate whether or not the RACH transmission was successful.

6.2.3 MAC Measurements

This component is responsible for performing traffic volume measurements. Traffic volume measurements involve measuring the buffer occupancy of a set of logical channels regularly over a period of time and reporting the measurements to the RRC. The buffer occupancy measurements can be represented in several different ways:

Current buffer occupancy. This is the buffer occupancy of a logical channel at a given instant in time.

Average buffer occupancy. This is the average buffer occupancy of a logical channel over a period of time.

Variance in buffer occupancy. This is the difference between the smallest and largest values of buffer occupancy of a logical channel over a period of time.

Depending on the implementation, the averaging and variance calculations may be grouped together with the RRC measurement functionality, in which case the MAC may simply report current buffer occupancy measurements on a periodic basis.

6.2.3.1 Periodic Measurements Periodic measurements involve reporting buffer occupancy to the RRC on a periodic basis. The period is set by the RRC.

6.2.3.2 Event Triggered Measurements Event triggered measurements involve reporting buffer occupancy measurements to the RRC only when a certain threshold is reached. Two events are defined for traffic volume measurements:

Event 4a. This event is triggered when the buffer occupancy exceeds a threshold TH_U. This is illustrated in Figure 6.5(a).

Event 4b. This event is triggered when the buffer occupancy falls below a certain threshold TH_L. This is illustrated in Figure 6.5(b).

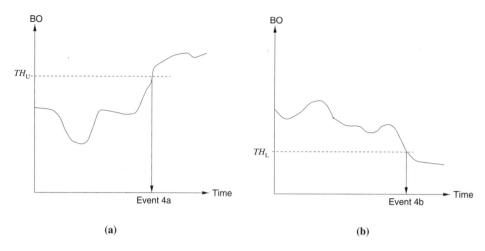

Figure 6.5. Trigger events for measurements. (a) Triggering of Event 4a; (b) Triggering of Event 4b.

The threshold values are provided by the RRC. Depending on the implementation, the event detection may be grouped together with the measurement functionality in the RRC. In this case the MAC may simply report periodic buffer occupancy measurements as instructed by the RRC.

6.2.3.3 *Communication Between the RRC and the MAC* Measurements are configured and reported via the exchange of two messages through the CMAC SAP:

CMAC_MEASUREMENT_REQ. This message is sent by the RRC to instruct the MAC to turn on or off the traffic volume measurements. It indicates which logical channels are to be measured and if the measurements are to be periodic or event triggered. If periodic, what the reporting period is, and if event based, the threshold values for event triggering are provided. It also indicates whether current, average or variance in buffer occupancy is to be reported.

CMAC_MEASUREMENT_IND. This message is sent by the MAC to inform the RRC of the traffic volume measurements as configured by the CMAC_MEASUREMENT_REQ message.

6.3 MAC RECEIVE FUNCTIONALITY

This section discusses the receive functionality of the MAC, namely the transfer of data from the physical layer to the RLC. The physical layer transfers received data to the MAC by sending a PHY_DATA_IND message. One PHY_DATA_IND is sent each TTI for each TrCH on which data is received. Each transport block received represents a single MAC PDU. By decoding the MAC header information in the PDU, the MAC is able to locate the relevant logical channel in the current configuration and hence the RLC entity to which the data is to be sent. The data is then sent to the RLC in a MAC_DATA_IND message, a

Table 6.3   TCTF values for FACH.

TCTF	Logical channel
00	BCCH
01000000	CCCH
10000000	CTCH
11	DCCH or DTCH

parameter of which identifies the target RLC entity. Only the MAC SDU is forwarded to the RLC, the MAC header is removed before the data is transferred. The following sections describe how the RLC routeing of data received on the different TrCHs is determined.

6.3.1 Reception on BCH

MAC PDUs received on the BCH are handled by the MAC-b and do not contain a MAC header as the BCH is only ever multiplexed to a BCCH. When the MAC-b receives a PDU on the BCH, it scans the current configuration to retrieve the BCCH information and identifies the RLC entity. The data is forwarded untouched to the identified RLC entity.

6.3.2 Reception on PCH

MAC PDUs received on the PCH are handled by the MAC-c/sh and also do not contain a MAC header. The process here is identical to that for BCH data, only this time the PCCH is used.

6.3.3 Reception on FACH

MAC PDUs received on the FACH are handled by the MAC-c/sh. The target logical channel is identified by examining the TCTF field of the MAC header. Table 6.3 illustrates the TCTF values used to identify the different logical channels that may be mapped to the FACH. TCTF bit values of 01000001 to 01111111 and 10000001 to 10111111 are reserved. Received PDUs with TCTF values in these ranges should be ignored.

6.3.3.1 *BCCH, CCCH or CTCH on FACH*   If the logical channel is identified as BCCH, CCCH or CTCH, there are no further MAC header fields to be read. The relevant logical channel may be retrieved from the current MAC configuration, the RLC entity identified and the data forwarded on to the RLC.

Figures 6.6 and 6.7 show the structures of MAC PDUs on each of these logical channels mapped to FACH.

6.3.3.2 *DCCH or DTCH on FACH*   If the logical channel is identified as DCCH or DTCH there is more information in the MAC header to read. First, the UE ID field must be read. The UE ID field is used to identify the target UE when dedicated data is transmitted

Table 6.4   UE ID Type field values.

Value in header	UE ID type
00	U-RNTI
01	C-RNTI (or DSCH-RNTI)

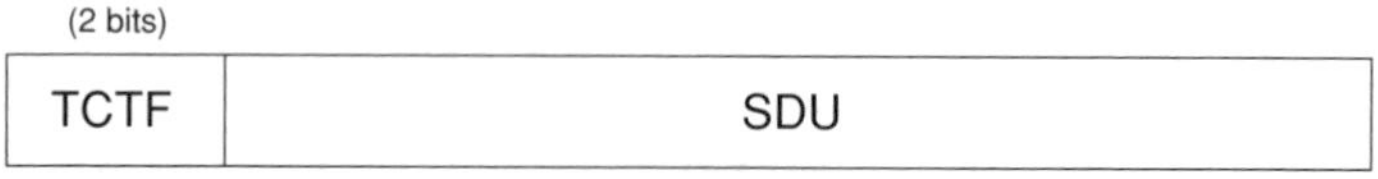

(2 bits)

TCTF	SDU

Figure 6.6. Structure of MAC PDU on BCCH mapped to FACH.

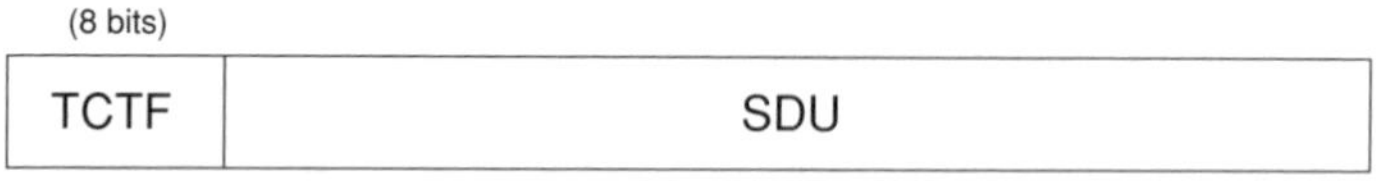

(8 bits)

TCTF	SDU

Figure 6.7. Structure of MAC PDU on CCCH or CTCH mapped to FACH.

using common TrCHs. MAC-c/sh reads the UE ID type field in the MAC header to determine what type of UE ID the header contains – these values are illustrated in Table 6.4. UE ID type values of 10 and 11 are reserved and PDUs using these values should be ignored. Based on the UE ID type field MAC-c/sh then reads either the 32 bit U-RNTI or the 16 bit C-RNTI field. If the UE ID does not match any currently configured UE ID in the MAC, the PDU is not intended for this UE and is ignored. If the UE ID does match one of the currently configured UE IDs then the data is intended for this UE. The PDU is now forwarded to the MAC-d entity to identify the logical channel. Unlike other logical channels, there may be multiple instances of DCCH or DTCH and more than one of those instances may be multiplexed to a single FACH. The next field of the MAC header, the C/T field, is used to identify the instance of logical channel if multiplexing is used. However, as the C/T field is not present if there is no multiplexing, i.e. only one instance of a DCCH or DTCH is mapped to the FACH, MAC-d must determine whether or not it needs to read the field. To do this, it scans the current configuration and retrieves the list of logical channels that are multiplexed on to the FACH. If only a single logical channel is found, then there is no multiplexing and therefore no C/T field to be read. If there are multiple logical channels, then the C/T field must be read from the header. The values of the C/T field, which is four bits long, are illustrated in Table 6.5. The logical channel can be identified in the list using the C/T field. Now the RLC entity may be determined and the data forwarded to the RLC.

Figure 6.8 shows the structure of a MAC PDU on either of these logical channels mapped to FACH, RACH and DSCH.

6.3.4 Reception on DCH

MAC PDUs are handled by the MAC-d block. PDUs received on the DCH are destined for the DCCH or DTCH and only contain MAC headers if several logical channels are multiplexed on to the DCH. On receipt of a PDU on the DCH, MAC-d scans the current

Table 6.5   C/T field to logical channel
instance mapping.

C/T field	DCCH/DTCH instance
0000	1
0001	2
0010	3
. . .	. . .
1101	14
1110	15

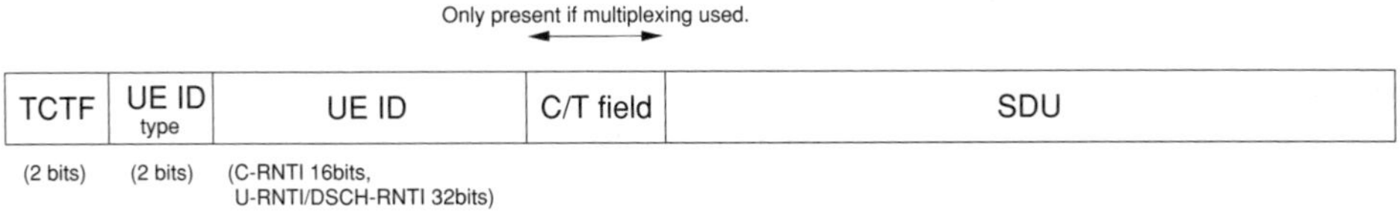

Only present if multiplexing used.

TCTF	UE ID type	UE ID	C/T field	SDU
(2 bits)	(2 bits)	(C-RNTI 16bits, U-RNTI/DSCH-RNTI 32bits)		

Figure 6.8. Structure of MAC PDU on DCCH or DTCH mapped to FACH, RACH and
DSCH.

configuration and retrieves the list of DCCHs and DTCHs for the DCH. If there is only
one logical channel, i.e. no multiplexing, then there is no C/T field to be read. If there is
more than one logical channel, the C/T field is read from the header and used to identify
the correct logical channel instance. From this, the RLC entity is identified and the data
forwarded to the corresponding RLC.

6.3.5 Reception on DSCH

PDUs received on the DSCH are initially handled by MAC-c/sh. As the DSCH is a shared
channel, the MAC headers of PDUs received on it contain a UE ID. The UE ID type field
and corresponding UE ID are read to determine if the PDU is intended for this UE. The UE
ID type is always DSCH-RNTI for PDUs received on the DSCH. If this does not match the
currently configured DSCH-RNTI, the PDU is ignored. If the UE ID does match, the PDU
is forwarded to MAC-d, which identifies the logical channel. The list of logical channels
for the DSCH is retrieved from the current configuration. If only one logical channel is
mapped to the DSCH, then there is no C/T field to be read. If multiple logical channels are
multiplexed, the C/T field is read from the header to identify the logical channel. From this
the RLC identity is identified and the data forwarded to the RLC.

6.3.6 Deciphering

MAC-d performs deciphering for the DCCHs and the DTCHs that are mapped to the TM
RLC entities. Once the MAC-d has identified the logical channel as a DCCH or DTCH, it
checks in the ciphering configuration, if the target RLC entity is in the list of entities that
require ciphering. If so, the SDU is passed on to the ciphering engine, along with the set of

stored ciphering parameters, for deciphering before it is forwarded to the RLC. Note that only the SDU part of the PDU is deciphered, not the header.

6.4 MAC TRANSMIT FUNCTIONALITY

This section discusses the transmit functionality of the MAC, namely the transfer of data from the RLC to the physical layer. The RLC indicates the buffer occupancy (the amount of data waiting to be transmitted) to the MAC. If there is data waiting, the MAC requests that the data be sent from the RLC by sending a MAC_STATUS_IND message to the RLC. The RLC sends the data to the MAC in a MAC_DATA_REQ message, a parameter of which is the transmitting RLC entity that is used to identify the logical channel to be used for transmission of the data. Before transferring the data to the physical layer, the MAC must construct and prepend the relevant MAC header to the data. The MAC sends the data to the physical layer by sending a PHY_DATA_REQ message via the PHY SAP.

A problem with this scenario is that the MAC does not implicitly know when it is able to transmit data, or rather when the physical layer is in such a state that it is able to send data. For the purpose of this text, it is assumed that a trigger exists that allows the MAC to determine that the physical layer is able to transmit. This trigger could be implemented by means of a message sent from the physical layer to indicate that it is ready, for example, a message PHY_TX_READY_IND could be sent. This message could be sent every TTI for transmission on the DCH and on receipt of an acknowledgement on the AICH after the preamble cycle for the RACH.

6.4.1 TFC Selection

TFC selection is the process of determining the most suitable TFC for transmission from the current set of valid TFCs [139]. A TFC is said to be a valid TFC if it fulfils the following criteria:

Belong to the TFCS. The TFC must belong to the set of currently configured TFCs.

Not be in the blocked state. Each TFC requires a different amount of transmit power due to the amount of data that may be transmitted with that TFC. It may be possible for a TFC to require more power than the currently configured UE transmit power. TFCs whose transmit power requirements exceed the currently configured transmit power levels may be blocked to prevent them being used for transmission. Although the decision to block a TFC could in theory be made in the MAC it may be a better idea for the physical layer to monitor the TFCs and inform the MAC of those that should be blocked due to excess power requirements.

Be compatible with the current RLC configuration. The TFC must be compatible with the current data size configurations in the RLC.

Not be capable of carrying more bits than can be transmitted in a TTI. This applies when compressed mode is configured. The physical layer may be responsible for signalling this restriction to the MAC.

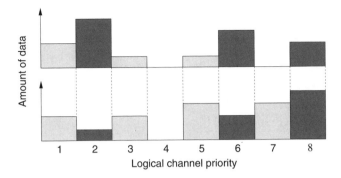

Figure 6.9. Transmitted data capability of two TFCs categorized by priority.

6.4.1.1 *Selecting the Best TFC*   The most suitable TFC is described as the one that can carry the greatest amount of highest priority data. The highest priority data is determined by examining the uplink logical channels in priority order, highest first, and identifying the highest priority channels with data to transmit. The TFC is selected applying the rules in the list below. If the first rule results in multiple possible TFCs, apply the second rule iteratively in decreasing priority order. The third and fourth rules are applied if multiple possible TFCs remain after the application of rules one and two.

1. Identify the TFC that permits the transmission of the greatest amount of highest priority data.

2. Identify the TFC that permits the transmission of the greatest amount of next-highest priority data.

3. Identify the TFC with the highest bit rate.

4. Identify the TFC which would cause the RLC to generate the least padding PDUs.

The two histograms in Figure 6.9 show the amount of data, categorized by priority, that could be transmitted by two TFCs. Rule one of the selection process would fail to identify a single TFC, as both TFCs could carry the same amount of priority one data. In this case, rule two would apply and the first TFC would be chosen as it is able to carry more priority two data.

Note that the choice must hold across a TrCH for a whole TTI, as the transport format for a TrCH can only change at a TTI boundary.

6.4.2 Transmission on DCH

For the DCH, a PHY_DATA_REQ is sent every TTI for each TrCH in response to the trigger. Once the MAC-d has retrieved the data from the RLC and selected a TFC, it must construct one MAC PDU for each transport block that is to be transmitted, as identified by the TFC selection algorithm. MAC PDUs carried on the DCCH or DTCH mapped to DCH only carry a MAC header if several DCCHs or DTCHs are mapped to the same DCH. In this case, the C/T field is prepended to the SDU to be transmitted.

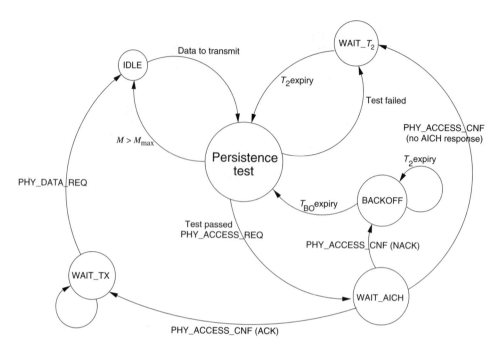

Figure 6.10. State machine for RACH access procedure.

6.4.3 Transmission on RACH

As the RACH is a contention-based channel, the MAC controls when RACH transmissions are executed and as such it must indicate to the physical layer that it wishes to transmit on the RACH, the physical layer will be unable to provide a regular trigger for the RACH case. The RACH access procedure [139], which is implemented in the MAC-c/sh, can be described by means of a state machine, as illustrated in Figure 6.10.

6.4.3.1 Initiating a RACH Access   The initiation of the RACH access attempt involves the following steps:

1. Set the value of M, the current preamble cycle, to zero.

2. Select the ASC. In the case where the ASC is not selected by the RRC, the MAC must perform the selection. The selection is based on the logical channel priorities and is made by selecting the lowest value between the maximum available ASC and the maximum logical channel priority. The ASC identifies the PRACH partition and a persistence value.

3. Perform persistency test. Increment the value of M. If $M > M_{max}$ then the maximum number of preamble ramping cycles has been reached and the RACH procedure has failed. Indicate this failure to the RRC in a CMAC_STATUS_IND. If $M \leq M_{max}$ then the following actions are performed:
   (a) start the 10 ms timer T_2. Should this persistence test fail, this timer is used to ensure that a subsequent persistence test is not performed within the same TTI.

(b) generate a random number $R \in \{0...1\}$. If $R \leq P_i$, where P_i denotes the persistence value for the selected ASC, then the persistence check succeeds and the RACH cycle starts. At this stage the MAC submits a PHY_ACCESS_REQ to the physical layer. If the persistence check fails, the MAC must wait for the timer T_2 to expire before it can start another persistence test. This procedure is repeated until the test succeeds. In the case where the PHY_ACCESS_REQ is sent, the MAC should wait for the PHY_ACCESS_CNF to determine the response on the AICH.

6.4.3.2 Acknowledgement ACK on AICH   If the PHY_ACCESS_CNF indicates that an *Acknowledgement* (ACK) was received on the AICH, the data may be sent to L1 by sending a PHY_DATA_REQ message. It may be that in the case of an ACK, the physical layer provides a trigger to the MAC to send the PHY_DATA_REQ. The fact that the RACH transmission was successful should be indicated to the RRC.

6.4.3.3 Negative Acknowledgement NACK on AICH   If the PhCH (RACH) is busy, the PHY_ACCESS_CNF message will indicate a *Negative Acknowledgement* (NACK) on the AICH. In this case the MAC should initiate the backoff procedure as follows:

1. If T_2 has not expired, wait for expiry.

2. Determine the value of the backoff timer T_{BO} by generating a random number in the range N_{BO1min} and N_{BO1max}. Start the T_{BO} and wait for it to expire. At this point, go back to Stage three in the procedure listed in Section 6.4.3.1.

6.4.3.4 No Response on AICH   If the PHY_ACCESS_CNF indicates that no response was received on the AICH, the MAC awaits the expiry of T_2 and goes back to Stage three of the procedure in Section 6.4.3.1. No backoff is performed in this case.

6.4.4 MAC Headers on RACH

The sections below outline the MAC headers used for PDUs transmitted on the RACH.

6.4.4.1 CCCH on RACH   MAC PDUs transmitted on the CCCH require a MAC header. In the case of CCCH only the TCTF is required in the header as shown in Figure 6.11. The TCTF value to use for CCCH on RACH is illustrated in Table 6.6.

6.4.4.2 DCCH or DTCH on RACH   MAC PDUs transmitted on the DCCH or DTCH require a MAC header. The logical channel must be identified by setting the TCTF value. The value to use for DCCH or DTCH on RACH is illustrated in Table 6.6. The UE ID fields

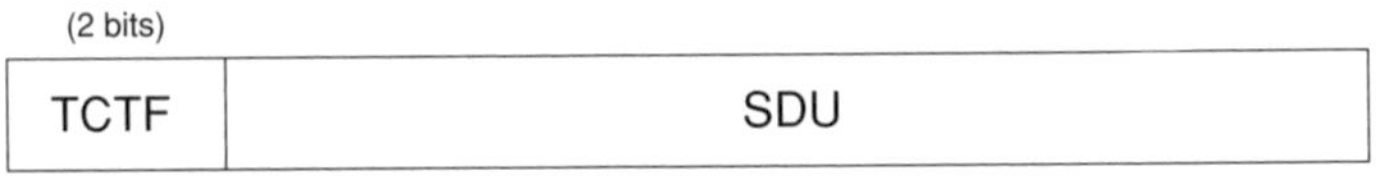

(2 bits)

TCTF	SDU

Figure 6.11. Structure of a MAC PDU on a CCCH mapped to the RACH.

Table 6.6   TCTF values for RACH.

TCTF	Logical channel
00	CCCH
01	DCCH or DTCH

in the MAC header must also be populated, only the C-RNTI is used in the uplink. In the case where multiple DCCHs or DTCHs are multiplexed on to the RACH, the C/T field must also be populated. The TCTF and C/T fields are populated by the MAC-d, while the UE ID fields are populated by the MAC-c/sh.

6.4.5 Ciphering

The MAC performs the ciphering for the DCCHs and the DTCHs that are mapped to the TM RLC entities. Once MAC-d has identified the logical channel as a DCCH or DTCH, it checks in the ciphering configuration if the source RLC entity is in the list of entities that require ciphering. If so, the SDU is forwarded to the ciphering engine, along with the set of stored ciphering parameters, for enciphering before it is passed on to the physical layer. Note that only the SDU part of the PDU is enciphered, not the header.

7

Radio Link Control

Seraj Ahmad, Ian Blair and Francoise Bannister

7.1 INTRODUCTION

The *Radio Link Control* (RLC) block is a sublayer within Layer 2. The purpose of the RLC layer is to provide radio link services for use between the UE and the network. At the transmitter, the higher layers (*Radio Resource Control* (RRC), *Broadcast/Multicast Control* (BMC), *Packet Data Convergence Protocol* (PDCP), or circuit switched voice or data) provide data on radio bearers in *Service Data Units* (SDUs). These are mapped by the RLC layer into *Protocol Data Units* (PDUs) which are sent on the logical channels provided by the *Medium Access Control* (MAC). The position of the RLC within the access stratum is shown in Figure 7.1.

The RLC is configured by the RRC via the RLC Control SAP, as illustrated in Figure 7.1, according to Section 8 in [139].

Inside the RLC block are discrete RLC entities: all radio bearers require RLC entities, as do the special purpose logical channels (BCCH, PCCH and CTCH). Within the UE, bidirectional radio bearers need both a transmit and a receive RLC entity, or a bidirectional RLC entity, whereas receive only channels use unpaired receive RLC entities. The total number of RLC entities configured at a given time depends on the RRC state, and also the configuration of radio bearers in use.

For each radio bearer, the RLC layer will provide one of three types of service, supplied by different kinds of RLC entity. Each service is suited to the needs of different types of data user:

Transparent Mode (TM). Transparent mode is used for simple, often time-critical, data transfer, such as might be required for example by a speech codec. Transparent data is

WCDMA – Requirements and Practical Design. Edited by R. Tanner and J. Woodard.
© 2004 John Wiley & Sons, Ltd. ISBN: 0-470-86177-0.

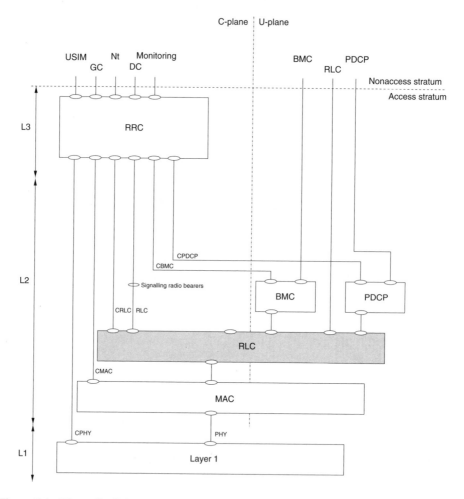

Figure 7.1. The radio link control block within the protocol (access stratum) structure.

passed through the RLC in PDUs which are unchanged, (except for one special case, in which data may be segmented or reassembled in a very simple way).

There is no signalling between peer TM RLC entities, and the TM RLC PDUs have to fit the sizes of the MAC PDUs exactly. (These in turn depend on the size of the transport blocks provided by Layer 1, after allowance is made for any header used by the MAC layer).

Transparent mode is not capable of detecting PDUs which have been lost in transmission from the peer RLC entity, the best it can do is pass on the results of any CRC check performed by Layer 1.

Unacknowledged Mode (UM). In unacknowledged mode, RLC signalling is added to the data to permit a variable relationship between the user of the data and the capabilities of the radio channel. An unacknowledged mode RLC entity can segment or concatenate data from the user SDUs to fit variously sized MAC PDUs.

Table 7.1 Mapping of radio bearers onto logical channels and RLC entities.

Radio bearer ID	Logical channel	RLC entity type	User
—	BCCH	TM (downlink only)	RRC
—	PCCH	TM (downlink only)	RRC
—	CTCH	UM (downlink only)	BMC
RB0 (SRB0)	CCCH	TM (UL) / UM (DL)	RRC
RB1 (SRB1)	DCCH	UM (UL and DL)	RRC
RB2 (SRB2)	DCCH	AM (UL and DL)	RRC
RB3 (SRB3)	DCCH	AM (UL and DL)	RRC (for NAS)
RB4 (SRB4)	DCCH	AM (UL and DL)	RRC (optional for NAS)
RB5-32	DTCH	TM, UM or AM	User plane (PDCP, or CS data/voice)

Unacknowledged mode PDUs have sequence numbers, and hence the UM RLC can indicate when data has been lost in transmission from the peer UM entity.

The flexibility of use of PDU sizes by UM may be used, for example, to take maximum advantage of a radio channel with varying capacity. Alternatively, if the relationship between SDU and PDU sizes is fixed, UM mode may also be used in time critical applications where the added feature over TM of lost PDU detection may be useful.

Acknowledged Mode (AM). In acknowledged mode, not only is RLC signalling added to the payload data, but additional PDUs are defined to permit bidirectional signalling between peer RLC entities. This signalling is used primarily to request the retransmission of missing data, and hence the AM RLC is particularly suited to applications which require reliability of data transmission, at the expense of a variable delivery time. Acknowledged mode RLC entities support segmentation and concatenation in a similar manner to UM, but, because of the need to identify missing PDU data simply, payload data is sent in fixed size PDUs. Note that AM is intrinsically bidirectional, UM and TM may be applied in a single direction at a time.

The RLC layer is also responsible for ciphering of UM and AM data. In both cases the associated transmit RLC entity ciphers the payload and length parts of each PDU; ciphering is not applied to the sequence number in the RLC header data.

Finally, the RLC maps radio bearers to the logical channels provided by the MAC. Some radio bearer and logical channel mappings are fixed, others are configured by the network when, for example, a radio bearer is set up. The mapping of different types of radio bearers to logical channels is shown in Table 7.1.

The user plane radio bearers from Layer 2 shown in Table 7.1 are presented to the user as *Radio Access Bearers* (RABs). A RAB is a radio bearer or group of radio bearers which are referenced by a RAB ID, and also the core network domain which the RAB is intended for. The mapping of the RABs to radio bearers is provided to Layer 2 by the RRC when the radio bearers are configured. For *Circuit Switched* (CS) services, RABs may contain multiple radio bearers; e.g. for an AMR codec (see Chapter 11), three radio bearers are used so that different sets of codec bits can be afforded different levels of error detection

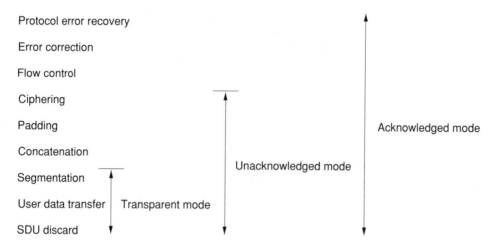

Figure 7.2. The RLC functionality at the transmitter.

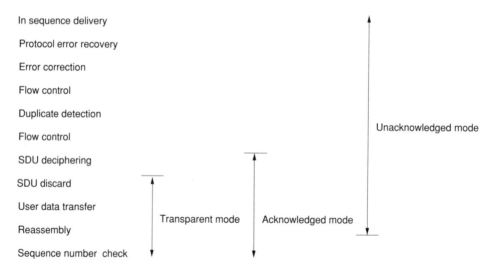

Figure 7.3. The RLC functionality at the receiver.

and correction. In contrast, *Packet Switched* (PS) RABs are mapped to single PDCP radio bearers.

Figures 7.2 and 7.3 summarize the available functionality of TM, UM and AM entities at the transmitter and receiver sides.

Each RLC entity can be viewed as an *Enhanced Finite State Machine* (EFSM) implementing the logic to provide the required data transfer services. The RLC provides a control unit to manage these RLC EFSM instances. The control unit services can be accessed through the *RLC Control* (CRLC) interface. This interface is also known as a *Service Access Point* (SAP).

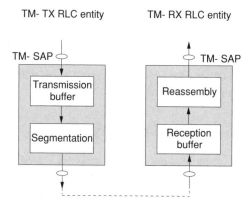

Figure 7.4. Overview of transparent mode data transfer.

Note that RLC entities share transmission and buffer occupancy conditions with the MAC, see Chapter 6. The MAC translates RLC buffer occupancy into the transport block occupancy needed during *Transport Format Combination* (TFC) selection. Therefore, for the purpose of efficiency, the RLC and MAC may be tightly coupled in implementation, blurring the responsibility of the MAC SAP shown in Figure 7.1.

In the following sections we describe in detail the RLC's operation for TM, UM and AM, based on the relevant 3GPP specifications:

- TS25.322 [140] specifies the RLC protocol;

- TS25.321 [139] specifies the MAC (lower layer) protocol;

- TS25.331 [141] specifies the *Radio Resource Controller* (RRC) protocol which configures the RLC entities.

7.2 TRANSPARENT DATA TRANSFER SERVICE

The *Transparent Mode* (TM) is used to transport streaming and real time data across the air interface. Figure 7.4 shows a transmit and receive TM RLC entity exchanging data and gives an overview of the operations carried out. Subsequent subsections describe the various aspects of the transparent mode entity.

7.2.1 Transparent Mode Transmit Entity (TM-TX)

The TM-TX entity provides data transfer without adding any header. The absence of a header in the TM-TX entity causes it to have no concatenation and only limited segmentation capabilities. Any segmentation is explictly configured by the RRC and is assisted by the MAC.

When segmentation is configured, only a single RLC *Service Data Unit* (SDU) can be transmitted within a *Transmission Time Indicator* (TTI). The RLC SDU is segmented into equal sized chunks, which must be transmitted in the same TTI. The size and number of segments depend on the allowed *Transport Format Set* (TFS) of the underlying logical channel onto which the TM-TX entity is mapped. The size of SDUs received by the TM-Tx RLC entity must be a multiple of one of the available sizes in the TFS.

When segmentation is not configured, one or more RLC SDUs can be transmitted each TTI. The size of SDUs received by the TM-TX RLC entity must equal one of the available sizes in the TFS. All the SDUs transmitted in the same TTI must be of equal size as dictated by the homogeneity of transport blocks.

7.2.2 Transparent Mode Receive Entity (TM-RX)

The TM-RX entity is the complimentary peer of a TM-TX entity. It receives data PDUs from downlink logical channels, performs reassembly if needed and delivers the resultant RLC SDUs to the higher layer. Reassembly is explicitly configured by the RRC to match the peer TM-Tx entity segmentation configuration. If reassembly is configured, then the TM-RX RLC entity delivers a single SDU every TTI. Otherwise, the TM-RX RLC entity can deliver one or more SDUs each TTI.

Sometimes a TM SDU may be erroneous if one or more of its constituent PDUs are corrupted. The RRC can configure the TM-RX RLC entity to either deliver these erroneous SDUs, or discard them. The TM-RX entity delivers the data to higher layers over the TM-RLC SAP.

7.2.3 TMD PDU Schema

Transparent Mode Data (TMD) PDUs are the basic units of data exchanged between peer TM RLC entities. Each PDU is formed from a higher layer SDU, possibly with segmentation of the SDU into several PDUs. As pointed out in the preceding sections, no RLC header is added to the SDU. The TMD PDU length is not constrained to be a multiple of eight bits but must be nonzero.

7.2.4 Transparent Mode Data Transmission

The TM-TX entity receives the data to be transmitted from the TM-RLC SAP. These TM SDUs received from higher layers are kept in a transmission queue until they are transmitted. The TM-TX entity provides the SDU buffer occupancy to the MAC depending on its segmentation configuration. In the absence of segmentation, the TM-TX entity informs the MAC of the size of SDU and the number of waiting SDUs. The MAC then uses this information to select a suitable TFC, see Section 6.4.1. When segmentation is configured, the MAC needs to know only the size of the SDU, as the number of SDU segments is decided by the MAC according to the available transport block sets.

After the TFC selection, the MAC informs the TM-TX entity of the number and size of the TMD PDUs it can transmit. The TM-TX entity uses this information to generate TMD PDUs and submit them to the MAC.

7.2.5 Transparent Mode Data Reception

The TMD PDUs received in one TTI are reassembled into one SDU when reassembly mode is configured. Otherwise, all TMD PDUs are treated as individual SDUs. Reassembly is assisted by the MAC, which labels each TM data PDU with a sequence number. This sequence number is utilized by the TM-RX entity to correctly reassemble the SDU. The TM SDUs are delivered to the upper layer within the same TTI. Each TMD PDU delivered to the TM-RX entity contains a CRC-pass/fail indication, which is computed in the physical layer. This CRC status may be used to filter erroneous TM SDUs according to the following configuration rules:

No Detect. All SDUs are delivered to the upper layer without any attempt to discover the error status.

Yes. All SDUs are delivered to the upper layer along with an error status.

No. All erroneous SDUs are discarded and not delivered to the upper layer.

7.2.6 Transparent Mode Discard

To prevent the transmission queue from overflowing, the TM-TX entity uses two different buffer management schemes. In the first scheme, the TM-TX entity tries to maintain only those SDUs in the queue that have arrived in the same TTI. Ideally, all SDUs that arrive in one TTI should have been transmitted in that TTI itself, but this is not always possible. So remaining SDUs may be transmitted in the next TTI when no further SDUs are received. However, the remaining SDUs are discarded as soon as a new SDU is received in the next TTI. This scheme is used when the discard mode is configured as No Discard.

The second scheme, where a timer-based discard is configured, schedules a discard timer as soon as it receives an SDU from the higher layer. If the SDU has not been transmitted by the expiry of this discard timer, it is discarded. However, an asynchronous discard may interfere with the TFC selection procedure within the MAC, which relies upon the integrity of the *Buffer Occupancy* (BO) reported by the RLC. So, to protect the BO integrity and to avoid a race condition, the expiry of the discard timer should be serialized with the TFC selection procedure.

7.3 UNACKNOWLEDGED DATA TRANSFER SERVICE

The unacknowledged mode is used to transport traffic which may need segmentation and/or concatenation capabilities, but does not need retransmissions. Examples of such services include measurement reporting and cell broadcast services. Unacknowledged mode entities are unidirectional in nature, and there can be more than one peer receiving entity for a single transmitting entity. Figure 7.5 shows a transmit and receive UM RLC entity exchanging data and gives an overview of the operations carried out. The following subsections describe the operation of unacknowledged mode entities.

7.3.1 Unacknowledged Mode Transmit Entity (UM-TX)

The UM-TX entity augments the TM-TX entity capabilities to support concatenation and unrestricted segmentation by introducing a header. It embeds segmentation and concatenation information in the header of each UM PDU transmitted to the peer entity. This header

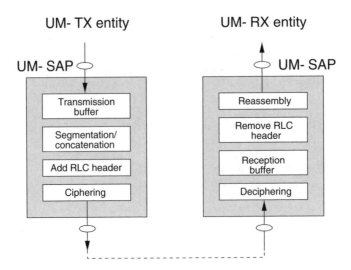

UM- TX entity   UM- RX entity

Figure 7.5. Overview of unacknowledged mode data transfer.

information takes the form of sequence numbers and *Length Indicators* (LIs), as described later in Section 7.3.3.

The UM-TX entity can transmit one or several equal sized PDUs each TTI. As the UM-TX entity does not support retransmissions, the PDU size can change every TTI. The PDU size is selected by the MAC according to the available transport block sizes. If a PDU is not filled by SDU data and header information, then the remaining space is padded. An especially defined padding LI is used in the RLC header to signal the padded space to the peer entity.

The UM mode segmentation and concatenation is performed on the fly as the PDU size may not be known beforehand. The UM-TX entity maintains a state variable VT(US), which holds the sequence number of the next *Unacknowledged Mode Data* (UMD) PDU to be transmitted. The VT(US) is initialized at zero, and incremented after submitting a UMD PDU to the MAC. The VT(US) can have a maximum value of 127 after which it wraps around to 0. The UM-TX entity receives SDUs through the UM-RLC SAP and transmits PDUs on a configured logical channel to the MAC.

7.3.2 Unacknowledged Mode Receive Entity (UM-RX)

The UM-RX entity receives PDUs from the downlink logical channels, performs reassembly, and delivers the completed SDUs to the upper layer. Since every incoming PDU contains a sequence number, the UM-RX entity attempts reassembly in all cases and no explicit configuration is required. The UM-RX entity keeps adding the received data to an assembly buffer until it obtains an LI which indicates that the current SDU is complete.

The reassembly procedure expects the PDUs to arrive in sequential order. An out of sequence UM PDU will cause the reassembly of the current SDU to be aborted. The reassembly procedure recovers from an aborted state only after it receives an LI in a UM PDU.

Table 7.2   The extension bit E.

E	Octet class
0	The next octet is data or padding
1	The next field is a LI and another E bit

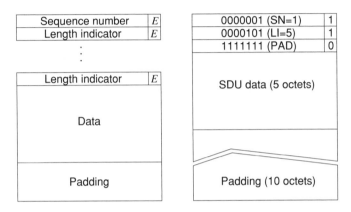

Figure 7.6. The UMD PDU.

7.3.3 UMD PDU Schema

UMD PDUs are the basic units for data exchanged between peer UM RLC entities. The layout of a general UMD PDU is shown on the left hand side of Figure 7.6, while the right hand side shows a specific example.

The UMD PDU header always has one octet, at the start of the PDU, which gives a seven bit sequence number and a one bit extension (E) field. It may also contain one or more LIs. As shown in Table 7.2, this extension to include LIs is signalled through the one bit E field. Each LI also contains an E bit to indicate if a further LI is also included in the header, and so using these E bits the header can grow recursively.

The size of each LI is either 7 or 15 bits, depending on the size of the largest PDU configured for the UM entity. Each RLC UM entity is configured to use either 7 or 15 bit LIs, and so both the receiving and transmitting entities will know what length LIs are used. With one E bit per LI, each LI takes one or two octets. The total length of a UMD PDU is always a multiple of eight bits.

The LIs are set to indicate the number of octets between the start of the SDU data in this PDU and the end of the SDU. If the SDU does not end in this PDU, no LI is included for this SDU. If the SDU does end in this PDU, an LI will be included unless the SDU exactly fills the remaining capacity of the PDU. In this case, there is no space for the LI in this PDU, and the first LI in the next UMD PDU is used to indicate the end of this SDU. The special value of zero is used as the LI to indicate this.

Various other special values of LI are defined – these are listed in Tables 7.3 and 7.4 for 7 and 15 bit LIs respectively. The special values of LI for AM PDUs, described in Section 7.4.3, are also described in these tables.

Table 7.3   Length indicators (LIs) with seven bits.

LI value	Description (UM)	Description (AM)
0 × 00	The last segment of an SDU exactly filled the previous PDU and there was no LI to indicate the end of the SDU	
0 × 7C(124)	The first data octet of the RLC PDU is also the first data octet of an RLC SDU	Reserved
0 × 7D(125)	Reserved	Reserved
0 × 7E(126)	Reserved	The RLC PDU includes piggybacked status after the last data octet
0 × 7F(127)	The RLC PDU contains padding after the last data octet. The padding length will fill the rest of the PDU, and can be zero	

Table 7.4   Length indicators(LIs) with 15 bits.

LI value	Description (UM)	Description (AM)
0 × 0000	The last segment of an SDU exactly filled the previous PDU and there was no LI to indicate the end of the SDU	
0 × 7FFB	The last segment of an SDU was one octet short of exactly filling the previous PDU and there was no LI to indicate the end of the SDU. The last octet in the previous PDU should be discarded	
0 × 7FFC	The first data octet of the RLC PDU is also the first data octet of an RLC SDU	Reserved
0 × 7FFD	Reserved	Reserved
0 × 7FFE	Reserved	The RLC PDU includes piggybacked status after the last data octet
0 × 7FFF	The RLC PDU contains padding after the last data octet. The padding length will fill the rest of the PDU, and can be zero	

LIs are packed sequentially in the order of the SDUs to which they refer. They can be visualized as pointers to the end of an SDU. Note that an LI whose value exceeds the length of the UMD PDU is invalid.

Figure 7.7 shows an example of the RLC TM-TX entity packing data from three input SDUs into four output PDUs in a given TTI. A few octets from SDU $n - 1$ remain to be transmitted at the start of this TTI. Two other SDUs are also available to be transmitted. The first RLC PDU contains the remaining data from SDU $n - 1$ and the start of the data from SDU n. The header for this first PDU contains a sequence number (SN) and an LI indicating the end of the data from SDU $n - 1$. Note that no LI is included for SDU n in this PDU, as this SDU does not end in the first PDU.

The next two PDUs contain just data from SDU n and the PDU sequence numbers. The data from this SDU is exhausted in the final PDU, which therefore contains an LI indicating the remaining length of the SDU. This final PDU is also large enough to contain all the data

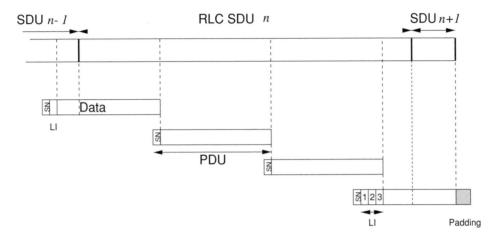

Figure 7.7. Example UM RLC packing of SDUs into PDUs.

from SDU $n + 1$, and so this data and an LI for it is also included. Finally, padding is used to fill the remaining space in the PDU, and so a third LI is used to indicate to the receiving entity that the remaining data in the PDU is padding.

7.3.4 Unacknowledged Mode Data Transmission

All UM SDUs are buffered in the transmission queue until they are transmitted. The UM-TX entity informs the MAC of the SDU buffer occupancy. The exact information provided in the UM buffer occupancy is implementation dependent, but it should be enough for the MAC to correctly compute the transport block occupancy for different transport block sizes. The MAC uses the TB occupancy to select a suitable TFC.

Upon completion of the TFC selection procedure, the MAC requests the UM-TX entity to transmit its UMD PDUs, specifying the PDU size and the number of PDUs. The VT(US) state variable is encoded in the sequence number field of the UMD header and is incremented by one for each PDU. The LIs are packed as described above, and finally the UMD PDUs are submitted to the MAC.

7.3.5 Unacknowledged Mode Data Reception

Upon receipt of a UMD PDU from the MAC, the UM-RX entity extracts the sequence number and length indicators, and performs an integrity verification. If the UMD PDU contains an invalid or reserved length indicator, then verification fails and the PDU is ignored. Otherwise, if this PDU completes an SDU, the SDU is delivered to the higher layer.

7.3.6 Unacknowledged Mode Discard

The UM-TX entity SDU buffer management scheme has the same two modes of SDU discard as described in Section 7.2.6 for the TM-TX entity. When a timer-based discard is configured the operation of the UM-TX entity is similar to that of the TM-TX entity. However, in addition to the steps taken during the TM discard procedure, the state variable

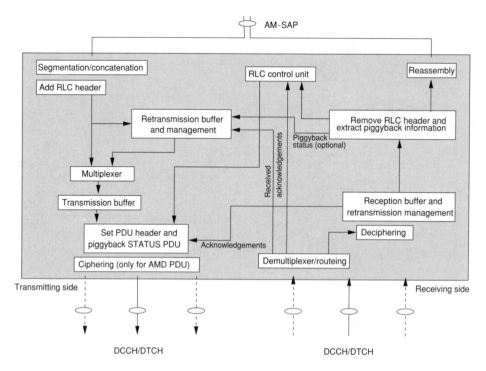

Figure 7.8. The AM entity.

VT(US) is here increased by one after an SDU discard. This introduces a gap in the sequence number which prevents the peer entity from reassembling the discarded SDU.

The UM SDU buffer management differs from transparent mode when the discard mode is configured as No Discard. In the UM-TX entity, the SDUs are not discarded as long as the buffer has space. When the buffer reaches its capacity, the UM-TX entity discards the oldest SDUs in the queue. If one or more segments of the discarded SDU have been submitted to the MAC, then the state variable VT(US) is incremented by one after the SDU discard. However, this is not necessary if no segments of discarded SDU have been submitted to the MAC.

7.4   ACKNOWLEDGED DATA TRANSFER SERVICE

The acknowledged mode is used to transfer reliably non real time data. Such a service may be used for example by traditional Internet applications. The acknowledged mode is by nature bidirectional, meaning that an acknowledged mode RLC entity contains both a transmitting and a receiving entity, as shown in Figure 7.8. High reliability is provided through the retransmission of erroneous PDUs, at the expense of a possible higher end-to-end delay than in both TM and UM operations.

7.4.1   Acknowledged Mode Transmit Entity (AM-TX)

The AM-TX entity further augments the UM-TX entity capabilities by incorporating an *Automatic Repeat Request* (ARQ) mechanism, including control and flow. The acknowledged mode defines several special PDUs to support these mechanisms. These PDUs are collectively referred to as control PDUs. Control PDUs may be transmitted together with data PDUs on a single logical channel or on a separate logical channel. The control PDUs have higher transmission priority than the *Acknowledged Mode Data* (AMD) PDUs, supporting error free and efficient operation of the protocol. To transmit (respectively receive) these control PDUs, an AM-RX (respectively AM-TX) entity must have an associated AM-TX (respectively AM-RX) entity, making the AM mode a bidirectional entity.

To support retransmissions, the upper layer configures the AM-TX entity with a data PDU size that remains fixed during the operation of the AM-TX entity and can be changed only through reestablishment. The size of the control PDU is the same as the size of the data PDU when one logical channel is active. However, the size for the control PDUs is not fixed if they are transmitted on a separate logical channel, and can be dynamically selected by the MAC, according to both the control buffer occupancy and availability of a suitable transport block size. The use of a separate logical channel for control PDUs saves radio bandwidth for large data PDUs and allows control PDUs to be sent at higher priority.

All AM SDUs are segmented or concatenated into fixed size *Acknowledged Mode Data* (AMD) PDUs which are defined by the RRC during the AM-TX entity configuration. If the length of an AMD PDU, due for transmission, is shorter than the configured length, it may be padded to match the configured PDU size.

All AMD PDUs are kept in the transmission buffer until their reception is positively acknowledged by the peer entity. AM entities use a dedicated structure known as STATUS_REPORT to inform the transmitting peer entity about receiver buffer conditions and AMD PDU reception status. A STATUS_REPORT may contain a positive and/or a negative acknowledgement for one or more AMD PDUs. Negative acknowledged AMD PDUs are retransmitted subject to a maximum transmission value configured by the higher layer. Retransmitted AMD PDUs have a higher priority than AMD PDUs that are transmitted for the first time, thereby reducing the average delivery time for an AM SDU.

A STATUS_REPORT can be split over one or more STATUS PDUs, which can be transmitted separately. To efficiently utilize radio bandwidth when only one logical channel is configured, STATUS PDUs can be piggybacked on to AMD PDUs providing enough space is available. No other PDU type can be piggybacked.

The AM-TX entity may poll the receiver to send a STATUS_REPORT, by setting a flag in the AMD PDU header. Which AMD PDU has the flag set is determined by poll triggers that are configured by the higher layer. There are eight different poll triggers and a combination of them may be used to ensure a timely request for the transmission of a STATUS_REPORT. The AM-TX entity receives SDUs from the AM-RLC SAP and transmits PDUs on one or two logical channels.

7.4.1.1 The AM-TX Entity State Variables
The AM-TX operation is governed by a certain number of state variables, such as:

VT(S). This state variable keeps the sequence number of the next AMD PDU to be transmitted for the first time and is incremented when the corresponding AMD PDU is transmitted.

VT(A). This state variable forms the lower end of the transmission window and contains the sequence number of the first AMD PDU not yet acknowledged by the receiver.

VT(MS). This state variable forms the upper end of the transmission window and contains the sequence of the first AMD PDU which cannot be transmitted owing to the fact that it will be rejected by the receiver.

VT(WS). This state variable contains the size of the transmission window. Its initial value is configured by the upper layer.

VT(DAT), VT(MRW), VT(RST). These are used to limit the number of times an AMD PDU *Move Receiving Window* (MRW) command or RESET PDU respectively is transmitted until successful reception.

7.4.2 Acknowledged Mode Receive Entity (AM-RX)

The AM-RX entity is similar to the UM-RX entity with respect to the SDU reassembly and delivery process, but it has a different receive buffer management. The AM-RX entity buffers out of sequence AMD PDUs as long as they are within the receiver window, and sends STATUS_REPORT acknowledgements using an associated AM-TX entity.

When a complete SDU has been received, it is reassembled and delivered to the upper layer through the AM-RLC SAP. The AM-RX entity can be configured to guarantee in-sequence SDU delivery. In that case, the reassembly function is only invoked upon arrival of the first in-sequence AMD PDU containing a length indicator. The AMD PDUs removed from the receiver buffer for reassembly are positively acknowledged in a STATUS_REPORT.

In addition to the generation of a STATUS_REPORT, in response to a poll request, the receiver may be configured by the RRC to automatically generate and transmit such a report.

7.4.2.1 *The AM-RX Entity State Variables*   The AM-RX operation is itself governed by a certain number of state variables, such as:

VR(R). This state variable contains the sequence number of the next expected in-sequence AMD PDU.

VR(MR). This state variable contains the upper end of the receive window. Any received AMD PDU with sequence number greater than or equal to VR(MR) will be discarded.

VR(EP). This state variable contains the number of AMD PDUs whose retransmission is still expected as a consequence of the transmission of the latest status report. It is used by the *Estimated PDU Counter* (EPC) mechanism.

7.4.3 AM PDU Schema

The AM mode defines and uses two different kinds of PDU, known as AMD PDUs and control PDUs, to transport respectively data and RLC control information. The first bit in an AM PDU, the *Data Control* (D/C) bit, is used to distinguish AMD PDUs from control PDUs, see Figure 7.9.

7.4.3.1 *Acknowledged Mode Data PDU*   AMD PDUs are the basic unit of data exchanged between peer AM RLC entities. In addition to data, they carry poll information and optionally piggybacked STATUS information. The length of the AMD PDUs is always

Table 7.5   The header extension.

HE value	Description
00	Octet following AMD header contains data
01	Octet following AMD header contains a length indicator and an *E* bit

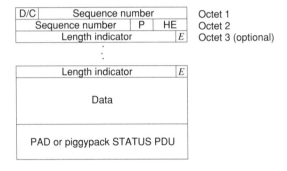

Figure 7.9. The AMD PDU.

a multiple of eight bits. The field *Header Extension* (HE) defined in Table 7.5 can be used to extend the AMD header to include a length indicator and an extension bit. The header extension field (2 bits) serves the same purpose as the extension field (1 bit) defined in Table 7.2.

The semantics of the UM and AM length indicators are almost identical, the only difference being that an AMD PDU can also carry a piggybacked STATUS PDU instead of padding bits. The size of the length indicator used within the AM-TX entity depends upon the configured AMD PDU size. The maximum AMD PDU size for a seven bit length indicator is 126 octets.

7.4.3.2 *Control PDUs*   The AM control PDUs are used to carry control information, such as AMD PDU acknowledgement (positive and/or negative), window update for SDU discard, flow control, reset request and reset acknowledgement. The first four bits constitute the control header which is divided into two fields, the D/C and control PDU type. The AM RLC defines and uses four different types of control PDU known as STATUS PDU, piggybacked STATUS PDU, RESET PDU and RESET ACK PDU. The control PDU type field is used to convey this information. The exact layout of the octets following the control PDU header depends on the type of control PDU being carried. The following subsections describe the different control PDUs mentioned here.

STATUS PDU. The status PDU is used to exchange status information between peer AM RLC entities. The basic unit of status information is a *SUper-FIeld* (SUFI), see Figure 7.10, which identifies a specific type of status information. A STATUS PDU can include super-fields of different types, encoded as *Type, Length, Value* (TLV) for variable length SUFIs or *Type, Value* (TV):

MRW. This SUFI is used to request the receiver to move its receive window in case either the timer or MaxDAT based AM SDU discard procedure has been invoked.

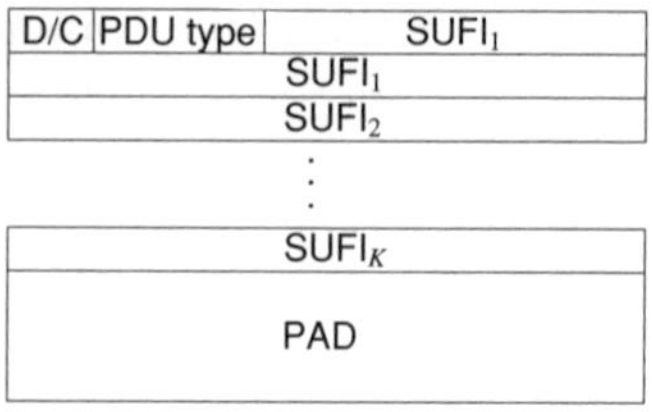

```
┌────┬────────┬──────────────────┐
│D/C │PDU type│      SUFI₁        │
├────┴────────┴──────────────────┤
│            SUFI₁               │
├────────────────────────────────┤
│            SUFI₂               │
└────────────────────────────────┘
                 ⋮
┌────────────────────────────────┐
│            SUFIₖ               │
├────────────────────────────────┤
│                                │
│             PAD                │
│                                │
└────────────────────────────────┘
```

Figure 7.10. The STATUS PDU.

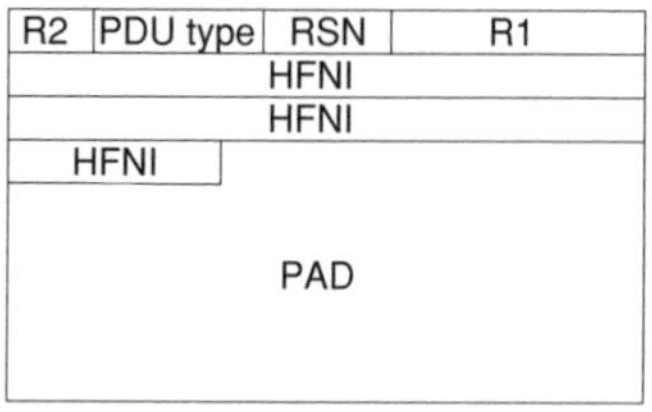

```
┌────┬────────┬──────┬───────────┐
│R2  │PDU type│ RSN  │    R1     │
├────┴────────┴──────┴───────────┤
│             HFNI               │
├────────────────────────────────┤
│             HFNI               │
├──────────────┬─────────────────┤
│    HFNI      │                 │
├──────────────┘                 │
│                                │
│             PAD                │
│                                │
└────────────────────────────────┘
```

Figure 7.11. The RESET and RESET ACK PDU.

MRW ACK. This SUFI acknowledges the reception of an MRW SUFI.

WINDOW. This SUFI is used when the receiver wishes to change the transmission window of the peer AM entity.

NACK. Used to indicate negative acknowledgement of a data PDU using one of LIST, BITMAP or RLICT SUFI, see Section 7.4.6.

ACK. This SUFI is used to acknowledge all AMD PDUs received up to a certain sequence number and is usually sent in response to a poll request or status trigger. It should be the last SUFI in the last STATUS PDUs generated from the STATUS_REPORT.

NO_MORE. This SUFI indicates the end of the data part of a STATUS PDU.

PIGGYBACKED STATUS PDU. The format of the piggybacked STATUS PDU is the same as STATUS PDU except that the D/C field is replaced by a reserved bit.

RESET PDU. The RESET PDU is used to request resynchronization with the peer AM RLC entities.

RESET ACK PDU. After receiving a RESET PDU, the peer entity responds with a RESET ACK PDU to complete the reset procedure. RESET and RESET ACK PDUs are paired using the *Reset Sequence Number* (RSN) field as shown in Figure 7.11.

7.4.4 Acknowledged Mode Transmission

Depending on the available bandwidth, the PDU selection for submission to the lower layer follows an ordered process:

1. Select any control PDUs awaiting transmission.

2. Select any AMD PDUs waiting in the transmission queue.

3. Generate new AMD SDUs from SDUs awaiting transmission.

7.4.4.1 *AMD PDU Transmission*   In addition to buffering, the transmission queue provides a mechanism to keep track of the segments that are derived from the same SDU. This is needed to both send confirmations to the upper layer and to support the discard procedure. The AM buffer occupancy reported to the MAC is used in the TFC selection as well as traffic volume measurement.

A VT(DAT) state variable is associated with every buffered PDU and is initialized at zero. At the time the corresponding AMD PDU is scheduled to be transmitted, its state variable is incremented by one, and a check is performed to ensure VT(DAT) < MaxDAT. If that is the case, the AMD PDU is submitted to MAC. Otherwise, a discard procedure or reset procedure is initiated depending on the discard mode configuration.

7.4.4.2 *AMD PDU Retransmission*   An AMD PDU requested for retransmission is added to a retransmission queue. The corresponding change in data buffer occupancy is reported to the MAC. The MAC selects a suitable TFC and indicates the number of PDUs that can be submitted for transmission.

7.4.4.3 *Control PDU Transmission*   The scheduling of control PDUs for transmission depends on several triggers:

- a RESET PDU is scheduled for transmission in order to initiate the reset procedure in case of a protocol error;

- a RESET ACK PDU is scheduled for transmission in order to acknowledge the receipt of a RESET PDU;

- the STATUS_REPORT transmission scheduling can be triggered by one or more events, such as a poll request, a status report trigger, explicit discard notification or flow control.

After a control PDU has been scheduled for transmission, the corresponding change in buffer occupancy is reported to the MAC. In cases where two logical channels have been configured, the control buffer occupancy is reported separately from the data buffer occupancy. The MAC selects a suitable TFC and indicates the number of PDUs, and optionally the size of the control PDU when two logical channels are configured. The AM RLC generates the appropriate control PDUs from the scheduling and transport format information provided by the MAC, and then submits these PDUs to the MAC.

7.4.5 Acknowledged Mode Reception

7.4.5.1 *AMD PDU Reception*   Upon receipt of an AMD PDU from the MAC, the AM-RX entity extracts the *Sequence Number* SN, length indicators and the poll bit. If the poll bit is set, a STATUS_REPORT is triggered. It then performs integrity verification of the received AMD PDU. If an AMD PDU contains an invalid or reserved length indicator or an invalid piggybacked STATUS PDU header, the verification fails and the AMD PDU is discarded.

If the AMD PDU has valid fields, is not a duplicate PDU and the sequence number satisfies VR(R) ≤ SN ≤ VR(MR), the piggybacked STATUS PDU, if present, is extracted and the AMD PDU is buffered at a position given by its sequence number until reassembly is possible. Otherwise, the AMD PDU is discarded. If, from the received AMD PDU sequence number, missing PDUs are newly detected, and MISSING PDU status trigger has been configured, a STATUS_REPORT transmission is triggered.

After updating the state variables, the AM-RX entity further attempts a reassembly if the received AMD PDU contains one or more length indicator. However, if in-sequence SDU delivery is configured and SN ≠ VR(R), the reassembly is deferred. Reassembled SDUs are delivered to the upper layer through the AM-SAP.

7.4.6 Control PDU Reception

The STATUS or piggybacked STATUS PDUs are processed depending on the SUFI contained within the received PDU.

7.4.6.1 MRW   This SUFI contains a LENGTH, a list of (LENGTH + 1) sequence numbers and an *Nlength* field. If the LENGTH field is zero and SN[0] > VR[R], all the buffered PDUs in the Rx RLC entity are discarded and the reassembly buffer is emptied. If the LENGTH field is not zero then segments of each SDU ending in a PDU whose sequence number is listed are discarded. If the PDU with SN[LENGTH] sequence number has already been received, then Nlength octets contained in it can be discarded immediately. After discarding the required data, the receiver window is moved to SN[LENGTH] and an acknowledgement for the MRW command is scheduled, sending a STATUS PDU containing MRW_ACK SUFI.

7.4.6.2 MRW_ACK   This SUFI contains an *N* and an SN_ACK field. The *N* field is equal to the Nlength field in the corresponding MRW SUFI, and The SN_ACK field indicates the updated value of VR(R) after the reception of the MRW SUFI. These two fields are used together to ensure that the MRW–ACK is not obsolete and that the associated MRW command has been correctly executed. Reception of a valid MRW_ACK SUFI terminates the SDU discard with an explicit signalling procedure.

7.4.6.3 WINDOW   This SUFI contains a *Window Size Number* (WSN) field after the SUFI header and is used to control the window size of the transmitting peer entity. It is worth noting that the transmitter window size (or VT(WS)) can only be changed between a minimum and maximum value configured for the AM-TX entity by the upper layer. A WINDOW SUFI has no effect if the WSN field is set to an invalid value.

7.4.6.4 LIST   This SUFI contains a LENGTH field and several (SN, L) pairs after the SUFI header. The SN field identifies the beginning of an error gap, while the associated

L field identifies the number of consecutively missing PDUs following that AMD PDU. Hence, a LIST SUFI can be used to convey error patterns having several small burst errors. Identified missing PDUs are scheduled for retransmission.

7.4.6.5 *BITMAP*   This SUFI contains a LENGTH field, a *First Sequence Number* (FSN) field identifying the sequence number of the PDU for the first bit in the bitmap, and a bitmap identifying the status (received/ not received) for each consecutive PDU, starting from FSN. Upon reception of this SUFI, PDUs identified as not received are scheduled for retransmission. A BITMAP SUFI is most suitable to encode random errors.

7.4.6.6 *RLIST*   This SUFI contains, after the SUFI header, a list length field (LENGTH), the FSN for the first erroneous AMD PDU in the RLIST and a list of LENGTH number of *Code Words* (CW). The number identified by the set of CWs represents the distance between the previous indicated erroneous AMD PDU up to and including the next erroneous AMD PDU. A special value of CW ($= 0001$) is used to indicate that the following concatenation of CWs represents the length of the error burst. Hence, the RLIST SUFI encodes blocks of consecutive correctly received PDUs, followed by an error burst length. Such a SUFI is most suitable for encoding sparse or large burst errors. Upon reception of this SUFI, PDUs identified as not received are scheduled for retransmission.

7.4.6.7 *ACK*   This SUFI contains a *Last Sequence Number* (LSN) field after the SUFI header. This SUFI is included as the last SUFI of the STATUS PDU and acknowledges the reception of all AMD PDUs with sequence number SN $\leq$ LSN and not identified as missing in the earlier part of the STATUS PDU. It is used to update the state variable VT(A), providing the SN value of the first erroneous PDU is greater than LSN.

7.4.6.8 *NO_MORE*   This SUFI indicates the end of the STATUS PDU. All octets following this SUFI are interpreted as being padding octets.

7.4.7 Error Correction

AM entities use an error control function to provide reliable data link services over an error prone transmission medium. The AM error control function is based on a combination of *Automatic Repeat reQuest* (ARQ) used in conjunction with *Forward Error Correction* (FEC), commonly referred to as Hybrid-ARQ. A series of actions involving both AM RLC and the physical layer is performed in order to provide link layer reliable services, which are described below:

- **Channel coding.** The physical layer introduces redundancy in the outgoing RLC PDUs, encoded within transport blocks, using the FEC mechanism, which allows the receiver to correct transmission errors up to a certain extent.

- **Error detection.** The physical layer introduces a *Cyclic Redundancy Check* (CRC) for outgoing transport blocks containing RLC PDUs. It enables the receiving physical layer to detect residual errors in the transport block, and indicate the status of each block to MAC. Such error status is forwarded to RLC along with an RLC PDU.

- **Integrity verification.** Upon receipt of an RLC PDU, an integrity verification is performed over the RLC header in order to detect any residual errors which may have affected the RLC header. The integrity verification involves the sequence number check, length indicators integrity check, reserved bits integrity check, super field format integrity, absence of conflicting status, etc.

- **ARQ.** The AMD PDUs which are discovered to contain errors sare discarded within the AM/ UM entity. In the case of AM, a STATUS_REPORT is sent to the peer asking it to retransmit the erroneous AMD PDUs.

7.4.8 Error Recovery

Peer AM RLC entities have their transmitted and received sequence numbers properly synchronized during normal operation. The sequence numbers, however, can lose synchronization due to residual errors in data or control PDUs. These errors manifest themselves in various forms at the transmitter, such as, for example, excessive repetition for certain PDUs. In order to recover from the effect of residual or other protocol errors, AM mode uses a resynchronization procedure known as the reset procedure. The reset procedure helps AM entities to resynchronize send and receive state variables as well as ciphering parameters.

When a reset condition is satisfied, the AM entity (master) detecting the error sends a RESET PDU to the peer. Upon receipt of the RESET PDU, the peer entity reinitializes the transmit and receive state variables and responds by sending a RESET_ACK PDU on the return channel. The reset procedure is completed when the master AM entity receives a RESET_ACK PDU for the RESET PDU.

7.4.9 Automatic Repeat Request Mechanism

AMD PDUs are buffered in the transmitter until they are acknowledged. If the acknowledgements are delayed, the upper edge of the transmission window may be reached, preventing further transmission. This results in a reduced throughput at the RLC level. The RLC provides mechanisms within the transmit and receive entity to trigger timely transmission or retransmission of a STATUS_REPORT.

7.4.9.1 Polling The transmitter driven STATUS_REPORT triggering is known as polling, where the poll bit within the AMD PDU header is set to request the receiver to transmit a STATUS_REPORT on the return channel. The frequency of polling depends on various triggers that can be configured by the higher layer:

Timer-based. The AM-TX entity polls the receiver periodically soliciting a STATUS_REPORT. The value of the timer is configured by the higher layer entity.

Window-based. The AM-TX entity triggers a poll when a percentage of the transmission window is equal to or exceeds the Poll_Window parameter configured by the higher layer entity.

PDU frequency-based. The AM-TX entity polls the receiver after having transmitted a specified number of PDUs, defined as Poll_PDU and configured by the higher layer entity.

SDU frequency-based. This scheme is similar to the previous one, except that it uses SDU-based counting. The poll is set in the PDU containing the last segment of the SDU target for polling. The polling frequency is defined as Poll_SDU and is configured by the higher layer entity.

Last PDU in transmission queue. The AM-TX entity triggers a poll in the last PDU transmitted from its transmission queue.

Last PDU in retransmission queue. The AM-TX entity triggers a poll in the last PDU retransmitted from its retransmission queue.

In addition to these triggers, two mechanisms can be used to optimize the polling performance:

Timer poll prohibit. This mechanism is used to prevent multiple poll requests being sent in a short span of time.

Timer poll. This mechanism is used to overcome the loss of a poll request upon the loss of an AMD PDU containing the poll bit set. If the timer expires while no corresponding STATUS PDU has been received, the receiver is polled once more.

7.4.9.2 Status Reporting

The acknowledged mode also provides receiver driven status report triggering, where the receiver either discovers the need for a STATUS_REPORT or just transmits it perodically:

Missing PDU. The AM-RX entity triggers the generation of a STATUS_REPORT when missing PDUs are detected. The missing PDU event is detected by keeping track of gaps in the receive window.

Timer-based. The AM-RX entity triggers periodically the generation of a STATUS_REPORT. Timer-based status reports also serve as 'keep a life' mechanisms.

In addition to these triggers, two mechanisms can be used to optimize the STATUS_REPORT transmissions and retransmissions:

Status prohibit. This mechanism is used to prevent the transmission of multiple status reports in a short span of time resulting from the arrival of too many polls or the detection of too frequent missing PDU events, or a combination of both.

EPC-based retransmission. This mechanism is used to retransmit a status report which may have been lost. A retransmission timer based on the average round trip delay may be used to trigger the retransmission of the status report if the requested PDUs don't arrive before the timer expires. $VR(EP)$ is then used to adapt the retransmission timer to the current data rate.

8

PDCP

Hardik Parekh

8.1 INTRODUCTION

This section outlines the responsibilities and operation of the PDCP sublayer. The *Packet Data Convergence Protocol* (PDCP), which is defined for *Packet Switched* (PS) domain only, sits above the *Radio Link Control* (RLC) in the *Access Stratum* (AS) of the protocol stack, see Figure 8.1.

The user plane is the vertical part of the protocol stack that deals with the user data (both the *Circuit Switched* (CS) and the packet switched data). The network layer or transport layer protocols of various packet data applications use the access stratum for the transmission of user data. The PDCP is the layer where various network layer, transport layer or upper layer protocols like TCP/IP or RTP/UDP/IP converge, see Figure 8.2. The PDCP converts all these packet data network packets into RLC *Service Data Units* (SDUs) and the RLC SDUs into packet network packets at the transmitter and the receiver side respectively.

Also, being a part of the PS domain in the user plane, highly reliable data transfer is expected from the applications, e.g. for data streaming applications. Though RLC provides reliable services for data transmission, reliability is not guaranteed during the *Serving Radio Network Subsystem* (SRNS) relocation. For this, PDCP maintains PDCP sequence numbers to avoid any data loss during an SRNS relocation. The basic functionality of the PDCP is:

1. Transfer of packet user data;

2. To perform header compression and decompression of IP data streams for the transmit and the receive entity respectively;

3. Maintain the sequence numbers for *Radio Bearers* (RBs) which are configured to support lossless SRNS relocation.

WCDMA – Requirements and Practical Design. Edited by R. Tanner and J. Woodard.
© 2004 John Wiley & Sons, Ltd. ISBN: 0-470-86177-0.

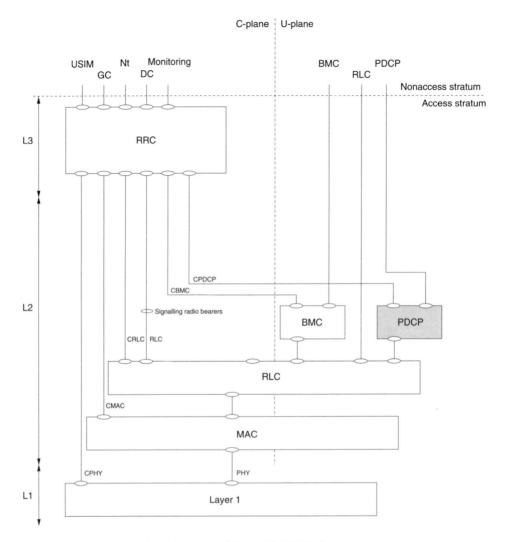

Figure 8.1. The protocol architecture with the PDCP block.

All relevant 3GPP specifications are listed below.

- TS25.323 [145] specifies the PDCP;
- TS25.331 [141] specifies the *Radio Resource Control* (RRC) protocol;
- TS25.322 [140] specifies the RLC protocol;
- TS25.303 [146] specifies the interlayer procedures in connected mode;
- RFC 2507 [147] specifies the IP header compression;
- RFC 3095 [148] specifies the *Robust Header Compression* (ROHC).

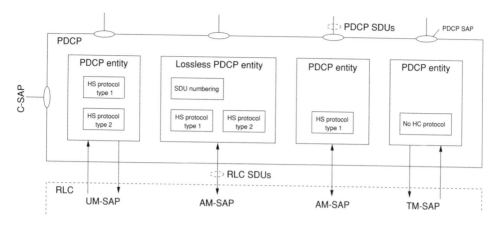

Figure 8.2. PDCP detailed architecture.

8.2 OVERALL ARCHITECTURE

Each of the PS domain *Radio Access Bearers* (RABs) is associated with one RB. This RB is in turn associated with one PDCP entity. Each PDCP entity is then associated with one or two RLC entities depending on the RLC mode. A PDCP entity can be associated with two *Unacknowledged Mode/Transparent Mode* (UM/TM) RLC entities (one of them in the uplink and the other in the downlink) or one *Acknowledged Mode* (AM) RLC entity. The PDCP entity can be configured to use none, one or several header compression algorithms[1].

Considering the PDCP functionality, the PDCP layer can be logically divided into mainly two basic submodules for the implementation:

Control plane part of the PDCP. Basically, the part of the PDCP which deals with the *PDCP-Control-Service Access Point* (C PDCP SAP). This part of the PDCP communicates with the *Radio Resource Control* (RRC) layer [141] through the C PDCP SAP and is responsible for the creation, reconfiguration or deletion of the PDCP entities. The RRC configures PDCP entities with all the required parameters for the header compression parameters and associates the PDCP entity with RABs and RBs.

User plane part of the PDCP. Receives uplink data from RBs on PDCP SAPs. In the uplink, the headers of the IP data packets are compressed (subject to PDCP configuration) and the *Protocol Data Units* (PDUs) are then forwarded to the respective RLC entity through RLC *Service Access Points* (SAPs). In the downlink, the PDCP receives the compressed header data from the RLC through the RLC SAP, which is then decompressed (subject to PDCP configuration) and forwarded to the upper layer through PDCP SAPs. Below is a list of the functions contained in this submodule:

---

[1] Release 99 supports only the RFC 2507 header compression algorithm.

- transfer of the packet data through PDCP and RLC SAPs;
- a PDCP entity could be configured to use any header compression algorithm. For each type of header compression, the PDCP will maintain a header compression space where all the compressor and decompressor context is stored;
- if a PDCP is configured for SRNS relocation, then the PDCP needs to maintain an SDU sequence numbering submodule.

Thus the PDCP submodule maintains various PDCP entities according to what the entities are configured for. Note, however, that firstly, the *Transparent Mode* (TM) RLC SDU should be an integer multiple of a valid TM PDU size. The RFC 2507 header compression type does not result in a fixed size of compressed data after header compression is performed. So, a PDCP entity which is configured to use the RFC 2507 header compression type cannot be associated with the TM RLC. Secondly, if a PDCP entity is configured to use the RFC 3095 header compression type with segmentation option, then it can be associated with the TM RLC.

8.3 PDCP INTERFACE

The PDCP interface with other layers can be divided into two parts according to the SAP used for the communication, *Control plane part of the PDCP* (CPDCP) and *User plane part of the PDCP* (UPDCP)[2].

8.3.1 CPDCP

This part interacts via the C-PDCP-SAP with the RRC layer and comprises four control plane primitives (messages), CPDCP_CAPABILITY_REQ, CPDCP_CONFIG_REQ, CPDCP_RELEASE_REQ and CPDCP_RELOC_REQ. Note that a primitive is a signal that carries information across interfaces. Next we discuss each primitive in turn.

8.3.1.1 CPDCP_CAPABILITY_REQ This primitive is actually not a 3GPP defined message but is used by the RRC to communicate the capability of the PDCP, see Figure 8.3.

The PDCP may support various header compression types and also could have limited capabilities for each header compression type. While replying to this primitive, PDCP includes all its capabilities (header compression types supported, parameters of header compressions supported, whether lossless SRNS relocation is supported or not), which is then forwarded to UTRAN by the RRC.

8.3.1.2 CPDCP_CONFIG_REQ This primitive is used by the RRC to configure or reconfigure a PDCP entity. The RRC configures/reconfigures the PDCP entity if one of the following cases is true [145]:

- when the radio bearer control message requests a configuration (creation) or reconfiguration of the PS RAB. The reconfiguration is used to indicate the

---

[2]CPDCP and UPDCP are not defined by 3GPP but are used here to simplify the discussion of the PDCP interface.

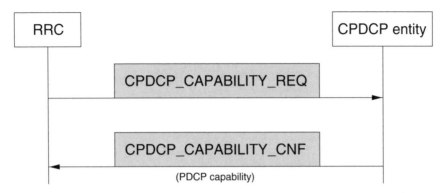

Figure 8.3. The CPDCP_CAPABILITY_REQ primitive.

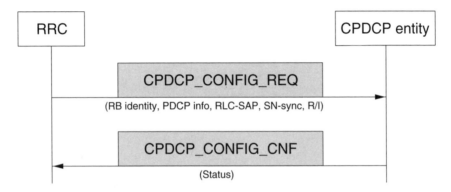

Figure 8.4. The CPDCP_CONFIG_REQ primitive.

reconfiguration of the PDCP entity along with the reconfigurations of header compression parameters;

- when the SRNS requests that all header compression algorithms should be reinitialized. At this point, a flag 'R/I' is set to a value which indicates that the header compression algorithm needs to be reinitialized;

- when an RLC entity, which is mapped to a lossless PDCP entity, is reset or reestablished not because of an SRNS relocation, then the RRC sends this primitive. The 'SN-sync' parameter is set such that it indicates the lossless PDCP entity to start a PDCP sequence number synchronization procedure.

The parameters shown in Figure 8.4 are interpreted as follows [145]:

RB identity denotes the connection between an RB and a PDCP entity.

PDCP info is received in the reconfiguration message that establishes or reconfigures a RAB. This message includes the header compression information such as the type of header compression algorithms to be used, parameters of the header compression type, i.e. whether lossless SRNS relocation has to be supported or not.

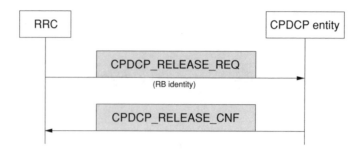

Figure 8.5. The CPDCP_RELEASE_REQ primitive.

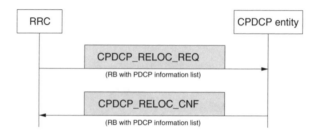

Figure 8.6. The CPDCP_RELOC_REQ primitive.

RLC-SAP denotes the type of RLC SAP to which the PDCP entity is associated, e.g. RLC-TM, RLC-UM or RLC-AM.

SN-sync indicates that the PDCP should start a PDCP sequence number synchronization procedure. SN-sync is set when the RLC is reset or reestablished and not because of an SRNS relocation.

R/I indicates when an SRNS relocation is performed. The RRC informs the PDCP to reinitialize all *Header Compression* (HC) algorithms with this parameter. Note that all the PDCP entities are informed to reinitialize the HC algorithm regardless of whether it is a lossless PDCP entity or not.

8.3.1.3 CPDCP_RELEASE_REQ The RRC sends this primitive to release a PDCP entity associated with an RB mapped to a PS RAB that is to be released, see Figure 8.5. The parameter is the identity flag:

RB identity declares which PDCP needs to be released.

8.3.1.4 CPDCP_RELOC_REQ This primitive triggers the SRNS relocation in lossless PDCP entities. The RRC sends this primitive upon the receipt of a peer RRC message that indicates an SRNS relocation (request). In response to this message, the PDCP sends a CPDCP_RELOC_CNF message to the RRC. The key parameter, also shown in Figure 8.6, means:

Figure 8.7. The PDCP_DATA_REQ primitive.

RB with PDCP info includes the next expected *Uplink* (UL) receive sequence number maintained by the PDCP at UTRAN for all lossless PDCP entities. The PDCP responds to the RRC with a CPDCP_RELOC_CNF message, which includes the locally maintained next expected DL receive sequence number for each lossless PDCP entity. The RRC will then send it back to the peer RRC in the completed message.

8.3.2 UPDCP

This part interacts through the RLC SAP (RLC-TM, RLC-UM or RLC-AM SAP) and PDCP-RABM-SAP with the RLC layer and the *Radio Access Bearer Management* (RABM) respectively.

8.3.3 PDCP_DATA_REQ

The data processing is outlined below upon reception of a PDCP_DATA_REQ at the PDCP SAP.

- If the PDCP is configured to use a 'NO PDCP header' (the PDCP entity is configured by default for neither HC nor to support lossless SRNS relocation), the PDCP will not attach any header before the data and send a 'PDCP-No-Header' PDU via the RLC SAP.

- If the PDCP entity is configured to use a PDCP header (PDCP entity is configured to use HC and not configured to support lossless SRNS relocation), then the PDCP will compress the received SDU and then fill the compressed data into a PDCP_DATA_PDU and pass it on to the RLC SAP.

- If the PDCP entity is configured to use a PDCP header, then the PDCP will compress or not compress the received SDU subject to the HC configuration and send a PDCP_DATA_PDU. If the sequence number synchronization procedure is triggered by the RRC, then the data shall be sent along with the 'send sequence number' parameter of the particular PDCP entity in a PDCP_SN_PDU message. A PDU type field in the beginning of the PDCP header indicates whether it is a PDCP_DATA_PDU or a PDCP_SN_PDU message.

When the RLC provides data to the PDCP, then the same process is applied but in reverse order.

8.4 HEADER COMPRESSION

In any network/transport layer protocols, the consecutive headers belonging to the same packet streams have a significant amount of redundancy. The header size can be significantly compressed by sending the static field information first, e.g. destination address, source address, and then later sending the full field followed by the delta transmission for the field and utilizing the dependencies and predictability of the other fields.

The *Header Compression* (HC) is specific to a particular network layer, transport layer or upper layer protocol combination, e.g. TCP/IP, RTP/UDP/IP. When a *Packet Data Protocol* (PDP) context is activated, the network layer protocol, e.g. IP or *Point-to-Point Protocol* (PPP), is known during this activation process. The PDCP configuration is defined by UTRAN and states whether or not to use HC, type of the HC and the parameters and the support of lossless SRNS relocation, based on the following requirements:

- quality of service requirement for the particular application. UTRAN decides whether to use the HC or not for a particular reliability and bit rate, also whether to support lossless SRNS relocation or not;

- link quality, e.g. poor link, low data rate, is also a parameter used to decide the HC type and its parameters;

- PDCP capability, which is indicated to the network by the UE capability information message during the RRC connection setup procedure in the RRC connection setup complete message. That includes all types of HC algorithm supported by the PDCP, the limitations of the parameters supported by the PDCP for each compression algorithm, whether SRNS relocation is supported by the UE or not.

The PDCP configuration is then sent to the UE in a radio bearer setup message which actually creates a PS RAB that also includes the configuration of the associated PDCP entity. The RRC then configures the PDCP layer according to the content of the received configuration message.

8.4.1 Packet Identifier Value PID

It is possible to have a single PDCP entity configured for multiple HC types. Each HC type will have various types of packet, and when a packet is received by the PDCP, it has to process it according to [145]:

- the type of HC protocol, e.g. RFC 2507, RFC 3095;

- the type of packet within that particular HC protocol.

As a result of this requirement, a new field called 'PID' is added in the PDCP PDU. Each PDCP maintains a *Packet Identified* (PID) table which is based on the configuration of the entity, see Table 8.1. The rules which apply for the PID are:

- each PDCP entity should maintain a separate PID table and each PDCP entity could have different packet types mapped to different PID values;

- PID value '0' indicates that the packet contained in the data field of the PDCP PDU is unchanged by the sender and it should not be decompressed by the receiver;

Table 8.1 PID values for the RFC 2507 table.

PID value	Compression type	Packet type
0	No header compression	Regular header
1	RFC 2507	Full header
2	RFC 2507	Compressed TCP
3	RFC 2507	Compressed TCP no delta
4	RFC 2507	Compressed nonTCP
5	RFC 2507	Context state
6	HC Type 1	Packet type 1 of 'HC type 1'
7	HC Type 1	Packet type 2 of 'HC type 1'
8	HC Type 1	Packet type 3 of 'HC type 1'
9	HC Type 2	Packet type 1 of 'HC type 2'

Table 8.2 PID values for the RFC 2507 table.

PID value	Packet type
$n+1$	Full header
$n+2$	Compressed TCP
$n+3$	Compressed TCP no delta
$n+4$	Compressed nonTCP
$n+5$	Context state

- PID values are assigned in ascending order starting from one. The first available PID value will be assigned to the first packet type of that particular algorithm;

- PID values can always be mapped to some other packet types after the reconfiguration of the header compression protocols for a PDCP entity.

8.4.2 IP Header Compression (RFC 2507)

This is discussed in detail in reference [147]. The PID values are fixed for this type of header compression. In 3GPP release 99, only one type of HC algorithm is supported per entity, which is the RFC 2507. The PID values for this algorithm will always have values from 1 to 5 as described in Table 8.1. Release 4 and higher also support RFC 3095 (robust header compression) [148]. The PID values for RFC 2507 are listed in Table 8.2: Value n denotes the number of PID values already mapped to other protocol packet types. The following aspects are important for the implementation.

- RFC 2507 states that the length of the data should be indicated by the link level protocols. In the UMTS stack, the length of the received packet is indicated by the RLC as the PDU length.

- RFC 2507 states that the type of packet being sent should again be identified by another mechanism. We already have a PID that will identify the type of packet being transmitted.

Table 8.3 The PID values for
RFC 3095A.

PID value	Packet type
$n + 1$	CID1
$n + 2$	CID2
...	...
...	...
$n + x$	CIDx

Table 8.4 The PID values for RFC 3095B.

PID value	Packet type
$n + 1$	RFC 3095 packet format

- RFC 2507 specifies certain error recovering algorithms that should be implemented by the algorithm.

- The expected reordering in the *Information Element* (IE) PDCP information field indicates whether a packet reordering shall be performed or not.

- Always perform a TCP checksum after the decompression of the TCP/IP header and then compare it with the checksum included in the received compressed packet.

8.4.3 Robust Header Compression (RFC 3095)

This HC type is discussed in detail in IETF RFC 3095. Relevant information from past packets of a packet data stream is maintained in a context. Each context is uniquely identified by a *Context Identifier* (CID). If these values are used then:

- if included in the PDCP PDU header itself, the PID values will be directly mapped to the CIDs; or

- included in RFC 3095 packet format, and the PIDs will indicate the packet format.

Which of the above methods is to be used is indicated by higher layers [141, 145].

8.4.3.1 *CIDs Included in the PDCP PDU Header* PID values for the RFC 3095 directly map to the CID values used by RFC 3095. In this case, CIDs shall not be introduced in the RFC 3095 packet format by PDCP. The maximum CID value is provided by higher layers.

8.4.3.2 *CIDs Used Within Robust Header Compression (ROHC) Packet Format* The PID values for the RFC 3095A map to the RFC 3095B packet format, and the CIDs shall not be introduced in the PDCP PDU header by PDCP [145]. Term n in Table 8.4 denotes the number of PID values already assigned to other protocol packet types.

8.5 SRNS RELOCATION

There exists an Iu-PS connection between the SRNC and the *Serving GPRS Support Node* (SGSN) during a PS call. This is also true when a UE is moved to a *Drift Radio Network Controller* (DRNC), because there exists an Iur connection between the SRNC and DRNC. When the load over the Iu-PS or Iur connection increases, then it is decided to relocate to the SRNC. During this SRNC relocation process, the following steps are carried out:

- new AM/UM RLC entities are created at the target SRNS. To have both peer entities, UTRAN and UE, in sync all the AM/UM RLC entities at the UE are also reestablished;

- all the AM/UM RLC entities at the source SRNS are deleted after the successful relocation;

- at the target SRNS, all the PDCP entities are created with the information provided by the source SRNS. The source SRNS sends the following information to the target SRNS:
  - UL receive PDCP sequence numbers;
  - DL send PDCP sequence number;
  - transmitted but not yet acknowledged PDCP SDUs with their respective DLs send PDCP sequence numbers;
  - nontransmitted PDCP SDUs.
  Although the newly created PDCP entities at the target SRNS are transferred, as explained above, the header compression context is not transferred. So it is necessary to reinitialize the header compression algorithm for all the PDCP entities at the UE side while maintaining the *Sequence Number* (SN) information to keep both UE and UTRAN PDCP entities in sync;

- the source SRNS sends the UL receive sequence number to the UE PDCP entities via an upper layer (RRC) message, whereas the PDCP entity at the UE indicates the DL receive sequence numbers to the target SRNS via an upper layer (RRC) message;

- the PDCP entities at the source SRNS are deleted after the successful relocation.

8.5.1 Interpretation of Lossless SRNS Relocation

If the RLC mapped to the PDCP entity is in the AM mode, then there should not be any data loss. However it is possible that some data loss may occur during the SRNS relocation process mentioned above, since the RLC entities are reestablished and established at the UE and UTRAN respectively. Data loss is undesirable for many applications in packet switched mode and it is the PDCP's responsibility to make the SRNS relocation lossless. Those PDCP entities that are serving such an application are configured to maintain sequence numbers. These are exchanged at the time of SRNS relocation and help make the relocation lossless. These PDCP entities are called the lossless PDCP entities [145].

8.5.2 Lossless SRNS Relocation

All the lossless PDCP entities should be mapped to the acknowledge mode, with sequential delivery for the downlink RLC entities only. As mentioned above, where the UE capability is transmitted to UTRAN, it is known by the UE PDCP if the lossless SRNS relocation is supported or not. The PDCP entities are configured by upper layers to support *Lossless SRNS Relocation* (LSR). All lossless PDCP entities keep sequence numbers which actually confirm the PDCP SDUs which are transmitted, but the confirmation of the faithful reception for these SDUs has not yet been received from the lower layer. The data transfer then begins with the unconfirmed PDCP SDU [145].

8.5.3 PDCP Sequence Numbering

The value of the PDCP sequence number ranges from 0 to 65 535. These sequence numbers are set to '0' when the PDCP entity is set up for the first time. The sequence numbers are always incremented and never decremented. Basically, the peer PDCP maintains four *Sequence Numbers* (SN) in total, two sequence numbers at the UE side and two at the SRNC side [145].

- At the UE, the following sequence numbers are maintained for each PDCP entity:
  - UL transmit PDCP SN shall be set to zero for the first PDCP SDU submitted to the lower layer and should be incremented by one for the next PDCP SDU submitted to the lower layer;
  - DL receive PDCP SN shall be set to zero for the first PDCP SDU received from the lower layer and should be incremented by one for the next PDCP SDU received from the lower layer.

- At the UTRAN, the following sequence numbers are maintained for each PDCP entity:
  - DL transmit PDCP SN shall be set to zero for the first PDCP SDU submitted to the lower layer and should be incremented by one for the next PDCP SDU submitted to the lower layer;
  - UL receive PDCP SN shall be set to zero for the first PDCP SDU received from the lower layer and should be incremented by one for the next PDCP SDU received from the lower layer.

8.5.4 SDU Management in a Lossless PDCP Entity

Only a lossless PDCP entity stores the received SDUs until the confirmation of reception is received. When the AM RLC confirms the reception of the SDU by the peer entity, the stored SDU is discarded.

The 'PDCP SN window size' parameter indicates the maximum number of transmitted SDUs, which are not confirmed to have been successfully transmitted to the peer entity by the RLC. The PDCP cannot have more than 'PDCP SN window size' SDUs transmitted but unconfirmed. At the time of SRNS relocation, the next expected receive sequence numbers for the peer lossless PDCP entity are received, which are routed through to upper layers. The PDCP then discards all the transmitted but unconfirmed PDCP SDUs with sequence numbers less than the 'next expected DL/UL sequence number' sent by the peer entity. After the successful relocation, the data transmission starts from the (first) unconfirmed SDU having a sequence number equal to the next expected sequence number by the PDCP entity.

8.5.5 Sequence Number Exchange During a Relocation

After an SRNS relocation has occurred, the network sends the 'UL receive' sequence numbers for all the lossless RBs in an RRC message, which are then passed on to the PDCP entity. Then the PDCP takes the following actions.

- All the PDCP SDUs with a sequence number up to the 'UL receive' sequence numbers are discarded. In this way, all the transmitted but not yet acknowledged SDUs, which are actually received by the peer PDCP entity, are acknowledged.

- If the received 'UL receive' sequence number is less than the 'UL send' sequence number of the first transmitted but not acknowledged SDU, or greater than the sequence number of the unsent SDU, then it is interpreted as an invalid sequence number, and the 'PDCP sequence number synchronization' procedure is triggered.

- After the relocation, if some transmitted SDUs are still left unacknowledged, the data transmission is resumed by retransmission of the SDU with the 'UL send' sequence number equal to the 'UL receive' sequence number. Otherwise, the data transmission is resumed with the transmission of the first unsent SDU.

8.5.6 PDCP SN Synchronization Procedure

When an RLC AM entity is reset or reestablished, it does not assure a reliable transmission of the data. At the time of an SRNS relocation, the data is made lossless through the aforementioned process. However, when the RLC entities mapped to a lossless PDCP entity are reset or reestablished for reasons other than SRNS relocation, then it is the PDCP's responsibility that the peer lossless PDCP entities do not go out of synchronization. For this reason, the PDCP has to trigger a 'PDCP sequence number synchronization procedure'. A PDCP sequence number synchronization procedure is triggered when either of these cases is true:

- the PDCP is given an invalid UL/DL receive sequence number during an SRNS relocation;

- the RLC entity mapped to a lossless PDCP entity is reset or reestablished but not because of an SRNS relocation.

When a PDCP sequence number synchronization is triggered, the PDCP will submit a 'PDCP sequence number' PDU to the lower layer. When the confirmation of the successful transmission of this PDU is received from the lower layer, this procedure terminates. If the UE PDCP entity receives a PDCP sequence number PDU, it sets the value of the 'DL receive' PDCP sequence number to the value indicated in the PDCP sequence number PDU. When the UTRAN PDCP entity receives a PDCP sequence number PDU it sets the 'UL receive' PDCP sequence number to the value indicated in the PDCP sequence number PDU [145].

8.6 PDCP HEADER FORMATS

PDCP PDUs should be multiples of eight bits when the RLC entity mapped to then is either in AM mode or UM mode. If the RLC entity is in the TM mode, then it can be bit aligned.

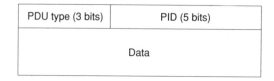

Figure 8.8. The PDCP data PDU format.

There are three types of PDCP PDU [145], PDCP no header PDU, PDCP_DATA PDU and PDCP sequence number PDU.

8.6.1 PDCP no Header PDU

This PDU, a single block of data, is used when the PDCP is configured to not attach any PDCP header before the received SDU. The receiver entity should also not process this PDU but directly forward it to the higher layer. The PDCP is configured to use this format when it serves an application that does not require header compression or lossless data transmission.

8.6.2 PDCP_DATA PDU

The data PDU is used when the PDCP is expected to apply header compression but no lossless data transmission is expected. The following information can be transferred in this PDU:

- uncompressed PDCP SDUs, e.g. regular header packets;

- header compression related control signalling, e.g. RFC 2507 packets like CONTEXT_STATE;

- compressed PDCP SDUs (RFC 2507/RFC 3095 packet types, e.g. compressed TCP).

The format is shown in Figure 8.8. The format includes a three bit field PDU type and a five bit field PID value plus the actual data block [145].

8.6.3 PDCP Sequence Number PDU

The PDCP sequence number PDU is used by a lossless PDCP entity to convey the PDCP sequence number to the peer entity. This PDU is mainly used when the sequence number synchronization is lost or expected to be lost between the transmit and receive entities [145]. The PDCP data PDU contains:

- PDCP sequence numbers;

- uncompressed PDCP SDUs;

- compressed PDCP SDUs.

Note that header compression related control signalling, e.g. RFC 2507 packets like CONTEXT_STATE, is not transmitted through this kind of packet since the control signalling information for header compression algorithms is generated dynamically at any

Table 8.5   The three bit PDU type field.

Bit	PDU type
000	PDCP data PDU
001	PDCP sequence number PDU
010–111	Reserved

Table 8.6   The five bit PID field.

Bit	Description
00000	No header compression
00001–11111	Dynamically negotiated header compression identifier

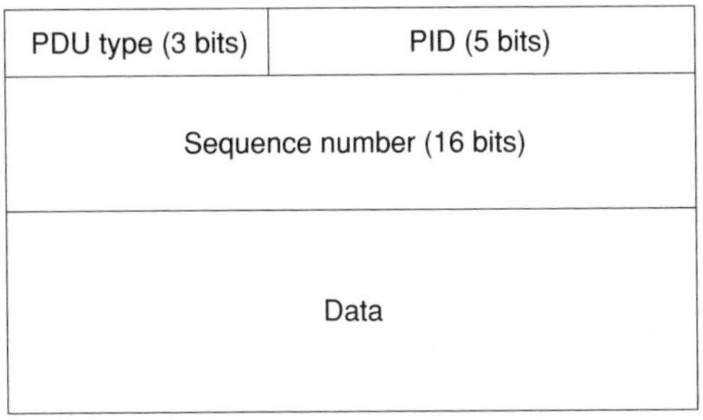

PDU type (3 bits)	PID (5 bits)
Sequence number (16 bits)	
Data	

Figure 8.9. The PDCP sequence number PDU.

time and not assigned any sequence number, e.g. CONTEXT_STATE. The control signalling information for the header compression algorithm cannot be included in a sequence number PDCP PDU. The PDU format is shown in Figure 8.9.

8.6.4 PDCP Header Fields

There are four PDCP header fields:

PDU type. This field indicates the PDCP data PDU type and consists of three bits, see Table 8.5

PID. This five bit field indicates the used header compression and packet type, or a context identifier, see Table 8.6 The receiving PDCP entity processes the received PDU after identifying the packet type.

Data. Data includes uncompressed PDUs, compressed PDUs or header compression feedback control information. Note that the data field does not include header compression control signalling for PDCP SDUs when the PDU type is a 'SeqNum PDU'.

Sequence number. The sequence number is a 16 bit field and has a range from 0 to 65 535.

8.7 HANDLING AN INVALID PDU TYPE AND PID

All received PDUs with a PDU type set to reserved bit values 010 to 111 should be discarded. Similarly, if a PDCP entity is received that is not configured to support lossless SRNS relocation, then it should be discarded once it receives a PDCP sequence number PDU. The PDU shall also be discarded when a PDCP entity receives a PDCP PDU with a PID value that is not mapped to a valid HC packet type [145].

9

Broadcast/Multicast Control

Joby Allen

9.1 INTRODUCTION

This chapter discusses the structure and operation of the BMC layer. The *Broadcast/Multicast Control* (BMC) layer sits above the RLC in Layer 2 of the *Access Stratum* (AS) and is only active in the user plane, which handles the user data (see Figure 9.1).

Operators may offer a variety of services to their subscribers, amongst others, for example, football results or the weather forecast. The user requests one type of information on his handset, for instance the latest football news, and the application will ensure that the information becomes available. If the user enables the football news, then the BMC will read the corresponding frame and forward the data to the application, to be processed and displayed. In essence, the BMC is responsible for receiving *Cell Broadcast Service* (CBS) traffic on the *Common Traffic Channel* (CTCH) logical channel routed via an *Unacknowledged Mode* (UM) RLC entity from a *Forward Access* (FACH) transport channel. The CBS traffic consists of the following:

CBS messages. Messages containing user data for broadcast to the *NonAccess Stratum* (NAS). The NAS configures the BMC with a list of the CBS messages it wishes to receive; only these messages are being sent to the NAS while others are discarded.

Schedule messages. Messages containing information regarding which CBS messages are broadcast in which radio frame. The BMC uses this information in conjunction with the list of messages requested by the NAS to determine which frames it needs to receive. This information is then forwarded to the *Radio Resource Controller* (RRC)

WCDMA – Requirements and Practical Design. Edited by R. Tanner and J. Woodard.
© 2004 John Wiley & Sons, Ltd. ISBN: 0-470-86177-0.

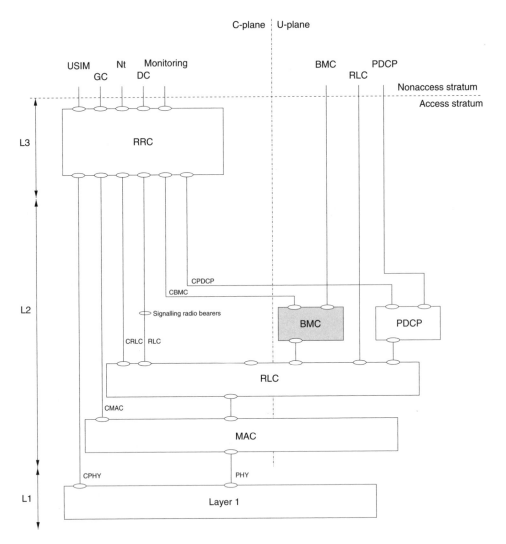

Figure 9.1. Physical location of the BMC in the access stratum.

which configures the physical layer for *Discontinuous Reception* (DRX) so that only the frames containing the CBS messages required by the BMC are received.

For this chapter, the following 3GPP specifications are most relevant:

- TS25.324 [149] defines the BMC;

- TS25.925 [150] defines the radio interface for broadcast and multicast services, and gives further information as to how the BMC is used;

- TS25.331 [141] specifies the RRC.

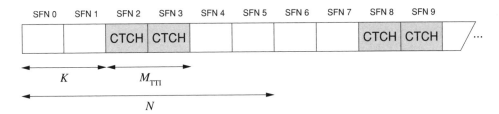

Figure 9.2. CTCH schedule example with $N = 6$, $K = 2$, $M_{TTI} = 2$.

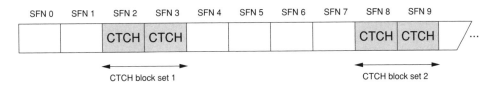

Figure 9.3. CTCH block sets.

9.2 CTCH SCHEDULING

CTCH traffic is scheduled on two levels, known as Level 1 and Level 2.

9.2.1 Level 1 Scheduling

CBS traffic is broadcast on the CTCH periodically, over the whole cell/sector, by the network. The frames that contain the CTCH data and the period at which CTCH data is broadcast are defined by the following parameters:

M_{TTI} is the number of frames carrying CTCH data each period;

N is the period in frames for which the CTCH must be allocated in order to ensure the reception of a complete set of CTCH data for that period. N must be equal to, or a multiple of, M_{TTI}.

K is the offset in frames of the first CTCH frame within the period of N frames. It must either be zero or a multiple of M_{TTI}.

The value of M_{TTI} is equivalent to the number of frames received each TTI on the FACH used. The values of N and K are defined in the system information *System Information Block* SIB5 received by the RRC, see also Chapter 10 or [141]. Figure 9.2 shows an example with $N = 6$, $k = 2$ and $M_{TTI} = 2$.

Each block of CTCH frames is called a CTCH block set, see Figure 9.3. The first CTCH block set starts at SFN K and is called block set 1, block set i starts at SFN $N(i-1) + K$.

9.2.2 Level 2 Scheduling

Level 2 scheduling is also known as in-band scheduling. A schedule message is transmitted to the BMC every Level 1 schedule period. The schedule message contains information

Table 9.1   Translation of a message type field.

Message type value	Type of message
1	CBS message
2	Schedule message

regarding which CBS messages are broadcast in which block set. This information allows the BMC to provide the RRC with a list of frames it wishes to receive so that the RRC may configure the physical layer for DRX in order to receive only the frames carrying the required messages. At other times the receiver is used to either receive other data or is inhibited to save power.

9.3 BMC OPERATION

On startup the BMC may be considered to be in an idle state, i.e. the list of CBS messages that the NAS has indicated it wishes to receive is empty, and the BMC receives no CBS traffic on the CTCH.

9.3.1 BMC Activation

The BMC is activated when it receives a BMC_ACTIVATE_REQ control command from the NAS. The message contains a list of the CBS messages that the NAS wishes to receive. On receipt of this message, the BMC takes the following actions:

1. Stores the list of requested messages;

2. Informs the RRC that the lower layers should be configured to receive the CTCH using Level 1 scheduling. This is done by sending a CBMC_RX_IND message to the RRC through the CBMC.

9.3.2 Message Handling

The BMC receives CTCH frames containing BMC messages. The first octet of a BMC message contains the message type field, which indicates the type of BMC message that has been received. The possible values of message type are defined in Table 9.1.

9.3.2.1 CBS Messages   The BMC maintains a list of the CBS messages requested by the NAS. This list is controlled by the NAS through use of the messages BMC_ACTIVATE_REQ and BMC_DEACTIVATE_REQ. The BMC_ACTIVATE_REQ is sent to add new messages to the list, while the BMC_DEACTIVATE_REQ is sent to remove CBS messages that are no longer required from the list. For each message, the message ID and the serial number of the last CBS message received with that ID (serial number) are stored in the BMC.

Message type (1 octet)	Message ID (2 octets)	Serial number (2 octets)	Data coding scheme (8 bits)	CBS data (n octets, where $0 < n <= 1246$)

Figure 9.4. Structure of a BMC CBS message.

The fields of the CBS message are shown in Figure 9.4 and are defined as follows [149]:

Message type identifies the type of BMC message and is always 1 for a CBS message;

Message ID is the identifier of the CBS message;

Serial number identifies the version (serial number) of the CBS message;

CBS data contains the actual user data, such as football results or news.

Upon receipt of a CBS message, the BMC extracts the message ID and the serial number. It then searches the requested message list to see if the NAS has requested that particular message ID. If the message ID exists in the list and the stored serial number is different to the most recently received serial number, the CBS message is forwarded to the NAS by means of a BMC_DATA_IND message. If the message ID is not stored in the table, then the NAS has not requested to receive that particular message and it will be discarded. Similarly, if the stored serial number matches the newly received serial number, the message is a repeat broadcast and is also discarded. This operation is illustrated in Figure 9.5.

9.3.2.2 Schedule Messages Schedule messages contain information regarding which CBS messages are broadcast over the air in which frames. Different applications may use a different frame. The structure of a schedule message is illustrated in Figure 9.6. The schedule message fields have the following meanings [149]:

Message type identifies the type of BMC message and is always 2 for a schedule message;

CTCH BS offset is the offset, relative to the CTCH block set index in which this schedule message was received, of the first CTCH block set index of the next schedule period;

CBS schedule period length is the number of CTCH block sets in the next scheduling period;

New message bitmap contains a bit representing each CTCH block set in the next schedule period, as illustrated in Figure 9.7. In Figure 9.7, i is the index of the first CTCH block set in the next CBS schedule period and j is the index of the last CTCH block set in the CBS schedule period. A bit value of 1 for a CTCH block set is used if at least one of the following conditions is true:

1. The CTCH block set contains a BMC message that was not sent in the last schedule period; or

2. It contains a BMC message that was sent but not scheduled in the last schedule period; or

3. Reading of the CTCH block set is advised; or

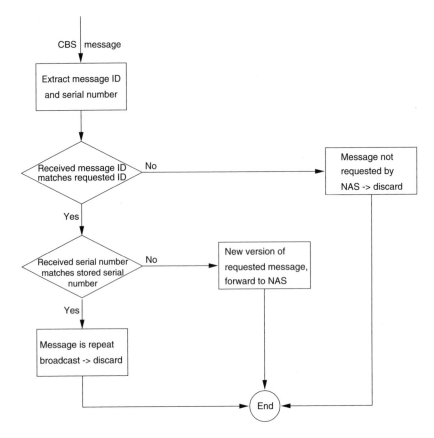

Figure 9.5. The CBS message handling process.

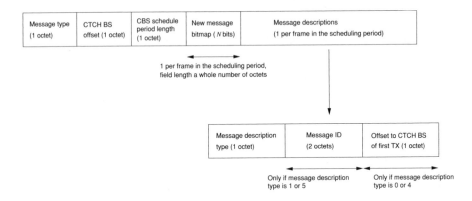

Figure 9.6. Structure of a BMC schedule message.

Figure 9.7. New message bitmap.

4. It contains a schedule message; or

5. It contains either the first transmission or a repeat transmission of a BMC message in the schedule period.

A value of 0 is used if none of the above conditions apply.

Message descriptions is a message description for each bit in the bitmap. Each message description contains fields to describe the nature of the message represented by that bit.

The BMC uses the new message bitmap and message description fields in conjunction with the requested message list to determine which frames in the next schedule period contain the messages requested by the NAS. This information is forwarded to the RRC in a CBMC_RX_IND message. The RRC uses the information to configure the physical layer to receive only the frames required.

9.3.3 BMC Deactivation

The BMC returns to the idle state when it receives a BMC_DEACTIVATE_REQ that causes all requested messages to be deleted from the table. Recall that the UE maintains a table of messages as requested by the NAS, and the table acts as the filter for the messages. BMC informs the RRC that it no longer needs to receive CBS traffic by sending a CBMC_RX_IND.

10

RRC

Dinesh Mittal, Rohit Bhasin, Hardik Parekh, Ian Blair, Ian Hunter and Joby Allen

10.1 INTRODUCTION

This chapter discusses the structure and operation of the RRC. The *Radio Resource Controller* (RRC) forms the core of the *Access Stratum* (AS), see Figure 10.1.

The RRC is responsible for coordinating the use of radio resources in the UE. To do this it must select a cell on a network, and then extract information about it so that the UE can communicate with the cell correctly. It can then be instructed to make connections so that control and user data can be exchanged with the network.

The RRC sits between the *Dedicated Control* (DC), *Notification* (NT), and *General Control* (GC) *Service Access Points* (SAPs), which connect with the *NonAccess Stratum* (NAS), and the internal components of the access stratum (RLC, MAC, PDCP and BMC, through the CRLC, CMAC, CPDCP and CBMC SAPs). It also communicates with the physical layer using the CPHY interface. It exchanges information with the network RRC through PDUs which are sent through the RLC (RLC SAPs). This is illustrated in Figure 10.2.

The primary 3GPP specifications relevant to the RRC are:

- TS25.331 [141] specifies the RRC;

- TS25.304 [151] describes the requirements of idle mode, which particularly concerns the RRC.

In general, most TS25.xxx specifications are also important. In particular:

- TS25.301 [142] describes the architecture of the AS;

WCDMA – Requirements and Practical Design. Edited by R. Tanner and J. Woodard.
© 2004 John Wiley & Sons, Ltd. ISBN: 0-470-86177-0.

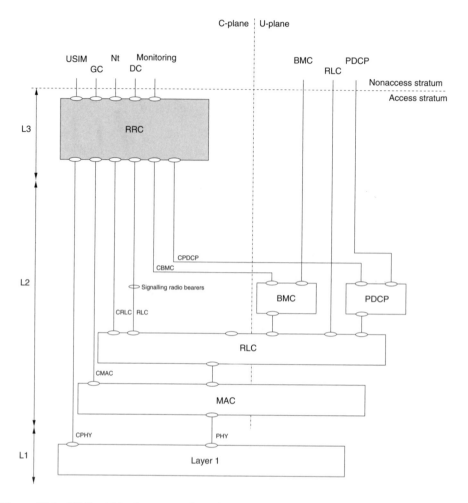

Figure 10.1. RRC within the protocol stack.

- TS25.302 [144] describes the functionality required by the AS from Layer 1;
- TS25.303 [146] describes the main AS scenarios;
- TS25.321 [139] describes the MAC, and its interface to the RRC;
- TS25.322 [140] describes the RLC, and its interface to the RRC;
- TS25.323 [145] describes the PDCP, and its interface to the RRC;
- TS25.324 [149] describes the BMC, and its interface to the RRC;
- TS25.402 [152] describes how Layer 1 timing is managed in the AS.

10.1.1 Architecture

The internal layout of the RRC is not prescribed in any detail in the 3GPP specifications: TS25.331 only defines entities to deal with broadcast, paging and dedicated control; hence it

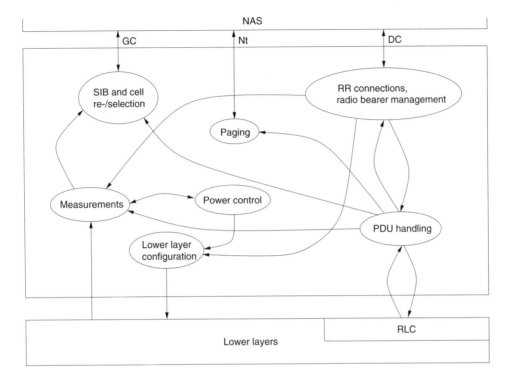

Figure 10.2. RRC architecture.

is for the designer to organize the RRC into a sensible arrangement of functional blocks. An example RRC architecture, showing these key functional aspects is illustrated in Figure 10.2. The different functional blocks will be discussed in the following sections, while the different features and their operation are explained in the remainder of this chapter.

10.1.1.1 *RRC Connection and Radio Bearer Management*
At the core of an RRC is an entity responsible for configuring and maintaining the RRC connection and associated radio bearers. This entity also contains a state machine which reflects the level of connectivity available at a given time. The RRC core state machine will change state as the UE encounters new radio environments, and also as the UE RRC is instructed to enter different states by the network. The key RRC states are:

- idle;
- Cell_FACH;
- Cell_DCH;
- Cell_PCH;
- Ura_PCH.

10.1.1.2 *Lower Layer Configuration*
As the RRC core changes state, it needs to reconfigue the lower layers. This aspect of functionality is delegated to the lower layer

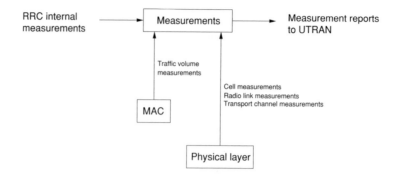

Figure 10.3. Measurements entity.

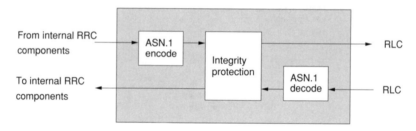

Figure 10.4. The PDU handling of the RRC.

configuration entity. This performs the detailed procedures necessary to configure the lower layers of the access stratum. It configures the physical layer and the MAC, RLC, PDCP and BMC sublayers on request from the RRC state machine, taking into account input from other RRC entities, such as power control.

10.1.1.3 *Measurements*   The measurements entity provides support for internal UE L3 RRC specific measurements and also measurement reporting to the UTRAN. It takes measurements from the MAC and physical layer as well as internal RRC measurements and reports them to UTRAN, see Figure 10.3.

10.1.1.4 *Paging*   The paging module handles incoming paging messages. Any received paging messages that contain paging information for this UE are forwarded to the NAS on the NT SAP.

10.1.1.5 *PDU Handling*   This entity is responsible for the reception and transmission of RRC PDUs. As RRC PDUs are sent over the air in ASN.1 encoded format, the PDU handler contains functionality to encode RRC PDUs (in the transmit direction) and decode (in the receive direction), as illustrated in Figure 10.4. This module is also responsible for implementing the integrity protection functionality.

10.1.1.6 Power Control The power control entity implements the open loop, inner loop and outer loop power control algorithms in the RRC.

10.1.1.7 SIB and Cell Selection/Reselection The SIB and cell selection/reselection entity is responsible for:

- reception and storage of system information received on the BCH;
- evaluation of cell measurements made by the measurements entity for the purpose of cell selection.

10.1.2 Functionality

RRC is required to perform a number of basic functions:

- cell selection;
- cell reselection;
- reception of broadcast system information;
- paging and notification;
- establishment, maintenance and release of an RRC connection between the UE and UTRAN;
- establishment, reconfiguration and release of *Radio Access Bearers* (RABs);
- assignment reconfiguration and release of radio resources for the RRC connection;
- RRC connection mobility functions;
- routeing of higher layer PDUs;
- control of requested quality of service (QoS);
- UE measurements;
- outer loop power control;
- arbitration of radio resources on uplink DCH;
- integrity protection;
- ciphering management;
- PDCP control;
- CBS control.

10.1.2.1 Cell Selection After startup, the RRC is responsible for finding a cell on which it can find service, using a *Public Land Mobile Network* (PLMN) as requested by higher layers. The search may utilize information from cells which were previously used for service, or from a sequential search of frequencies which have detectable cells. When a valid cell has been found the RRC will start to collect system information from the cell.

10.1.2.2 *Cell Reselection*   As the UE moves, new cells become better for communication. The RRC is responsible for monitoring cells and for controlling the selection of new cells when appropriate.

10.1.2.3 *Reception of Broadcast System Information*   The RRC is provided with basic information about the network from information which is broadcast by the cells it sees. This information details the channels which may be used on the cell, how measurements are to be made when using the cell, and also includes some information for the NAS applicable when using the cell. System information is transmitted to the UE by UTRAN on a BCH transport channel. This channel distributes small units of information which are built up into more complex entities called *System Information Blocks* (SIBs). SIBs control the general behaviour of the RRC when it is idle, and provide additional information for use when the UE is connected.

10.1.2.4 *Paging and Notification*   Paging provides the functionality needed so that a UE can be alerted that the network wishes to connect to it. This is performed using special techniques to minimize the power required by the UE. Paging is also used to ask a UE which is in Cell_PCH or URA_PCH states to change to Cell_FACH so that data may be exchanged with it. The same method is also used to notify idle UEs of updates to NAS information, or changes to broadcast system information, without requiring an RRC connection.

10.1.2.5 *Establishment, Maintenance and Release of An RRC Connection Between the UE and UTRAN*   To use a network, a UE must be able to exchange dedicated signalling messages between itself and the network. This is carried out by establishing an RRC connection to the network, i.e. exchanging information so that the UE RRC can communicate with its peer. An RRC connection is itself used to effect signalling connections for packet services, circuit switched services or both. Such signalling connections are maintained until released by the network, provided that radio conditions are sufficient.

10.1.2.6 *Establishment, Reconfiguration and Release of Radio Access Bearers*   When a user wishes to convey data, the RRC uses the signalling connection with its peer to configure data services. These are provided in the form of *Radio Access Bearers* (RABs). A single RAB might provide, for example, a voice connection or a packet service context.

10.1.2.7 *Assignment Reconfiguration and Release of Radio Resources for the RRC Connection*   As the RRC establishes connections or sets up RABs, different demands are placed on lower layers. The required PHY, MAC, RLC and other lower layer resources are normally configured by the RRC in accordance with the instructions it receives from UTRAN, although the RRC may also act autonomously, e.g. if radio coverage is lost.

10.1.2.8 *RRC Connection Mobility Functions*   As the user moves, an RRC connection will require routeing through different cells. Connection mobility functions ensure that the UE can be contacted as it moves through the network. This includes handover between cells as the user passes closer to their transmitters.

10.1.2.9 *Routeing of Higher Layer PDUs*   The RRC provides specific services to transfer NAS data to and from its network peer. This facility is also used for the transfer of *Short Message Service* (SMS) data.

10.1.2.10 *Control of Requested QoS*   The RRC provides features that permit the network to manage the quality of service (QoS) of connected radio bearers. This is partly achieved by providing sufficient information to the UTRAN to permit it to adjust the configuration of radio bearers as required. The RRC may also perform outer loop power control to adjust the QoS of a given service (see below).

10.1.2.11 *UE Measurements*   The RRC will make a range of measurements on cells, on the UE itself and on active channels. Sometimes these are reported to the UTRAN as instructed by the network, at other times they are used for cell selection/reselection, or for open or outer loop power control.

10.1.2.12 *Outer Loop Power Control*   The RRC will adjust the inner loop power control target to achieve a desired (block) error rate for a service.

10.1.2.13 *Arbitration of Radio Resources on Uplink DCH*   In high data rate mobiles, the RRC is responsible for applying a dynamic resource allocation procedure to the uplink DCH.

10.1.2.14 *Integrity Protection*   The RRC marks its PDUs in such a way as to determine whether a third party is tampering with the signalling between the network and mobile. The process of marking transmitted PDUs, and checking received PDUs, is termed integrity protection.

10.1.2.15 *Ciphering Management*   Ciphering parameters are also managed by the RRC. In particular, the RRC ensures that the use of ciphering is synchronized between the UE and UTRAN.

10.1.2.16 *PDCP Control*   For packet service bearers, the RRC is responsible for configuring and reporting parameters for PDCP entities.

10.1.2.17 *CBS Control*   The RRC is also responsible for configuring the BMC, and for setting up Layer 1 DRX for the reception of *Common Transport Channel* CTCH PDUs at the correct times.

10.2 CELL SELECTION AND RESELECTION

10.2.1 PLMN Selection and Reselection

On request from the user (manual mode) or periodically (automatic mode), NAS can request the RRC to search for all the PLMNs that are available. The RRC shall search on all the carriers available in its frequency band and find the strongest cell on each carrier. The RRC shall read the master information block on each cell to find out the PLMN that cell belongs to and shall report each found PLMN on each carrier to NAS. If the signal strength for a cell (CPICH_RSCP in the case of UTRA FDD cells) is below a minimum level [152], the RRC should include the received signal strength level of the cell. NAS can, at any time, stop or continue the PLMN search on the rest of the carriers.

NAS requests the RRC to camp on a selected PLMN and optionally includes a list of equivalent PLMNs. The PLMN shall be specified by different types of identity, either PLMN-identity for GSM type PLMNs or SID for ANSI-41 type PLMNs.

10.2.2 Cell Selection and Reselection in Idle Mode

On receiving a PLMN search request from NAS, the RRC shall try to find a 'suitable cell' on the requested PLMN and camp on it. Once camped on such a cell, UE is said to be in normal service state. For a cell to be suitable, it should satisfy the following criteria (henceforth called suitability criteria):

- cell should satisfy the S-criterion;

- cell should belong to one of the PLMNs as requested by NAS in the selected PLMN and equivalent PLMN list;

- cell should not be barred or reserved or belong to forbidden LAs as indicated in the system information.

Details of the above three criteria shall be duly described in the following text. To make the first cell search faster, the RRC can use the stored cell information from the *Universal Subscriber Interface Module* (USIM). This information can include the carrier frequency, primary scrambling code, measurement and timing related information of the cells. The RRC shall first try to find a suitable cell based on the stored cell information. This procedure is aptly named stored cell selection. If not successful, the RRC shall try on all the carriers available in its UTRA band (called initial cell selection). If this also fails, the RRC shall camp on any strong enough cell available, irrespective of its PLMN identity. In this state, the whole UE is said to be in limited service state and is only allowed to make emergency calls. This cell should also at least satisfy the first and third suitability criteria.

10.2.2.1 Cell Selection Process   After getting the measurements from the physical layer, the RRC shall collect the master information block for the strongest cell available and compare its PLMN identity with that requested by NAS. If it matches, the RRC shall try to receive other system information blocks. If the essential system information blocks (SIB 1, 3, 5 and 7) are not scheduled in a cell, the cell is considered to be barred. Once the necessary system information blocks are collected, the RRC shall check for the cell status according to information available in SIB 1 and SIB 3. The following checks need to be carried out to consider whether a cell is barred:

- the cell should not belong to a forbidden location area. This information is contained in the NAS system information received in SIB 1;

- the cell should not be barred or reserved as indicated in SIB 3.

The RRC shall evaluate the cell for S-criterion, which is explained in Section 5.2.3.1.2 of [151]. Parameters *Qualmin* and *Qrxlevmin* are as received in SIB 3. If the RRC finds the current strongest cell to be unsuitable for any of the above reasons, the RRC shall follow the same procedure for the next best cell found through measurements. Once the cell is evaluated for suitability, the UE is said to be in camped state. As explained above, if the RRC is not able to find any suitable cell to camp on, it should camp on any cell and will enter into the limited service state. If the RRC is not able to camp on any cell, it is said to be in any cell selection or no service state. In the latter state, the RRC shall keep on attempting periodically to camp on any cell.

10.2.2.2 Cell Reselection Process

Once camped on a cell, the RRC shall keep on continuously monitoring the neighbouring cells to find any better cell. The cell reselection procedure involves each cell being evaluated for the S-criterion first. All the cells satisfying the S-criterion are then ranked according to the H-criterion [151]. This ranking shall not be done if the current serving cell does not belong to a hierarchical cell structure as indicated in the system information block 11. Cells are then ranked according to the R-criterion. Cell ranking and the cell reselection criteria make it implicit that the RRC can only reselect a 'monitored' cell. A monitored cell is one which is specified in the neighbour cell information IE in system information type 11 or 12 (valid in connected mode only). If the best ranked cell is different from the current serving cell, the RRC shall reselect to the new cell if the new cell remains the best cell for *Treselection* time. The cell reselection procedure is specified such that the current serving cell gets somewhat of an advantage (by means of parameters *Qhyst, Qoffset, Treselection*) over other cells for being reselected. This is done so as to avoid rapid and sometimes false cell reselections. If, after the cell reselection procedure, none of the cells has been selected, then the RRC shall repeat the cell selection process.

10.2.3 Cell Reselection in Connected Mode

In connected mode, the RRC can do a cell reselection for the following reasons:

- cell reselection on reception of cell measurement results from the physical layer;

- cell reselection triggered by received messages from UTRAN.

The cell reselection procedure is the same as for the idle mode because of the first point. The only difference is that the RRC shall preferably use the information available in SIB 4 and SIB 12 for the cell reselection procedure if these SIBs are received. UE RRC can receive a message from a peer RRC asking for cell reselection to be performed. These messages may include the UTRAN preferred cell (indicated by the primary scrambling code) or even the frequency information, on which the cell can also be reselected. In this case, the RRC needs to perform a cell reselection only on the UARFCN stated in the messages. If the

RRC is not able to reselect to the same cell, as requested by UTRAN, the RRC updates the UTRAN RRC with its reselected cell (see Section 10.8 for details).

10.3 RECEPTION OF BROADCAST SYSTEM INFORMATION

The RRC is provided with basic information about the network from the information broadcast by each cell in the system. This information is called system information. This broadcast information describes the common channels in the cell, measurement and cell selection/reselection related information for the transmitting cell and its neighbouring cells, some NAS-related information, information used in intersystem handovers and also location services related information. Broadcast information is transmitted to the UE by UTRAN using the *Broadcast Control Channel* (BCCH) logical channel which is mapped onto the BCH transport channel in all states of UE except when in CELL_DCH state. System information is also essential for the UE initially to establish the timing relationship with the network.

10.3.1 Reception of System Information

The system information is broadcast in small units called *System Information Blocks* (SIBs). Each system information block encapsulates the same type of network information. This is required so as to allow UEs to capture only the required system information. Moreover, different kinds of information have different requirements on scope (cell scope, PLMN scope), validity and frequency of updates.

10.3.1.1 *Structure of System Information*
System information blocks are organized as a tree. The master information block gives the status of other system information blocks, and refers to where these may be found. This may be given explicitly or via further blocks known as scheduling blocks, see Figure 10.5.

The *Transport Block* (TB) size of the BCH transport channel is 246 bits. Therefore, the maximum length of a system information message is restricted to this size. Different system information blocks are contained within a generic system information message. Every system information message contains the P-CCPCH SFN (system frame number) apart from the SIBs. To fit the maximum information into one message, the system information blocks are either segmented or concatenated. A big system information block can be divided into more than one block and, similarly, several small system information blocks can be combined together. The part of the system information message containing the system information blocks is termed a segment. Depending upon the requirement, there can be nine different kinds of segment combination. The nine combinations are:

1. Complete;

2. First;

3. Last;

4. Subsequent;

5. Last + complete;

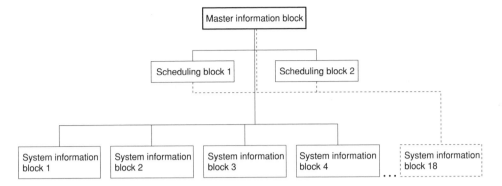

Figure 10.5. Organization and structure of SIBs.

   6. Last + complete list + first;

   7. Last + first;

   8. Complete + first;

   9. No segment.

Each segment contains a header which gives the UE an indication of what type of SIB it contains, the segment index for a subsequent segment and the total number of segments for a first segment. The segments are themselves ASN.1 encoded before being fit into the system information message. As system information messages use the TM RLC mode (BCCH logical channel), it is the peer RRC's responsibility to supply the whole SDU equal to the size of a TB. This requires the RRC to do some padding if a system information message is not filled fully with information. The padding bits shall be zeroes.

The RRC layer in a UE reassembles the segments belonging to a system information block in ascending order, and when finally all the segments are received, the whole system information block can be decoded. The RRC layer of a UE shall discard a SIB, reread the scheduling information and reread the SIB if:

- out of sequence segments are received;

- duplicate segments are received;

- segment index in a subsequent or last segment is equal to or larger than the SEG_COUNT in the first segment of the SIB.

10.3.1.2 Scheduling of System Information
The network keeps on broadcasting these blocks regularly with a certain repetition period, but a UE that has already read the system information once, is prevented from continually updating this information by the use of value tags or SIB validity timers. Scheduling of the SIBs is explained often in terms of parameters:

SEG_COUNT. Number of segments in a SIB. There can be 1 to 16 segments for a SIB.

SIB_REP. SIB repetition period in frames. This value is applicable to every segment of a SIB. Allowed values are 4, 8, ..., 4096 frames.

SIB_POS. Defines the position of the first segment of the SIB. This parameter can take all multiple values within the SIB_REP period, i.e. 0 to Rep-2 The steps have to be in multiples of 2 frames as a TTI for BCH is 20 ms.

SIB_OFF. This is an array of numbers and defines the offset for the subsequent segments for a SIB. Valid values are 2 to 32 frames, again with a step size 2. Position of the subsequent segment can be calculated as $SIB\_POS(i) = SIB\_POS(i-1) + SIB\_OFF(i)$.

Since a UE needs to first read the master information block to get the aforementioned scheduling information for other SIBs, scheduling parameters for a MIB are fixed. In the FDD mode, a MIB has $SIB\_POS = 0$, $SIB\_REP = 8$ and $SIB\_OFF = 2$.

SIBs are considered valid from the time they are received, unless:

- SIB_REP timer expires for a SIB whose validity is defined by expiration timers;

- for SIBs with modification as value tags, the SIB change can be explicitly indicated to a UE with the help of the BCCH modification information contained in paging type 1 or system information change indication messages;

- the cell or PLMN is reselected (note that PLMN scope SIBs, e.g. SIB 1, can still remain valid in the reselected cell);

- six hours have elapsed since a SIB was received and stored.

Modification of the SIBs is either achieved through the use of value tags or SIB expiration timers. Value tag information is received along with the scheduling information of the SIBs. Value tags can be of several types: PLMN value tag (valid for PLMN scope SIBs like SIB 1 and SIB 15.3); cell value tag (for cell scope SIBs like SIB 2, 3, 4 . . .); SIB occurrence value tag (for SIBs 15.2 and 15.3 for which there can be multiple occurrences); predefined configuration value tag (for SIBs type 16). Modification for SIBs 7, 9, 10, 14, 17, which contain the fast changing parameters of network information, is achieved through the use of validity timers. For SIBs 7 and 14, another related parameter is given in the respective SIB information, which allows the network to control the rate of change of these SIBs dynamically without forcing UE to reread the scheduling information.

10.3.2 System Information Blocks

Below is a more detailed list of information blocks.

MIB provides the scheduling of other system information blocks. It also contains the PLMN identity of the cell.

Scheduling blocks 1 and 2 contain the scheduling information for the SIBs which could not be provided in the MIB.

SIB 1 is a PLMN scope SIB, which means that some parts of SIB 1 information can be retained even after the cell reselection (if PLMN is the same in the new cell). SIB 1 contains the GSM-MAP cell NAS information for each CN domain. It also contains the timers and constant values to be used by the RRC.

SIB 2 contains the UTRAN registration area list that the current cell belongs to. This information is used by UE in URA_PCH state to evaluate whether a URA_UPDATE procedure must be initiated or not.

SIB 3 contains the cell selection and reselection parameters for the serving cell. SIB 3 also contains the cell access restriction information for the current cell. SIB 3 information is used by the RRC in idle mode and connected mode (only if SIB 4 is not available), to perform cell selection and reselection.

SIB 4 contains the same information as SIBs 3, but SIB 4 is used in connected states of the the RRC and if not scheduled in a cell, the RRC shall use SIB 3 instead.

SIB 5 contains parameters for common physical channels such as PRACH, SCCPCH, etc. These parameters are used in the establishment of channels in idle mode or connected states, except CELL_DCH.

SIB 6 contains the same parameters as SIB 5 but is to be used in connected mode states only.

SIB 7 contains fast changing parameters, particularly relating to RACH usage.

SIB 8 contains static CPCH parameters.

SIB 9 contains dynamic CPCH parameters.

SIB 10 contains DRAC parameters. This SIB is read only by the UEs with specific capability, i.e. able to read SCCPCH and DPCH simultaneously. This SIB is received on FACH TrCh rather than BCH.

SIB 11 contains measurement control information and the neighbour cell information list.

SIB 12 contains the same information as SIB 11 but is only to be read and used in connected mode.

SIB 13 contains the ANSI-41 version of SIB 1 and SIBs 13.1, 13.2, 13.3, 13.4 (ANSI-41 specific parameters).

SIB 14 is used for TDD.

SIB 15, 15.1, 15.2, 15.3, 15.4 contain UE positioning parameters.

SIB 16 contains predefined configuration parameters (for handover to UTRAN).

SIB 17 is used for TDD.

SIB 18 contains neighbouring PLMN IDs.

10.3.3 Modification of System Information

Modification of the system information is indicated to UE using the BCCH modification information, which is contained in either the paging type 1 messages (to reach the UEs in idle, CELL_PCH or URA_PCH states) or the SYSTEM_INFOR-MATION_CHANGE_INDICATION message (to reach UEs in CELL_FACH state). BCCH modification information contains the MIB value tag and optionally the BCCH modification time. The RRC shall compare the value tag received in the BCCH modification information with the one stored in the UE and if it is different, read the MIB again. If the received MIB has the same value tag as indicated in the BCCH modification information, then the RRC shall store the new MIB and take actions on the received MIB. If the received MIB has the same value tag as that stored in the UE, the RRC shall reread the MIB in the next scheduled period of MIB. The RRC shall store the new MIB and take actions on the received MIB if the received MIB has a different value tag compared to that indicated in the BCCH modification information and the one stored in UE. Also in this case, if ((MIB value tag in the BCCH modification.information − MIB value tag in received MIB) mod 8) is less than four, then the RRC shall reread the MIB on the next scheduled occasion. The BCCH modification time indicates to the UE the exact SFN at which the new MIB will be transmitted. If this is all true, then the RRC shall perform the aforementioned actions at this SFN only.

10.4 PAGING AND NOTIFICATION

Paging is a procedure by which the network can notify or page a selected number of UEs in the cell. After power on, the UE is registered with the network, so the network knows the cell in which the UE can be paged. There are two kinds of paging message that the network uses in different states of the RRC. Paging type 1 messages can be sent to UE in idle mode or in connected mode (URA PCH/CELL PCH), and paging type 2 messages for connected mode (CELL DCH/CELL FACH).

10.4.1 Paging in Idle Mode

A paging type 1 message is sent via the *Paging Control Channel* (PCCH) of the cell and is used to transmit paging information to a number of users in the cell. This is done by sending a number of paging records with a different UE identity from the dedicated UE. The message can direct the RRC to send a notification to NAS about an incoming call or direct the RRC to read the MIB in the cell by changing the value tag, which would imply change of some information in the SIB in the cell. The notification meant for NAS also contains the domain for which the terminating call has been requested.

10.4.2 Paging in Connected Mode

10.4.2.1 *Paging in CELL_DCH/CELL_FACH* A paging type 2 message is sent on the DCCH in connected mode CELL_FACH or CELL_DCH states of the RRC. This contains a notification for NAS for a mobile terminating call. This in turn triggers another signalling connection to be set up for a new domain, depending on the domain information in the message.

10.4.2.2 *Paging in CELL_PCH/URA_PCH* In the connected mode, the RNC, in addition to the tasks it can perform in idle mode, can also trigger a cell update procedure in the UE by sending a paging record for a UE without the paging information from the core network in connected mode. Paging in CELL_PCH would imply that the UE is known at the cell level so the paging is only in that cell. Paging in *UTRAN Registration Area* (URA) PCH would imply that the UE is moving quickly between cells so the UE is paged in the URA.

10.4.3 Paging in Idle Mode

In UMTS, the idle mode and the CELL_PCH/URA_PCH states support *Discontinuous Reception* (DRX). While in this mode, the UE can power down most of its normally used functionalities. So it is important that it is possible to receive the paging messages while in DRX mode. To achieve this, the RNC provides the DRX cycle length coefficient, which defines the periodicity of the DRX procedure.

The DRX cycle length is calculated as $\max(2^k, \text{PBP})$ frames, where k is the DRX length coefficient and the *Paging Block Periodicity* (PBP) is one for FDD.

The UE can receive the CN domain specific DRX cycle length coefficient for each domain. When the UE is connected to multiple domains, the UE should select the shortest cycle length. For CS domain this information is available from the SIBs but for PS the value is negotiated in NAS procedures. If not negotiated the value from the SIB is used instead. There is another DRX cycle length that is defined by the UTRAN in connected mode. In this case the UE will select the shortest of the following two options:

- UTRAN DRX cycle length;

- any of the stored CN domain specific DRX cycle lengths for the CN domains the UE is attached to with no signalling connection established.

Whilst in the DRX mode, the UE uses the *Page Indicator* (PI) to indicate the paging message. A PI is a short indicator transmitted on the *Paging Indicator Channel* (PICH) to indicate that a paging message is being transmitted on the *Paging Channel* (PCH) by the SCCPCH. For FDD mode, the number of PIs per radio frame (Np) can be 18, 36, 72 or even 144.

$$\text{PI} = \text{DRX Index} \bmod Np, \text{ where DRX Index} = \text{IMSI div } 8192 \qquad (10.1)$$

The UE shall monitor its paging indicator in the PICH frame with SFN given by the paging occasion. The paging occasion defines the SFN for the frame on which the UE must monitor the PICH to see if the paging message is being sent or not.

$$\text{Paging occasion} = [(\text{IMSI div } K) \bmod (\text{DRX cycle length div PBP})] *$$
$$\text{PBP} + n * \text{DRX cycle length} + \text{Frame Offset} \qquad (10.2)$$

where $n = 0, 1 \ldots \text{Max SFN}(4095)$. If the PI is received on the PICH in the paging occasion, the UE will read the paging message on the PCH on the associated SCCPCH. If the UE has no *International Mobile Subscriber Identifier* (IMSI), i.e. when making an emergency call without a USIM, then the UE shall use as default numbers: IMSI $= 0$ and DRX cycle length $= 256$.

10.5 ESTABLISHMENT, MAINTENANCE AND RELEASE OF AN RRC CONNECTION BETWEEN THE UE AND UTRAN

10.5.1 RRC Connection

The purpose of establishing an RRC connection is to make the UE known to the network so that it can exchange both user and signalling data with the network. To establish a connection, the RRC submits an RRC_Connection_Setup_Request message on the CCCH channel to the cell it is camped on. The UTRAN will then respond with an RRC_Connection_Setup message, returned on a FACH. The connection setup will assign the UE *Radio Network Temporary Identifiers* for use within the cell (C-RNTI) and across UTRAN (U-RNTI), and it will set up dedicated radio bearers to carry RRC signalling. Four or five radio bearers (depending on the network) are used for RRC signalling during a connection:

RB0 is the CCCH logical channel. It uses TM RLC and RACH in the uplink and UM RLC and FACH in the downlink, and is used, e.g. to establish a connection.

RB1 is a DCCH logical channel using UM RLC in both directions. It is used for peer RRC messages.

RB2 is a DCCH using the AM RLC. It is used for peer RRC messages which require *Acknowledged Mode* (AM)

RB3 is a DCCH using AM RLC. It is used for NAS messages carried through the RRC.

RB4 is an optional DCCH using AM RLC. It is used for low priority NAS messages. If this RB is not configured then RB3 is used.

10.5.2 Signalling Connections

When an RRC has a connection to the network it will support signalling connections for the user. There are two types:

1. CS domain signalling connections;

2. PS domain signalling connections.

These differ in the nature of the connection. CS domain connections provide a continuous channel for user traffic which must be filled or read at a more or less constant rate. PS domain connections are more suited to the bursty nature of packet data, although real-time packet services are also possible. Signalling connections are established using an initial direct transfer sent from the UE following the RRC connection establishment, or using the existing RRC connection if a second signalling connection is being established.

10.6 ESTABLISHMENT, RECONFIGURATION AND RELEASE OF RADIO ACCESS BEARERS

Once a signalling connection has been established, the user needs to establish bearers to carry traffic for the services they wish to use. These are termed *Radio Access Bearers* (RABs). Radio access bearers are provided for both packet and circuit domain connections. Nominally, they are treated in the same way, but there are significant differences between the two domains.

The properties of the RAB are determined by NAS exchanges. The network sets up RABs to match the quality of service demanded by the user, taking into account the available resources of the network. The outcome is that the RRC is configured with a radio bearer, or set of radio bearers, with characteristics chosen by higher layers to match the requirements of the service that they are supporting. RABs use RBs numbered from 5 to 32 inclusive, using DTCH logical channels.

PS RABs map to single radio bearers, but CS RABs can include sets of radio bearers. In particular for a voice call, data for a codec can be assigned to different parallel radio bearers, which are afforded differing levels of forward error correction and CRC protection. Radio access bearers are set up, modified or released by means of (respectively) radio bearer setup, reconfiguration or release messages sent from the network.

10.7 ASSIGNMENT, RECONFIGURATION AND RELEASE OF RADIO RESOURCES FOR THE RRC CONNECTION

The functionality of an idle UE is limited to receiving broadcast data from cells, and selecting cells on the basis of this information and the signal strength and interference level it detects. To actually communicate with a network, and to perform useful work for the user, a UE must establish a connection. This means establishing a defined set of physical, transport and logical channels which the UE can use to carry out dedicated data transfers with the network. The nature of the connection which is established is determined by the RRC state. 3GPP defines four basic states for a connected mobile:

1. Cell_FACH;

2. Cell_DCH;

3. Cell_PCH;

4. URA_PCH.

The principal states are Cell_FACH and Cell_DCH. Cell_PCH and URA_PCH are used to retain a logical connection with the network when there is little activity, but in many respects the behaviour of the UE in these states is similar to that of the idle state.

10.7.1 Cell_FACH

The Cell_FACH state is used when the user traffic is light and bursty. In the Cell_FACH state, the UE uses a RACH for uplink traffic, and a FACH for the downlink. The selection of the RACH/PRACH and FACH/S-CCPCH being made according to information contained in SIBs sent from the network. Whilst in the Cell_FACH state, the UE may also set up BCH channels to read SIB information to assess neighbouring cells. The UE also adopts a similar configuration when connecting, or when performing updates from the PCH states.

10.7.2 Cell_DCH

The Cell_DCH state is used when a continuous dedicated connection is required, or when the required data rate is too high to be accommodated by RACH and FACH. In particular, it is required when a typical voice channel is needed. In the Cell_DCH state, the UE uses DCH transport channels on both the downlink and uplink, and may use a DSCH (for more efficient use of downlink), or a FACH/S-CCPCH, if the DRAC procedure is used on the DCH. The Cell_DCH state also applies when the DSCH is used. The Cell_DCH state is the state used for high rate data transfers and takes the most resources from the network.

10.7.3 Cell_PCH

Cell_PCH state is used when the UE wishes to register as connected to the network, but takes the absolute minimum of resources from it. In this case, the UE drops into a state which is very similar to idle, but the NAS layer remains connected. As in idle mode, the UE monitors the BCH to assess cells, and listens to the PICH/PCH to detect when it is paged. When in the Cell_PCH state the access stratum can do very little other than update the network when it changes cells (and even this requires a brief return to CELL_FACH). Following a page, or if uplink data needs to be sent, the UE must move to the Cell_FACH state.

10.7.4 URA_PCH

URA_PCH is similar to Cell_PCH, except that the network is only updated when the user changes URAs. Paging a UE in this situation requires more in the way of network resources; a page has to be signalled across the URA rather than across a single cell. It therefore uses a minimum of resources from the network, but extra effort is required to page the mobile. A page must be sent on all cells in the URA if the UE is to be guaranteed a chance to receive it. UTRAN may use a number of procedures to change the UE state. It may perform a radio bearer, transport channel or physical channel reconfiguration, or it may change the configuration using the contents of a Cell_ or URA_ update confirmation.

10.8 RRC CONNECTION MOBILITY FUNCTIONS

As the UE moves, the connection and any associated RABs, are moved to new cells. This is performed in one of three ways:

1. Cell reselection;

2. *Hard Handover* (HHO);

3. *Soft/Softer Handover* (SHO/SOHO).

Cell reselection applies when there is no continuous data connection in place, i.e. the RRC is in Cell_FACH, Cell_PCH or URA_PCH states. The UE reports its cell reselections to the network using cell or URA updates. Also, the Cell_FACH state presents tighter constraints on when measurements may be made if downlink messages for the UE are not to be missed. Measurements, and decisions, for cell reselection are performed autonomously by the UE. Hard handovers are used when the RRC is in the Cell_DCH state. Existing radio links are torn down, and then established again using a new set of cells, on instruction from the network, using information originally provided by the UE in measurements.

Whilst cell reselections and HHOs have analogues in many other radio access technologies, SHOs are a particular feature of CDMA. In this case, the signals from a number of radio links (termed the *Active Set* (AS)) are combined so that interference or fading affecting one link is compensated by other links. The RRC is responsible for adding and removing links to particular cells for this purpose, in accordance with active set update messages received from the network, and also for reporting measurements of the cells involved, for the network to decide which cells to use. Softer handovers are basically a special case of soft handover whereby some of the cells are in the same physical location. This leads to economies in that the signal paths are the same; power control bits sent from such cells can be combined, reducing the power and hence interference to other UEs.

10.9 ROUTEING OF HIGHER LAYER PDUs

The RRC is responsible for routeing the NAS messages to and from the RNC, which in turn routes them to the core network. To do this the NAS has to establish a signalling connection to the domain to which it wants to transfer the data (CS/PS domain).

These messages are encapsulated in direct transfer RRC messages, on both uplink and downlink, using RB3 or 4. The RRC is also responsible for the transfer of SMS messages, using the same method. The messages to the different NAS entities are routed within NAS using information from the protocol discriminator field in the NAS message.

10.9.1 Direct Transfer Messages

There are three different types of direct transfer message, two in the uplink and one in the downlink.

10.9.1.1 *Initial Direct Transfer Message*   This is the message used to transfer the first signalling initiation message from the NAS. This message tells the RNC about a request for a new connection, which in turn informs the CN that a new signalling connection is being set up. The information contained in this is the NAS message itself, the domain the signalling connection is to be established and something known as the intradomain NAS node selector, which contains information about the core network node the message is to be routed to. All this information is utilized by the RNC to cause a successful connection to the requested domain.

10.9.1.2 *Uplink Direct Transfer Message*   After the transmission of the initial direct transfer message, any message sent by the NAS in the uplink for that domain will be sent in the uplink direct transfer message. This message just encapsulates the NAS message and contains the domain for which the message is meant.

10.9.1.3 *Downlink Direct Transfer Message*   The downlink direct transfer message contains the NAS message sent on the downlink. This message also contains the domain from which the message has originated so that the message can be routed to the correct domain in the NAS in the UE.

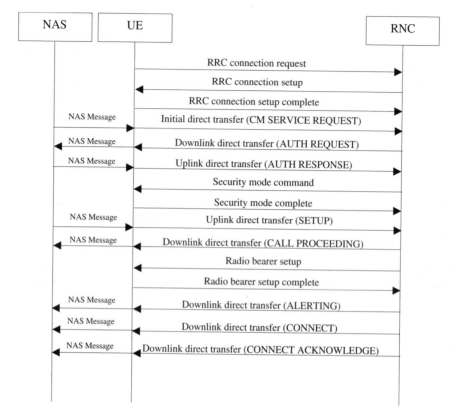

Figure 10.6. Sequence for mobile originated CS domain call.

10.9.2 Example for Mobile Originated CS Domain Call

The routeing of higher layer PDUs for a mobile originated CS call is illustrated in Figure 10.6.

10.10 CONTROL OF REQUESTED QoS

The RRC is charged with managing control of the requested QoS of the bearers provided, but in reality, the actions the RRC in the UE can take are limited, most of the complexity required for this resides in the network. The UE RRC provides support for QoS control by means of the following mechanisms:

Measurements. The UE can be requested to supply block error rate measurements to the network. Also, these can be used for outer loop power control;

Outer loop power control. The UE may be provided with target block error rates for transport channels.

Radio bearer configuration. The UTRAN can control the service provided at the UE by configuration of different channel configurations, e.g. switching between Cell_FACH and Cell_DCH states.

10.11   UE MEASUREMENTS

The measurement component provides support for internal UE L3 RRC specific measurements and also measurement reporting to the UTRAN. The RRC component uses cell RSCP and E_c/N_0 measurements for cell selection and reselection purposes and for open loop power control. It requires block error rate measurements for outer loop power control. The UTRAN requires measurements on a periodic or event driven basis for handover management, radio bearer control and UE positioning. The measurement component tracks timing measurements of cells, updating the relative cell timings used by the RRC as the measurements change. The measurement component maintains the information on cells to be measured, shared with the cell selection/reselection component and information provided to the UE in SIB 11, SIB 12 and measurement control, used to specify which measurements to make.

10.11.1   General

The measurement control and report procedure is used by the UTRAN to control the measurements being taken by the UE. The process is very versatile and allows many measurements with different characteristics to be set up, modified and released concurrently in all RRC states. The measurements are managed via SIB 11, SIB 12 and measurement control messages received by the RRC. In idle mode, measurement control information is read from SIB 11; in FACH and paging modes, SIB 12 is used. Measurement control messages can be used in all connected modes. Each measurement has a unique identity, type, object, measurement quantity, reporting quantity, reporting criteria, validity, mode and additional identities.

The following measurement types are supported:

1. Intrafrequency measurements;

2. Interfrequency measurements;

3. InterRAT measurements;

4. Traffic volume measurements;

5. Quality measurements;

6. UE internal measurements;

7. UE positioning measurements;

8. RACH measurement reporting.

These types are described in more detail in the following paragraphs.

The measurement object is the object on which the measurements are taken. This may be a list of cells or transport channels to take measurements on. The cells that measurements have to be taken on are stored locally within the RRC variable CELL_INFO_LIST by collating the IEs containing '..cell info list' received from all SIB 11, SIB 12 and measurement control messages. The cells fall into three mutually exclusive sets:

1. The active set is defined as the set of cells used for the current connection.

2. The monitored set is defined by the set of cells which the UE is requested to measure by the UTRAN for a particular measurement. This information is relayed in the IE cells for measurement and stored internally to the RRC in the variable CELL_INFO_LIST. These cells do not include members of the active set.

3. The detected set is defined as the cells detected by the UE which are not in the active or monitored sets.

The IE 'cells for measurement' contains a list of cells from CELL_INFO_LIST, these determine which cells each measurement type will read. If the optional IE is missing, the full contents of CELL_INFO_LIST will be used.

The measurement quantity defines the quantity to measure and the filtering to be applied to that object. Filtering is applied to measurements included in the IE 'measured results' and used for event generation, it is not applied to the IE 'measured results on RACH', or for cell reselection in idle or connected modes. The reporting quantities are the quantities that have to be included in any generated measurement reports. These can be further split into different sets of quantities for the active, monitored and detected sets, and include all of the measurement types described in the following paragraphs. The reporting criteria define whether the measurement reports are generated periodically, as a result of a specific event, or only as an additional measurement to be added to other measurement reports. The validity defines which RRC state the measurement report is to be generated in. If this optional IE is not included in the MC message, the measurement should be deleted on transition to any other state. The reporting mode determines whether the UE uses AM or UM RLC to transmit the measurement report. The additional measurement identities define the additional 'measured results' IEs which need to be generated when the reporting criteria are satisfied.

10.11.2 Physical Layer Measurements

The measurement component reads the following raw measurements from the physical layer, prior to processing and passing them on to the RRC in UE or UTRAN.

10.11.2.1 *CPICH RSCP*   This measurement is for handover evaluation, DL outer loop power control, UL open loop power control and for the calculation of pathloss. It is defined as *Received Signal Code Power* (RSCP), the received power on one code measured on the Primary CPICH of a cell. The reporting range for CPICH_RSCP is −115 to −40 dBm.

10.11.2.2 *UTRA Carrier RSSI*   This measurement is for interfrequency handover evaluation. It is defined as the received wideband power, including thermal noise and noise generated in the receiver, within the bandwidth defined by the receiver pulse shaping filter. The reporting range for UTRA carrier RSSI is −101 to −25 dBm.

10.11.2.3 *GSM Carrier RSSI*   This measurement is for handover between UTRAN and GSM. It is defined as the *Received Signal Strength Indicator* (RSSI), the wideband received power within the relevant channel bandwidth, performed on a GSM BCCH carrier.

10.11.2.4 CPICH_E_c/N_0

This measurement is for cell selection/reselection and for handover evaluation. The received energy per chip is divided by the power density in the band.

$$\text{CPICH\_}E_c/N_0 = \frac{\text{CPICH\_RSCP}}{\text{UTRA carrier RSSI}} \qquad (10.3)$$

Measurement is to be performed on the primary CPICH. The reporting range for CPICH_E_c/N_0 is -24 to -0 dB.

10.11.2.5 Transport Channel BLER

This measurement is for the estimation of the transport channel *Block Error Rate* (BLER). The BLER estimation is to be based on evaluating the CRC of each transport block associated with the measured transport channel after RL combination. The BLER shall be computed over the measurement period as the ratio between the number of received transport blocks resulting in a CRC error and the number of received transport blocks.

10.11.2.6 UE Transmitted Power

This measurement is the total UE transmitted power on one carrier. The reference point for the UE transmitted power is to be the antenna connector of the UE. The reporting range for UE transmitted power is -50 to $+33$ dBm.

10.11.2.7 Cell Synchronization Information

This measurement is used for handover timing purposes to identify time differences between active and neighbouring cell types in various units from chips to frames. The RRC reports Tm, OFF and COUNT-C-SFN frame difference to the network, where Tm is defined as the difference between the DL DPCH on the serving cell to the start of *System Frame Number* (SFN) on the neighbouring PCCPCH, given in chip units with the reporting range of 0 to 38 399 chips, see Figure 10.7.

OFF is defined as SFN on the neighbouring cell minus the *Connection Frame Number* (CFN) read from the DL DPCH on the serving cell (SFN-CFNTx)mod 256, giving the number of frames difference, with the reporting range of 0 to 255 frames, see Figure 10.7.

COUNT-C-SFN frame difference is defined as the four most significant bits of the difference between the 12 least significant bits of the RLC transparent mode COUNT-C in the UE and the SFN of the measured cell.

10.11.2.8 SFN–SFN Observed Time Difference

SFN–SFN observed time difference is the time difference of the reception times of frames from a serving cell and a target cell measured in the UE and expressed in chips. There are two different types, 1 and 2, where type 2 applies if the serving and target cell have the same frame timing. For type 1, the SFN–SFN observed time difference is defined as OFF 38 400 + Tm

Type 2 is defined as the difference between cell j and cell i when the UE receives one primary CPICH slot from cell j and receives the primary CPICH slot from cell i, that is closest in time to the primary CPICH slot received from cell j.

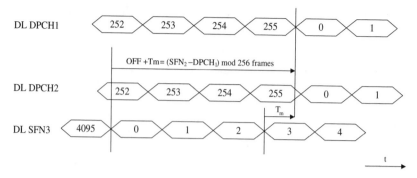

Figure 10.7. Definition of OFF and Tm according to [152].

10.11.2.9 UE Rx–Tx Time Difference Type 1   This measurement is used for call setup purposes to compensate propagation delay of DL and UL. The reference Rx path is to be the first detected path (in time) amongst the paths (from the measured radio link) used in the demodulation process.

10.11.2.10 UE Rx–Tx Time Difference Type 2   This measurement is used for UE positioning purposes. It contains the time difference in chips for each RL included in the active set, and is reported in UE internal measurements. The reference Rx path is to be the first detected path (in time) amongst all paths (from the measured radio link) detected by the UE.

10.11.2.11 Observed Time Difference to GSM Cell   This measurement is used to determine the system time difference between UTRAN and GSM cells. This is defined as the difference between the time at the beginning of the P-CCPCH frame with SFN $= 0$ from cell i and the time at the beginning of the GSM BCCH 51-multiframe from GSM frequency j received, closest in time after the P-CCPCH above.

10.11.2.12 Pathloss   This measurement is used to report the RSCP with respect to PCPICH Tx power. It is calculated as:

$$Pathloss = PCPICH \text{ Tx power} - CPICH\_RSCP \qquad (10.4)$$

The reporting range is from 46 to 158 dB.

10.11.3 Measurement Types

10.11.3.1 RACH Measurement Reporting   Measurements of the reporting quantity identified in SIB 11 or SIB 12 are broadcast on the BCH by the UTRAN. The IE 'reporting quantity' is used to define which of the following unfiltered measurements to report:

1. $CPICH\_E_c/N_0$;

2. $CPICH\_RSCP$;

3. Pathloss.

The results have to be ordered with the first cell being the 'best'. The total number of cells is dictated by the optional IE 'maximum number of reported cells on RACH'. The IE 'measured results on RACH' can be transmitted on many uplink PDUs including cell update, initial direct transfer and RRC connection request. These allow the UTRAN to take basic unfiltered cell measurements of a UE whilst it is not in the DCH state, where it would normally send measurement report PDUs to report cell status.

10.11.3.2 *Intrafrequency Measurements*

Intrafrequency measurements are defined as measurements on downlink physical channels on the current active set frequency. In DCH these measurements are reported to the UTRAN via a measurement report PDU, and in FACH are reported in the IE 'measured results in RACH' which can be transmitted in several UL PDUs. The IE 'report criteria' determine what mode the measurement will report in: periodic, event or additional measurement for adding to another measurement report. All types share the same cell list, measurement quantity, reporting quantity and validity. The IE 'measurement quantity' is used to define which measurements to evaluate:

1. CPICH E_c/N_0;
2. CPICH RSCP;
3. Pathloss;
4. ISCP measured on a timeslot basis.

The IE 'reporting quantity' is used to determine which measurements are reported in the measurement report. This IE stipulates specific quantities to be reported for active, monitored and detected sets. Any combination of the following quantities can be chosen:

1. SFN–SFN observed time difference, type 1 and type 2;
2. Cell synchronization information;
3. CPICH E_c/N_0;
4. CPICH RSCP;
5. Pathloss.

The IE 'measurement validity' dictates which RRC states the measurement is valid in. The IE 'reporting cell status' is used to restrict the overall number of cells which can be reported. Periodic measurements have a reporting amount and interval, defined in the IE 'periodic reporting criteria'. A measurement report containing the relevant cell measurements is generated every interval until the correct amount has been generated. Event driven measurements are configured by the IE 'intrafrequency measurement reporting criteria'. The following list defines the different events which can be configured:

Event 1a. A primary CPICH enters the reporting range.

Event 1b. A primary CPICH leaves the reporting range.

Event 1c. A nonactive primary CPICH becomes better than an active primary CPICH.

Event 1d. Change of best cell.

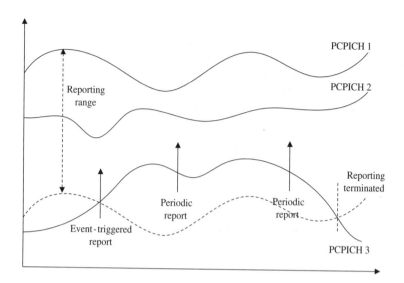

Figure 10.8. Periodic reporting triggered by Event 1a according to figure 14.1.4.1-1 [141].

Event 1e. A primary CPICH becomes better than an absolute threshold.

Event 1f. A primary CPICH becomes worse than an absolute threshold.

For the events defined above, a single cell is compared against a set of cells defined by the type of event using given formulae, an example is shown below. An event-triggered report will be generated, as marked in Figure 10.8, when a triggering equation is satisfied, and terminated when the leaving equation is satisfied, marked by 'Reporting terminated' in the figure. No measurement reports will be generated between the triggered report and the report being terminated, with the exception of event periodic reports for Events 1a and 1c. The triggering equation for Event 1a is (see Section 14.1.2.1 in [141]):

$$10 \cdot \mathrm{Log} M_{\mathrm{New}} + \mathrm{CIO}_{\mathrm{New}} \geq W \cdot 10 \cdot \mathrm{Log}\left(\sum_{i=1}^{N_A} M_i\right) +$$
$$(1 - W) \cdot 10 \cdot \mathrm{Log} M_{\mathrm{Best}} - (R_{1a} - H_{1a}/2) \quad (10.5)$$

while the leaving equation for Event 1a is:

$$10 \cdot \mathrm{Log} M_{\mathrm{New}} + \mathrm{CIO}_{\mathrm{New}} < W \cdot 10 \cdot \mathrm{Log}\left(\sum_{i=1}^{N_A} M_i\right) +$$
$$(1 - W) \cdot 10 \cdot \mathrm{Log} M_{\mathrm{Best}} - (R_{1a} - H_{1a}/2) \quad (10.6)$$

Parameters that affect these equations are $\mathrm{CIO}_{\mathrm{New}}$, W, M_{New}, M_{Best}, H_{1x}, R_{1x} and time to trigger, where:

- cell individual offset ($\mathrm{CIO}_{\mathrm{New}}$) is used to affect the measured result of individual cells and can be used to weight individual cells in the equations, both up and down;

- weighting (W) is used as a weighting parameter to bias the reporting range towards either the best or sum of measured results. If W is set to 0 the reporting range is defined relative to the best primary CPICH. If W is set to 1, then the reporting range is defined relative to the sum of cells M_i not in the forbidden list. Values in between set the range between the max and sum using a linear range;

- M_{new} is the current cell measurement;

- M_{Best} is the best cell measurement;

- hysteresis (H_{1x}) is used to limit the amount of reporting by offsetting the result;

- reporting range (R_{1x}) is used to set the trigger level relative to the weighted calculated sum against best before triggering;

- time to trigger determines a delay when the event should be active before the event is triggered;

- in addition, specific cells can be forbidden from affecting the reporting range.

For Events 1a and 1c only, event period reporting can be made during the time when the report is active, this is marked as 'periodic report' on Figure 10.8.

10.11.3.3 *Interfrequency Measurements*
Interfrequency measurements are defined as measurements on specified downlink physical channels on frequencies in addition to the current active set frequency. In DCH these measurements are reported to the UTRAN via a measurement report PDU. Interfrequency measurement processing is based on calculating a frequency quality estimate for 'used' and 'nonused frequencies to be compared against thresholds supplied by the UTRAN and also against each other to find the best frequency. A 'nonused' frequency is defined as a frequency that the UE has been instructed to measure but is not used for the current connection (i.e. not in the active set). A 'used' frequency is defined as a frequency that the UE has been instructed to measure and is also currently used for the connection. A 'virtual active set' is maintained for each nonused frequency, this is used in the evaluation of the frequency quality estimate for a 'nonused frequency. This is controlled via the IE 'interfrequency SET UPDATE' based on the mode parameter as follows. If the mode is set to 'Off' the UTRAN supplies the list of scrambling codes to be used in the virtual active set, subsequent measurement control messages can be sent to modify these lists.

If the mode is set to 'On' or 'On with no reporting' any interfrequency measurement must maintain a set of intrafrequency event measurements for that nonused frequency. These are either supplied by the UTRAN when the measurement is set up, or extracted from intrafrequency measurements being taken at the time the interfrequency measurement was set up. In addition, if the '..with no reporting' is set then an interfrequency measurement will not generate a measurement report when an intrafrequency measurement is satisfied on its nonused frequency. The IE 'report criteria' determine what mode the measurement will report in: periodic, event or additional measurement. All types share the same cell list, measurement quantity, reporting quantity, validity and interfrequency set update information.

The IE 'measurement quantity' is used to define the quantity to take measurements on. If the measurement is configured to update the virtual active set, then a separate measurement quantity of the same type as an intrafrequency measurement is stored for evaluating Events 1a, 1b and 1c for the nonused frequency. The following define the measurement quantity for the interfrequency measurement:

1. CPICH$\_E_c/N_0$;

2. CPICH_RSCP.

The IE 'reporting quantity' is used to determine what to report in a measurement report, this can include:

1. UTRA carrier RSSI;

2. Frequency quality estimate;

3. SFN–SFN observed time difference, type 1 and type 2;

4. Cell synchronization information;

5. CPICH$\_E_c/N_0$;

6. CPICH_RSCP;

7. Pathloss.

The IE 'measurement validity' dictates which RRC states the measurement is valid in. The IE 'interfrequency SET UPDATE' is used to define the mode of autonomous operation as described above. Event driven measurements are configured by the IE 'interfrequency measurement reporting criteria'. The following list defines the different events which can be configured:

Event 2a. Change of best frequency.

Event 2b. The estimated quality of the currently used frequency is below a certain threshold and the estimated quality of a nonused frequency is above a certain threshold.

Event 2c. The estimated quality of a nonused frequency is above a certain threshold.

Event 2d. The estimated quality of the currently used frequency is below a certain threshold.

Event 2e. The estimated quality of a nonused frequency is below a certain threshold.

Event 2f. The estimated quality of the currently used frequency is above a certain threshold.

The frequency quality estimate $Q_{carrier_j}$ used for interfrequency events is defined as (Section 14.2.0b.1 in [141])

$$Q_{\text{carrier}j} = 10 \cdot \text{Log} M_{\text{carrier}j} =$$

$$W \cdot 10 \cdot \text{Log} \left(\sum_{i=1}^{N_{Aj}} M_{ij} \right) + (1 - W_j) \cdot 10 \cdot \text{Log} M_{\text{Best}j} \qquad (10.7)$$

Parameters that affect this computation are $M_{\text{carrier}j}$, W_j, M_{ij}, $M_{\text{Best}j}$, where:

- $M_{\text{carrier}j}$ is the current measurement on the given frequency j;

- W_j is used to weight the result and pertains to frequency j;

- M_{ij} is the ith cell which is used to calculate the sum on frequency j;

- $M_{\text{Best}j}$ is the best cell measurement on frequency j;

- hysteresis (H_{2x}) is used to limit the amount of reporting by offsetting the result;

- time to trigger determines a delay when the event should be active before the event is triggered.

In order to take measurements on a nonused frequency the UE needs to use one of the following methods:

FACH measurement occasion. This is used to define a set of frames in FACH where the UE can take measurements on another frequency or RAT.

Compressed mode. This is a mechanism which is used in DCH, allowing measurements to be taken on nonused frequencies. It defines a set of gap patterns in which the physical channel will stop transmitting data, and allow it to take measurements on another frequency or RAT. These patterns are transmitted in various DL messages used by the UE via the IE 'DPCH compressed mode info'.

The UE performs checks on these patterns prior to performing measurements to ensure that similar resources are not trying to perform the same measurements at the same time. If these checks fail, then a measurement control failure is sent to the UTRAN to identify the error.

The physical layer is responsible for performing real time checks to ensure that the gap patterns do not overlap during run. If there are overlaps, the UE must send a physical channel reconfiguration failure PDU to inform the UTRAN of the error.

10.11.3.4 *InterRAT Measurements*   These are measurements of downlink physical channels on another RAT. In DCH these measurements are reported to the UTRAN via a measurement report PDU. The IE 'report criteria' determine what mode the measurement will report in: periodic, event or additional measurement. All types share the same cell list, measurement quantity, reporting quantity and reporting cell status. The IE 'measurement quantity' for UTRAN is used to compute the frequency quality estimate for the active set and can be:

1. Downlink E_c/N_0;

2. Downlink RSCP.

The measurement quantity for GSM can be GSM carrier RSSI.

Measurements on GSM cells can be requested with BSIC verified or BSIC nonverified. Different gap patterns are provided for GSM RSSI measurement, for initial BSIC, or for BSIC reconfirmation. Initial BSIC identification is defined as searching for the BSIC and decoding the BSIC for the first time when there is no knowledge about the relative timing between the FDD and GSM cell. BSIC reconfirmation is defined as tracking and decoding the BSIC of a GSM cell after initial BSIC identification is performed.

The IE 'interRAT reporting quantity' defines the quantities that the UE will report to the UTRAN when the event is triggered and can be:

1. Observed time difference to the GSM cell;

2. GSM carrier RSSI.

Periodic measurements have a reporting amount and interval, defined in the IE 'periodic reporting criteria'. A measurement report containing the relevant cell measurements is generated every interval until the correct amount has been generated. The IE 'reporting cell status' is used to restrict the overall number of cells which can be reported. Event driven measurements are configured by the IE 'interRAT measurement reporting criteria'. The following list defines the different events which can be configured:

Event 3a. The estimated quality of the currently used UTRAN frequency is below a certain threshold and the estimated quality of the other system is above a certain threshold.

Event 3b. The estimated quality of the other system is below a certain threshold.

Event 3c. The estimated quality of the other system is above a certain threshold.

Event 3d. Change of best cell in the other system.

The frequency quality estimate Q_{UTRAN} used for interRAT events is defined as (Section 14.3.0b in [141])

$$Q_{UTRAN} = 10 \cdot Log M_{UTRAN} =$$

$$W \cdot 10 \cdot Log \left(\sum_{i=1}^{N_A} M_i \right) + (1 - W) \cdot 10 \cdot Log M_{Best} \qquad (10.8)$$

Parameters that affect this computation are M_{UTRAN}, W, M_i, M_{Best}, where:

- M_{UTRAN} is the current cell measurement;
- W is used to weight the result;
- M_i is the ith cell which is used to calculate the measured sum;
- M_{Best} is the best cell measurement;
- hysteresis (H_{3x}) is used to limit the amount of reporting by offsetting the result;
- time to trigger determines a delay when the event should be active before the event is triggered.

10.11.3.5 *Traffic Volume Measurements*
In DCH and FACH, measurements of uplink traffic volume are reported to the UTRAN via a measurement report PDU. The IE 'report criteria' determine which mode the traffic volume measurements will report in: periodic, event or as an additional measurement. All types share the same object, measurement quantity, reporting quantity and validity. The IE 'measurement object' defines a list of transport channel identities of type DCH, RACH or USCH on which the traffic volume measurement will be performed. The IE 'measurement quantity' defines the time interval in the range 10 to 260 ms in steps of 20 ms and the quantity to be measured, which can be one of the following:

- buffer occupancy;
- average buffer occupancy;
- variance of buffer occupancy.

The reporting quantity indicates which of the buffer occupancy, average and variance are to be reported when a measurement report is generated. The UTRAN can configure all of

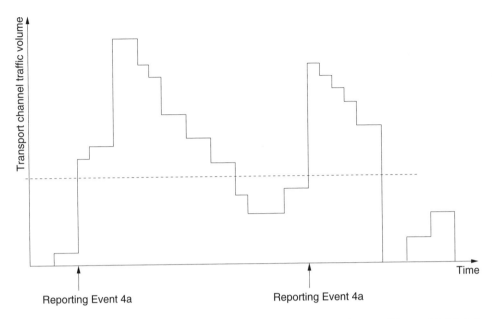

Figure 10.9. Periodic reporting triggered by Event 4a according to Figure 14.4.2.1-1 in [141].

the quantities to be reported, or any combination. Periodic measurements have a reporting amount and interval as defined in IE 'periodical reporting criteria' above. Event driven measurements are configured by the IE 'traffic volume measurement reporting criteria'. The following list defines the different events that can be configured:

Event 4a. Transport channel traffic volume becomes larger than an absolute threshold (see Figure 10.9).

Event 4b. Transport channel traffic volume becomes smaller than an absolute threshold.

Parameters that affect the generation of events are as follows:

- reporting threshold is the threshold value in bytes used to generate Event 4a and Event 4b. The range is 8 kB to 768 kB;

- time to trigger is used to ensure that the event is satisfied for a specific length of time prior to sending a measurement report. The range is 0 to 5 s;

- pending time after trigger is used to prevent another measurement report being generated for a given time period after an initial report. The range is 250 ms to 16 s;

- Tx interruption after trigger is used to prevent the UE from transmitting on the RACH for a short period after a measurement report has been generated. The range is 250 ms to 16 s.

10.11.3.6 *Quality Measurements* Downlink BLER is reported for each requested transport channel. The IE 'report criteria' determines which mode the traffic volume measurements will report in: periodic, event or as an additional measurement. All types share

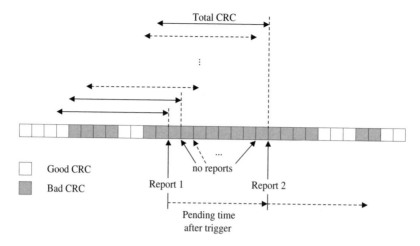

Figure 10.10. Periodic reporting triggered by Event 5a according to Figure 14.5.2.1-1 in [141].

the same reporting quantity. The IE 'reporting quantity' gives a list of downlink transport channel identities to be measured. This IE is optional, and if not present the default is to report BLER on all downlink transport channels. Periodic measurements have a reporting amount and interval as defined in IE 'periodical reporting criteria' above. Event driven measurements are configured by the IE 'quality measurement reporting criteria'. The following defines the event that can be configured:

Event 5a. A predefined number of bad CRCs is exceeded (see Figure 10.10).

Parameters that affect the generation of the event are as follows:

- transport channel ID, used to define the transport channel to report;

- total CRC, the length of the sliding window that event verification is to be performed upon. The range is 1 to 512 CRCs;

- bad CRC, the number of bad CRCs in the total CRC sliding window that are allowed before a quality event measurement report will be generated. The range is 1 to 512 CRCs;

- pending after trigger, the length of time, defined in terms of received CRCs, before another quality measurement report can be generated. The range is 1 to 512 CRCs.

10.11.3.7 UE Internal Measurements   UE internal measurements are measurements on transmission power and received signal strength. In DCH these measurements are reported to the UTRAN via a measurement report PDU. The IE 'report criteria' determine which mode the UE internal measurements will report in: periodic, event or as an additional measurement. All types share the same measurement quantity and reporting quantity. The IE 'measurement quantity' is defined as one of:

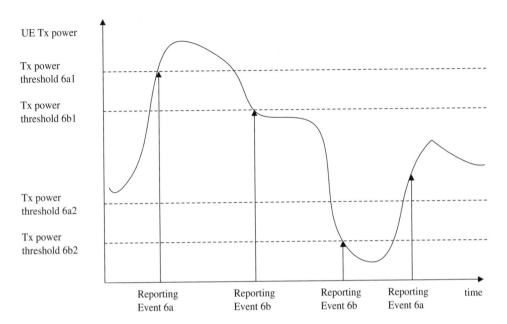

Figure 10.11. Periodic reporting triggered by Events 6a and 6b according to Figure 14.6.2.1-1 in [141].

1. UE transmitted power;

2. UTRA carrier RSSI;

3. UE Rx–Tx time difference.

The IE 'reporting quantity' defines the measurements to be reported. These can be the UE transmitted power and the UE Rx–Tx time difference. Periodic measurements have a reporting amount and interval as defined in the IE 'periodical reporting criteria' above. Event driven measurements are configured by the IE 'UE internal measurement reporting criteria'. The following list defines the different events that can be configured:

Event 6a. The UE Tx power becomes larger than an absolute threshold (see Figure 10.11).

Event 6b. The UE Tx power becomes less than an absolute threshold (see Figure 10.11).

Event 6c. The UE Tx power reaches its minimum value.

Event 6d. The UE Tx power reaches its maximum value.

Event 6e. The UE RSSI reaches the UE's dynamic receiver range.

Event 6f. The UE Rx–Tx time difference for an RL included in the active set becomes larger than an absolute threshold.

Event 6g. The UE Rx–Tx time difference for an RL included in the active set becomes less than an absolute threshold.

Parameters that affect the generation of events are as follows:

- time to trigger determines a delay when the event should be active before the event is triggered;

- UE transmitted power Tx power threshold, used in Events 6a and 6b, with a range of 50 to 33 dBm;

- UE Rx–Tx time difference threshold, used in Events 6f and 6g, with a range of 768 to 1280 chips.

10.11.3.8 *UE Positioning Measurements*   In DCH and FACH, measurements relating to the UE's position are reported to the UTRAN via a measurement report PDU. The IE 'report criteria' determines which mode the UE positioning measurements will report in: periodic, event or as an additional measurement. All types share the same measurement quantity, reporting validity, and can optionally contain *On Time Difference of Arrival* (OTDOA) assistance data for UE-assisted, OTDOA assistance data for UE-based or GPS assistance data. This assistance data can also be read from the SIBs broadcast on the BCH by the UTRAN. The IE 'UE positioning reporting quantity' defines both the method to be used (as UE-assisted, UE-based, UE-assisted preferred or UE based preferred) and the positioning type (as OTDOA, GPS or cell ID). It also provides specific control information for use in taking the measurement. In the cell ID based method, the position of a UE is estimated with the knowledge of its serving Node B. The cell coverage based positioning information can be indicated as the cell identity of the used cell, the service area identity or as the geographical coordinates of a position related to the serving cell. The position information will include a QoS estimate.

When geographical coordinates are used as the position information, the estimated position of the UE can be a fixed geographical position within the serving cell, the geographical centre of the serving cell coverage area, or some other fixed position within the cell coverage area. The OTDOA–IPDL method involves measurements made by the UE of the UTRAN frame timing. These measures are then sent to the UTRAN and the position of the UE is calculated by the UTRAN. The Node B may provide *Idle Periods in the Downlink* (IPDL) in order potentially to improve the hearability of neighbouring Node Bs. The support of these idle periods in the UE is optional. Support of idle periods in the UE means that its OTDOA performance will improve when idle periods are available. Network-assisted GPS makes use of UEs which are equipped with radio receivers capable of receiving GPS signals. The quantity to measure for UE positioning is dependent on the method and type requested in the IE 'UE positioning reporting quantity':

1. SFN–SFN observed time difference type 2, mandatory;

2. Rx–Tx time difference type 2, optional;

3. GPS timing of cell fames, optional.

The IE 'measurement validity' dictates which RRC states the measurement is valid in. Periodic measurements have a reporting amount and interval as defined in the IE 'periodical reporting criteria' above. Event driven measurements are configured by the IE 'UE positioning reporting criteria'. The following list defines the different events that can be configured:

Event 7a. The UE position changes more than an absolute threshold (UE-based methods only).

Event 7b. SFN–SFN measurement changes more than an absolute threshold.

Event 7c. GPS time and SFN time have drifted apart more than an absolute threshold.

Parameters that affect the generation of events are as follows:

- threshold position change shows how much the position should change compared to the last reported position before Event 7a is triggered. The range is enumerated from 0 to 1 000 000 m;

- threshold SFN–SFN change shows how much the SFN–SFN measurement of any measured cell is allowed to change before Event 7b is triggered. The range is enumerated from 0.25 to 5000 chips;

- threshold SFN–GPS TOW when the GPS TOW and SFN timer have drifted apart more than the specified value, Event 7c is triggered. The range is enumerated from 1 to 100 ms.

10.11.3.9 *Cell Selection/Reselection*   Unfiltered measurements are taken on cells identified in the CELL_INFO_LIST. These are passed on to the cell selection/reselection component. The measurement quantities to be stored for selection/reselection purposes are:

1. CPICH_E_c/N_0;

2. CPICH_RSCP.

The measurement component is only responsible for maintaining the CELL_INFO_LIST, taking the raw measurements on these cells and passing them on to the cell selection/reselection component.

10.12 POWER CONTROL

One of the more important roles the RRC performs is the management of *Power Control* (PC). There are three types of power control:

- open loop power control;
- inner loop power control;
- outer loop power control.

The simplest form of power control is open loop power control. In open loop power control, the RRC simply calculates the initial power of an uplink channel based on parameters sent by the network, and by the measurement of the CPICH_RSCP of the serving cell. Open loop power control is therefore only applied to uplink channels, i.e. the initial RACH preamble power, and to the power used for the DPCCH of a DCH when the channel is started.

Inner loop power control is the fast control of power signalled within Layer 1. As the UE moves, or the signal fades, the powers of both the UE and network transmitters are adjusted up and down. Inner loop power control requires signalling in both directions and therefore only applies to DCH and CPCH channels. Of course, inner loop power control requires a target value for the Layer 1 algorithm to work against, and it is this process that is termed *Outer Loop Power Control* (OLPC). OLPC is much more complex than the other forms of power control, and is also one place where the RRC designer is given leeway to implement his own algorithm in an area which considerably affects the performance of the system. OLPC is applied to DCH (and CPCH) channels.

Outer loop power control requires that the RRC sets up the Layer 1 target in accordance with the demands of the service that is being offered, and also in accoordance with the environment in which the UE sits. For example, if the UE is moving, it will encounter fading at a different rate from if it were stationary, and this might require a different target value for the inner loop power control if the UE is still to achieve the same performance. The exact nature of the algorithm used for outer loop power control is unconstrained by the specifications, and it is therefore left to the implementer to design the algorithm to achieve this. By way of an example, the UE might monitor the block error rate of the received channel for a given inner loop target, and use the performance of previously chosen inner power control targets to guide the selection for a new configuration.

10.13 ARBITRATION OF RADIO RESOURCES ON UPLINK DCH

This functionality is only required for UEs which support the *Dynamic Resource Allocation procedure* (DRAC), and it only applies to the Cell_DCH state. When it is configured, the RRC must monitor specified FACH channels on each cell in the active set, to read the SIB 10 sent by the active set cells. The SIB 10 contents provide random backoff parameters for particular DCH transport channels. When the UE wishes to transmit data on these channels the RRC must first compute the backoff parameters to determine whether the transport channel can be used.

10.14 INTEGRITY PROTECTION

Integrity protection is the process used to verify that messages received by the RRC are bona fide messages sent by the peer RRC. Messages received from illegitimate sources will fail the integrity check, and will be ignored. The integrity protection algorithm uses the following parameters:

- integrity key;
- count-I;
- direction (i.e. uplink/downlink);
- fresh;
- message;
- RB ID – 1.

The algorithm (F9) used by integrity protection is defined in [153, 154]. The initial integrity protection parameters are set up during connection establishment. The integrity key is obtained from the USIM, and the count-I value is initialized to the value of the start parameter saved in the USIM, multiplied by 16. Note that two sets of these values are used, CS and PS, depending on whether the connnection is for a CS or PS signalling connection. 'Fresh' is configured from UTRAN during security mode configuration. 'Message' is the contents of the RRC message to be integrity protected.

When integrity protection has started, the parameters are used to calculate a value called MAC-I, which is included in the PDU, and the value of count-I is then incremented. The receiving RRC recalculates MAC-I, and if the calculated MAC-I differs, the PDU is discarded. Integrity protection is updated each time the security parameters are changed, and also if a new signalling connection is made; the integrity protection algorithm always uses the parameters for the latest configured connection. Also, at the end of a connection the values of count-I are checked against threshold values stored in USIM. If the count exceeds the threshold, the key in USIM is deleted, forcing a new key to be created during authentication by the NAS when the next connection is established.

10.15 CIPHERING MANAGEMENT

Ciphering is used to prevent eavesdroppers from gaining access to the contents of messages sent between the UE and UTRAN. In some respects the ciphering process is similar to integrity protection, except that in this case it is performed by Layer 2. However, overall control of the process remains the responsibility of the RRC. The ciphering algorithm uses an algorithm termed F8, which requires the following parameters:

- ciphering key;
- count-C;
- direction (i.e. uplink/downlink);
- length (i.e. length of data to be encrypted);
- message;
- RB ID – 1.

Some of these parameters are either fixed or are the responsibility of Layer 2, the RRC is only concerned with the keys and count-C values required and also with the type of RLC affected (which determines whether ciphering is performed by MAC or RLC). The keys and count values operate similarly to the keys used for integrity protection, except that user RABs use the parameters defined for that CN-domain; only the signalling bearers use the parameters for the most recently connected signalling domain.

10.15.1 RRC Ciphering Actions

To manage ciphering, the RRC performs three activities:

- activating, reconfiguring or deactivating ciphering parameters;
- adding bearers with appropriate values of count-C;
- synchronizing count-C values for SRNS relocation.

When activating, reconfiguring or deactivating ciphering, the RRC needs to suspend and resume RLC entities to ensure that ciphering is applied at the activation times given by UTRAN, if the affected PDUs are to be decoded correctly. Bearers which are added require count-C values using the values which have been reached by other RBs for the same domain – hence the RRC must interrogate Layer 2 to calculate the count values to be used by the new bearers. For synchronization, the RRC is required to read the count-C values from Layer 2 and report them to the UTRAN. It may also be given new count-C values which Layer 2 should use after the relocation.

10.16   PDCP CONTROL

When radio bearers are configured for packet services, they are mapped through the PDCP entity in Layer 2. The PDCP entity may also be configured to perform header compression, and PDCP PDUs may be numbered to permit lossless reconfiguration. The RRC performs four roles in respect of the PDCP sublayer:

- configuration;
- reconfiguration;
- relocation;
- release.

10.16.1   Configuration

The RRC must initialize the PDCP entity for each PS radio bearer. The information for this is provided in the RRC configuration messages (such as radio bearer setup) which are sent by UTRAN to configure the required radio bearers.

10.16.2   Reconfiguration

The RRC will modify the PDCP entities for each PS radio bearer, e.g for reinitialization after an SRNS relocation. Again, the information to do this is provided in PDUs sent from UTRAN as part of the reconfiguration procedure. Depending on implementation, an RRC may also need to reinitialize PDCP entities in the event of an RLC reset, although this is more straightforwardly signalled directly between PDCP and RLC.

10.16.3   Relocation

In some PDCP modes lossless SRNS relocation may be supported. In this case the RRC must interrogate the PDCP entity after a reconfiguration, and report the next expected PDCP sequence numbers of the last PDUs sent by the peer, and the numbers of the PDUs last sent by the UE. Also, it will be signalled with numbers of the PDUs next expected by the peer, for configuration into the UE PDCP entity.

10.16.4   Release

Lastly, the RRC is responsible for releasing PDCP resources when the PS radio bearer is released.

10.17   CBS CONTROL

The RRC controls the configuration of the *Broadcast/Multicast Control* (BMC) sublayer of L2, which facilitates the reception of *Cell Broadcast Service* (CBS) messages that are used to provide user information to all receivers in a particular cell, a list of cells or geographical area. These messages contain information about sports updates, news updates, stock market updates, etc.

10.17.1   Configuration for the BMC

The RRC has to configure the L2 sublayers to receive CBS messages in an unacknowledged mode RLC on the CTCH logical channel mapped on the FACH transport channel via the S-CCPCH physical channel. For release 99 only, a single CTCH, which can be mapped only on FACH, will be created and the FACH will have a fixed TB size and fixed TTI.

10.17.2   Initialization Procedure

The user, i.e. a CBS application, triggers the BMC by activating some message IDs, which correspond to the type of message the user is interested in receiving. These message IDs are provided by the service provider (list of all the types of message he supports). On receiving this trigger the BMC requests the RRC to start reading the FACH data for CTCH. The RRC then waits for the CBS related system broadcast information (SIB 5), which contains the data for configuring the CTCH on FACH, and also the CTCH allocation period and frame offset for CTCH.

10.17.3   Scheduling of CBS Messages

There are two kinds of scheduling required to read the CBS messages efficiently at the physical layer at the time scheduled by the RNC.

10.17.3.1   Level 1 Scheduling   This is the scheduling information received by UE in the SIB 5 for the FACH, which supports the CTCH logical channel. This information contains the CTCH allocation period and CBS-frame offset, which help the UE to configure the CTCH on FACH at regular intervals and read the CBS messages if being transmitted on that radio frame by the RNC. This scheduling is essential at the beginning when the BMC has just been activated by the user as the UE does not know the radio frame at which the BMC will receive the CBS message which will contain the message ID activated by the user. So the UE will wait for the next SIB 5 to be transmitted in the cell, get the CBS related

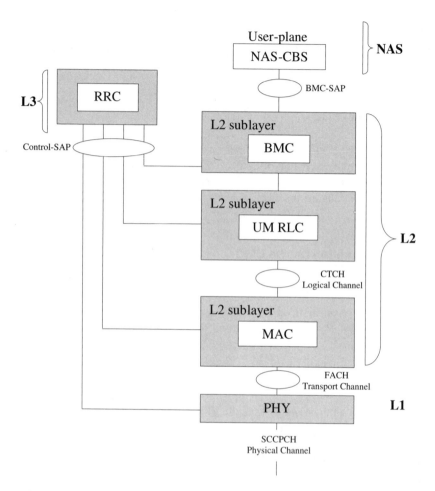

Figure 10.12. CBS control on UE side.

information and then read all the messages on the FACH for CTCH in order to read the CBS schedule message, which will provide the information about the messages scheduled during the next DRX schedule period.

If M is the number of radio frames in one TTI of the FACH used for CTCH, and N is the CTCH allocation period in number of radio frames:

$$M \leq N \leq \text{Maximum SFN} - K \tag{10.9}$$

where N is a multiple of M, and Maximum SFN is 4096.

If K is the CBS frame offset, an integer number of radio frames:

$$0 \leq K \leq N - 1 \tag{10.10}$$

where K is a multiple of M.

An example is given in Figure 10.13.

Table 10.1 Level 2 scheduling example.

All CTCH-BS index	SFN	Selected by BMC
1	3,4	1
2	9,10	0
3	15,16	1
4	21,22	1
5	27,28	0
6	33,34	0

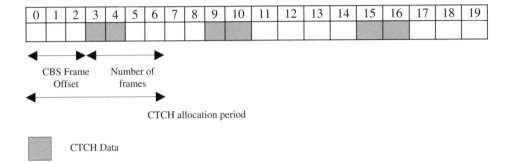

Figure 10.13. A level 1 scheduling example for $N = 6$, $K = 3$ and $M = 2$.

10.17.3.2 Level 2 Scheduling The BMC receives a CBS schedule message every DRX schedule period, which contains all the information on all the CBS messages that will be transmitted in the cell during the next schedule period. BMC evaluates this information based upon the message IDs that the user has activated it to receive and generates a bitmap for the messages it has to receive. Once the bitmap has been generated, the BMC sends this information to the RRC layer, so that the RRC can configure the lower layers to read the FACH for CTCH data only at the time the CBS message required by the user is scheduled. This saves a lot of unwanted scheduling at the UE. Values assumed for Table 10.1 below are $N = 6$, $K = 3$ and $M = 2$. The UE will read the CBS data only at the SFNs directed by the BMC.

10.17.4 Design Issues

There are two ways of looking at how the UE can control the scheduling. The first and the most obvious is that the RRC can configure the lower layers on initial activation from BMC and read the CTCH blocks whenever they arrive. The support of level 1 scheduling requires minimal design as the CTCH allocation and the CBS frame offset are fixed until such time as new parameters are not received. But support of the level 2 scheduling brings more complex designs as the BMC chooses the CBS messages it wants the UE to read and ignores the rest. In this case the list of the CTCH block set indices provided by BMC will require the UE to maintain a list and calculate the corresponding SFN at which the CBS message is to be accepted. This would require the scheduling to be taken care of by individual layers at the correct SFN, since the idea of configuration being controlled by L3 (RRC) would lead to an unnecessary overhead of configuring all the layers repetitively.

11

Speech Coding for UMTS[1]

Hee Tong How and Lajos Hanzo

11.1   INTRODUCTION – THE ADAPTIVE MULTIRATE (AMR) SPEECH CODEC

As described earlier, the access stratum in UMTS is capable of carrying a wide variety of services with different data rate, delay and error rate requirements. Hence, a wide variety of different services can be offered by operators to their users. However, even with this new flexibility, speech will remain a central application. In this chapter we give a brief description of the *Adaptive Multirate* (AMR) speech codec used for narrowband speech coding in UMTS terminals.

Speech encoders/decoders (known as *codecs*) have long been used in digital communication systems to remove redundancy from the speech signal, and so reduce the required data rate, before transmission. Typically, the lower the data rate of the speech codec used, the more users can be supported by the system. However, lower data rate speech coding usually results in lower quality, more synthetic sounding speech at the receiver. Therefore, much research has been carried out to reduce the bit rate needed for so-called 'toll quality' speech. In this chapter we give only a brief overview of these techniques and of the AMR speech codec. More details of speech coding generally can be found, for example, in [156–158].

Traditionally, second generation systems use a speech coding scheme which encodes at a fixed rate, such as 12.2 kbit/s for the GSM *Enhanced Full Rate* (EFR) codec. The AMR codec [159] can operate in any one of eight modes, offering eight different encoded bit rates

---

[1] This chapter is based on H.T. How, T.H. Liew, E.L. Kuan and L. Hanzo: An AMR Coding, RRNS Channel Coding and Adaptive JD-CDMA System for Speech Communications, submitted to *IEEE Journal on Selected Areas in Communications*, 2000 and was based on collaborative research with the co-authors [155].

WCDMA – Requirements and Practical Design. Edited by R. Tanner and J. Woodard.
© 2004 John Wiley & Sons, Ltd. ISBN: 0-470-86177-0.

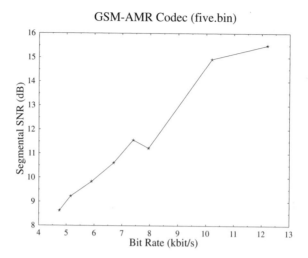

Figure 11.1. Segmental SNR performance of the AMR codec, operating at bit rates in the range between 4.75 kbit/s and 12.2 kbit/s.

from 4.75 to 12.2 kbit/s. This 'multirate' feature of the AMR codec allows the codec's rate to be lowered, for example to allow greater error protection for users with poor channel conditions, or increased capacity when the network is busy. Also, three of the AMR codec's modes correspond to existing standards, which renders communication systems employing the new AMR codec interoperable with other systems. Specifically, the 12.2 kbit/s mode is identical to the GSM EFR standard [160], the 7.4 kbit/s mode [157] corresponds to the US1 and EFR (IS-641) codecs of the TDMA (IS-136) system, and the 6.7 kbit/s mode is equivalent to the EFR codec of the Japanese PDC system [161].

The codec operates with narrowband speech sampled at 8 kHz[2]. It operates on a 20 ms frame of 160 speech samples, and generates encoded blocks of 95, 103, 118, 134, 148, 159, 204 and 244 bits for each 20 ms frame. This leads to bit rates of 4.75, 5.15, 5.9, 6.7, 7.4, 7.95, 10.2 and 12.2 kbit/s, respectively. The quality of the reconstructed speech from a speech codec can be approximately measured by its *Segmental Signal to Noise Ratio* (SEGSNR). The segmental SNR of the AMR speech codec is shown in Figure 11.1 for each of its eight bit rates.

In this chapter we first give an introduction to the general structure used for the AMR codec, before giving more details of individual blocks used and the bit allocation for different modes of the codec. We finally discuss the codec's sensitivity to errors in the received bits.

11.2 AMR STRUCTURE

The AMR codec employs the *Algebraic Code Excited Linear Predictive* (ACELP) model [163, 164] shown in Figure 11.2. Here we provide a brief overview of the AMR codec following the approach of [165–167].

---

[2]The inclusion of a wideband version of the AMR codec for speech sampled at 16 kHz and therefore offering more natural sounding quality, has also been standardized [162] for Release 5.

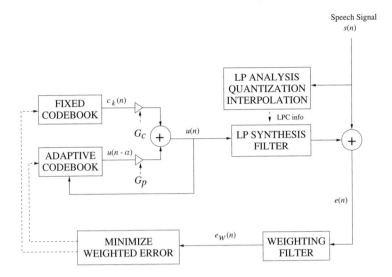

Figure 11.2. Schematic of ACELP speech encoder.

As shown in Figure 11.2, the ACELP encoder operates on the sampled input speech signal $s(n)$. For each 20 ms frame, *Linear Prediction Coding* (LPC) coefficients are found for a synthesis filter which describes the spectral envelope of the speech segment [161, 168]. The 20 ms frame is then split into smaller subframes, and for each subframe an excitation signal $u(n)$ is determined for the synthesis filter. An *Analysis by Synthesis* (AbS) procedure is employed, in order to find the particular excitation that minimizes the weighted *Mean Square Error* (MSE) between the reconstructed speech $\hat{s}(n)$ and the original speech signal $s(n)$. The MSE is weighted by a filter derived from the LPC synthesis filter. This allows the AbS procedure to take into account the fact that the quantization noise in the spectral neighbourhood of the speech formants[3] is less perceptible to listeners than noise in other spectral regions [161, 168].

The excitation signal $u(n)$ is given by the sum of weighted entries from two codebooks, the fixed and adaptive codebooks shown in Figure 11.2. The adaptive codebook consists of time shifted versions of past excitation sequences and describes the long term characteristics of the speech signal [161, 168]. In the original CELP codec [163] the fixed codebook contained pseudorandom noise sequences. In order to simplify the search of this fixed codebook, an *algebraic* structure is used in the ACELP codec instead. In order to reduce the complexity of the AbS search of the codebooks further, they are searched sequentially in order to find the perceptually best codebook entries, first for the adaptive codebook contribution, and then for the fixed codebook entry.

Once the encoder has determined the LPC filter coefficients and the excitation signal $u(n)$, parameters representing these are quantized and sent to the decoder. The decoder uses these parameters to form a replica of the synthesis filter and the excitation signal, and hence the reconstructed speech $\hat{s}(n)$.

---

[3] Formant is the terminology used to refer to a peak in the frequency response of the LPC filter modelling the vocal tract.

We now discuss each of the blocks in the ACELP structure shown in Figure 11.2, giving details for the 4.75 and 10.2 kbit/s modes.

11.3 LINEAR PREDICTION ANALYSIS

A 10th order LPC analysis filter is employed for modelling the short term correlation of the speech signal $s(n)$. Short term prediction, or linear predictive analysis, is performed once for each 20 ms speech frame using the Levinson–Durbin algorithm [168]. The ten LPC coefficients are transformed into *Line Spectrum Frequencies* (LSFs) for quantization and interpolation. The employment of the LSF [169] representation for quantization of the LPC coefficients is driven by their advantageous statistical properties. Specifically, within each speech frame, there is a strong intraframe correlation due to the ordering property of the neighbouring LSF values [168]. This vector quantization is described in Section 11.4.

The interpolated quantized LSFs are converted back to LPC filter coefficients, in order to construct a synthesis filter for each subframe. Without channel errors the same synthesis filter is used at the decoder for producing the reconstructed speech signal from the received excitation signal $u(n)$. As the weighting filter is not used at the decoder, the unquantized LSFs are used to produce this filter for each subframe.

11.4 LSF QUANTIZATION

In the AMR codec, the LSFs are quantized using interframe LSF prediction and *Split Vector Quantization* (SVQ) [165]. The SVQ aims to split the ten-dimensional LSF vector into a number of reduced dimension LSF subvectors, which simplifies the associated codebook entry matching and search complexity. Specifically, the SVQ scheme minimizes the average *Spectral Distortion* (SD) [170] achievable at a given total complexity. Predictive vector quantization is used [165] and the ten component LSF vectors are split into three LSF subvectors of dimension three, three and four. The bit allocations for the three subvectors will be described in Section 11.8 for the 4.75 and 10.2 kbit/s speech coding modes.

11.5 PITCH ANALYSIS

Pitch analysis using the adaptive codebook approach models the long term periodicity, i.e. the pitch, of the speech signal. It produces an output which is an amplitude scaled version of the previous excitation. The excitation signal $u(n) = G_p u(n - \alpha) + G_c c_k(n)$ seen in Figure 11.2 is determined from its G_p-scaled history after adding the G_c-scaled fixed algebraic codebook vector c_k for every 5 ms subframe. The optimum excitation is chosen on the basis of minimizing the weighted MSE E_w over the subframe.

In an optimal codec, the fixed codebook index and codebook gain, as well as the adaptive codebook parameters, would all be jointly optimized in order to minimize E_w [171]. However, in practice this is not feasible because of the associated excessive complexity. Hence, a sequential suboptimal approach is applied in the AMR codec, where the adaptive codebook parameters are initially determined under the assumption of zero fixed codebook excitation component, i.e. $G_c = 0$, since at this optimization stage no fixed codebook entry has been determined.

Given this assumption, the adaptive codebook delay α and gain G_p which minimize E_w can be determined. A so-called 'closed loop' search using the AbS approach over the full range of delays α is computationally expensive. Therefore, most CELP codecs employ both open loop and closed loop estimation of the adaptive codebook delay parameters, and this is the case in the AMR codec. The open loop estimate of the pitch period is used to narrow down the range of the possible adaptive codebook delay values and then the full closed loop AbS procedure is used for finding a high resolution delay around the approximate open loop position [164].

11.6 FIXED CODEBOOK WITH ALGEBRAIC STRUCTURE

Once the adaptive codebook parameters are found, the fixed codebook is searched by taking into account the now known adaptive codebook vector. The fixed codebook is searched by using an efficient AbS technique [172], minimizing the MSE between the weighted input speech and the weighted synthesized speech.

The fixed, or algebraic, codebook structure is specified in Table 11.1 and Table 11.2 for the 4.75 kbit/s and 10.2 kbit/s codec modes, respectively [165]. The computational complexity of the fixed codebook search is substantially reduced when the codebook entries $c_k(n)$ used are mostly zeroes. Hence, a structure based on the so-called *Interleaved Single Pulse Permutation* (ISPP) code design [171] is used. This algebraic structure of the fixed codebook allows for a fast search procedure. The nonzero elements of the codebook are equal to either $+1$ or -1, and their positions are restricted to a limited number of excitation pulse positions, as portrayed in Tables 11.1 and 11.2.

Table 11.1 Pulse amplitudes and positions for the 4.75 kbit/s AMR codec mode [165].

Subframe	Subset	Pulse: positions
1	1	i_0: 0,5,10,15,20,25,30,35 i_1: 2,7,12,17,22,27,32,37
	2	i_0: 1,6,11,16,21,26,31,36 i_1: 3,8,13,18,23,28,33,38
2	1	i_0: 0,5,10,15,20,25,30,35 i_1: 3,8,13,18,23,28,33,38
	2	i_0: 2,7,12,17,22,27,32,37 i_1: 4,9,14,19,24,29,34,39
3	1	i_0: 0,5,10,15,20,25,30,35 i_1: 2,7,12,17,22,27,32,37
	2	i_0: 1,6,11,16,21,26,31,36 i_1: 4,9,14,19,24,29,34,39
4	1	i_0: 0,5,10,15,20,25,30,35 i_1: 3,8,13,18,23,28,33,38
	2	i_0: 1,6,11,16,21,26,31,36 i_1: 4,9,14,19,24,29,34,39

Table 11.2 Pulse amplitudes and positions for the
10.2 kbit/s AMR codec code [165].

Track	Pulse	Positions
1	i_0, i_4	0,4,8,12,16,20,24,28,32,36
2	i_1, i_5	1,5,9,13,17,21,25,29,33,37
3	i_2, i_6	2,6,10,14,18,22,26,30,34,38
4	i_3, i_7	3,7,11,15,19,23,27,31,35,39

In the 4.75 kbit/s codec mode, the excitation codebook contains two nonzero pulses, denoted by i_0 and i_1 in Table 11.1, for each 40 sample subframe. The 40 positions in a subframe are divided into four so-called tracks. Two subsets of two tracks each are used for each subframe, with one pulse in each track. Different subsets of tracks are used for each subframe, as shown in Table 11.1, and hence one bit is needed for encoding the subset used. The two pulse positions i_0 and i_1 are encoded using three bits each, since both have eight legitimate positions in Table 11.1. Furthermore, the sign of each pulse is encoded using one bit. This gives a total of nine bits for the algebraic excitation encoding in a subframe.

In the 10.2 kbit/s codec mode of Table 11.2 there are four tracks, each containing two pulses. Hence, the excitation vector contains a total of $4 \times 2 = 8$ nonzero pulses. All the pulses can have the amplitudes of $+1$ or -1 and the excitation pulses are encoded using a total of 31 bits.

For the quantization of the fixed codebook gain, a gain predictor is used in order to exploit the correlation between the fixed codebook gains in adjacent frames [165]. The fixed codebook gain is expressed as the product of the predicted gain based on previous fixed codebook energies and a correction factor. The correction factor is the parameter which is coded, together with the adaptive codebook gain, for transmission over the channel. In the 4.75 kbit/s mode, the adaptive codebook gains and the correction factors are jointly vector quantized every two subframes, while this process occurs every subframe in the 10.2 kbit/s mode.

11.7 POST PROCESSING

At the decoder, an adaptive post filter [173] is used for improving the subjective quality of the reconstructed speech. The adaptive post filter consists of a formant based post filter and a spectral tilt compensation filter [173]. *Adaptive Gain Control* (AGC) is also used, in order to compensate for the energy difference between the synthesized and the post-filtered speech signals.

11.8 THE AMR CODEC'S BIT ALLOCATION

The AMR speech codec's bit allocation is shown in Table 11.3 for the 4.75 kbit/s and 10.2 kbit/s modes of the codec.

For the 4.75 kbit/s speech mode, 23 bits are used for encoding the LSFs using split vector quantization. As already stated, the LSF vector is split into three subvectors of dimension 3, 3 and 4, and each subvector is quantized using eight, eight and seven bits, respectively.

Table 11.3   Bit allocation of the AMR speech codec at 4.75 kbit/s and 10.2 kbit/s [165]. The bit positions for the 4.75 kbit/s mode, which are shown in round brackets assist in identifying the corresponding bits in Figure 11.3.

Parameter	1st subframe	2nd subframe	3rd subframe	4th subframe	Total per frame
LSFs					$8 + 8 + 7 = 23$ (1–23)
Pitch delay	8 (24–31)	4 (49–52)	4 (62–65)	4 (83–86)	20
Fixed CB index	9 (32–40)	9 (53–61)	9 (66–74)	9 (87–95)	36
Codebook gains	8 (41–48)		8 (75–82)		16
Total					95/20 ms = 4.75 kbit/s
LSFs					$8 + 9 + 9 = 26$
Pitch delay	8	5	8	5	26
Fixed CB index	31	31	31	31	124
Codebook gains	7	7	7	7	28
Total					204/20 ms = 10.2 kbit/s

4.75 kbps Speech mode

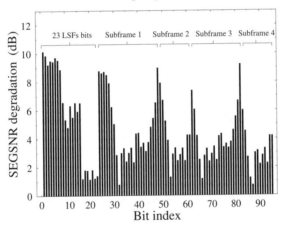

Figure 11.3. The SEGSNR degradations due to 100 % bit error rate in the 95 bit, 20 ms AMR speech frame. The associated bit allocations can be seen in Table 11.3.

This gives a total of 23 bits per frame for the LSF quantization. The pitch delay is encoded using eight bits in the first subframe and the relative delays of the other subframes are encoded using four bits each. The adaptive codebook gain is quantized together with the above-mentioned correction factor of the fixed codebook gain for every 10 ms using eight bits. As a result, a total of 16 bits per frame are used for encoding both the codebook gains. As described earlier, nine bits are used to encode the fixed codebook indices for every subframe, which gives a total of 36 bits per frame for these indices. This gives a total of 95 bits per 20 ms frame, or 4.75 kbits/s.

For the 10.2 kbit/s mode, the three LSF subvectors are quantized using eight, nine and nine bits respectively. Thus, a total of 26 bits per frame are used for quantizing the LSF vectors. The pitch delay is encoded using eight bits in the first and third subframes, and the relative delay of the other subframes is encoded using five bits each. The adaptive codebook gain is quantized together with the correction factor of the fixed codebook gain using a seven bit nonuniform vector quantization scheme for every 5 ms subframe. The fixed codebook indices are encoded using 31 bits in each subframe, or 124 bits per frame. This gives a total of 204 bits per 20 ms frame, or 10.2 kbits/s.

11.9 SPEECH CODEC'S ERROR SENSITIVITY

In this section, we demonstrate that some bits are significantly more sensitive to channel errors than others. These more sensitive bits need to be better protected by the channel codec than other bits [171]. A commonly used approach in quantifying the sensitivity of a given bit is to invert this bit consistently in every speech frame and evaluate the associated *Segmental SNR* (SEGSNR) degradation. The error sensitivity of various bits of the AMR codec determined in this way is shown in Figure 11.3 for the bit rate of 4.75 kbit/s, with the corresponding bit allocations shown in Table 11.3.

It can be observed from Figure 11.3 that the most sensitive bits are those of the first two subvectors of the LSFs, seen at positions 1–16. The least sensitive bits are related to the fixed codebook pulse positions, shown for example at bit positions 53–61 in Figure 11.3. This is because if one of the fixed codebook index bits is corrupted, the codebook entry selected at the decoder will differ from that used in the encoder only in the position of one of the nonzero excitation pulses. Therefore, the corrupted codebook entry will be similar to the original one. Hence, the algebraic codebook structure used in the AMR codec is inherently quite robust to channel errors.

The information obtained from techniques such as these can be used for designing a bit mapping procedure in order to assign different channel encoders according to the different bit error sensitivities. Typically, three classes of bit, known as Class A, Class B and Class C, are defined. The *Bit Rate Processing* (BRP, see Chapter 4) is configured to give most protection to the bits in Class A, and least protection to those in Class C.

Note that although appealing in terms of its conceptual simplicity, the above approach used for quantifying the error sensitivity of the various coded bits does not take into account the error propagation properties of different bits over consecutive speech frames. Better results can be demonstrated using different techniques – see [171].

11.10 CONCLUSIONS

In this chapter we have given a brief overview of the techniques used in the AMR speech codec. Narrowband speech was the only application for early mobile phones, and remains the main application for second generation systems. Although it will be an important application for third generation systems, it is likely to be joined by others such as downloads of video clips, video telephony and web browsing, etc. An overview of some of these other applications can be found for example in Chapter 2 of [6].

12

Future Developments

Rudolf Tanner and Carlo Luschi

12.1 INTRODUCTION

This chapter differs from the previous chapters in that its emphasis is not on the implementation of 3GPP functionality. In this chapter, we introduce capabilities which have either been recently standardized, e.g. 3GPP Release 5, or which are on the roadmap for future 3GPP releases. Future 3GPP releases, i.e. beyond Release 99, consider:

- a new high speed packet access system;
- new features for the support of location-based services;
- use of an interference cancellation scheme;
- multiple antenna systems;
- interworking schemes with other air interfaces.

The chapter also describes some principles and techniques which are believed to impact future implementations. The principal aim of all the techniques is to increase the network capacity by exploiting methods recently developed by the information theory and communications community. The selected techniques comprise:

- equalization for WCDMA;
- *Multiuser Detection* (MUD);
- *Space Time Coding* (STC);
- multiantenna systems, i.e. *Multiple Input Multiple Output* (MIMO).

WCDMA – Requirements and Practical Design. Edited by R. Tanner and J. Woodard.
© 2004 John Wiley & Sons, Ltd. ISBN: 0-470-86177-0.

12.2 3GPP RELEASE 5: HSDPA

Unlike the previous chapters, this chapter will not aim to address the implementation aspects. Before we proceed with discussing the 3GPP specification for HSDPA in detail, we expand the previous description of HSDPA by means of an example.

A number of substantially new features and performance enhancement techniques are specified in 3GPP Release 5 (R5). The objective of these features is to exploit the air interface in an even better way with respect to achievable system capacity. The key novelty in 3GPP R5 is the introduction of a downlink high speed packet access scheme.

Similar high speed packet technologies have been considered for the high speed evolution of WCDMA in 3GPP and cdma2000 in 3GPP2. In 3GPP2 1x[1] the set of new features is called *1x Evolution for Data and Voice* (1xEV–DV), while in 3GPP R5 it is referred to as *High Speed Downlink Packet Access* (HSDPA). The focus of the HSDPA specifications [174–176] is on streaming for both interactive and background services. 3GPP Release 4 already enables provision of services through an all IP core network. In this context, HSDPA will provide higher downlink data rates enabling effective wireless multimedia capabilities.

12.2.1 Introduction to HSDPA

HSDPA exploits a technique which is known in the literature as multiuser diversity in order to increase the downlink cell capacity. The principle of multiuser diversity is illustrated in Figure 12.1 and is explained next.

Figure 12.1 shows a cell with a *Base Station* (BS), i.e. Node B, and four mobile users, mobiles 1 to 4, whose propagation channels are independent of each other. The channel of mobile one appears to be rather stable and highly used, e.g. the user is sitting on a park bench with a strong line of sight component and may need to download lots of data. Mobile two's channel is fading slowly while user three has a fast fading channel. Mobile four may stay somewhere where the environment causes the occasional deep channel fade. All four users are being served through a high speed packet service. Given the limited resources, i.e. one shared downlink packet channel for all users, the BS needs to control the serving process. This is achieved with a scheduler in the BS. Obviously, serving a mobile when the channel is faded is not sensible. For that purpose, the BS needs to know the quality of each user's channel. On the other hand, every user should be served from time to time even though the channel quality may be poor. The different strategies of serving and scheduling are discussed later in Section 12.2.8. The channel state information is contained in the *Channel Quality Indicator* (CQI), which each user needs to provide. Each user frequently sends his CQI to the BS. The BS scheduler then sends a data packet to a user when his channel quality appears to be good, of course provided that the user has requested data. Figure 12.1 shows that user one can be served almost at any time, while user three can only be served occasionally with short packets. The packets depicted in Figure 12.1 vary in duration, the amount of transported bits and signal power. Bandwidth and the available downlink capacity is better utilized because of the randomness of the fading channel and the data requests from the users. Yet many complications arise from this flexibility and have

---

[1] 1x refers to a single carrier (with 1.25 MHz bandwidth), while 3x denotes the multicarrier air interface.

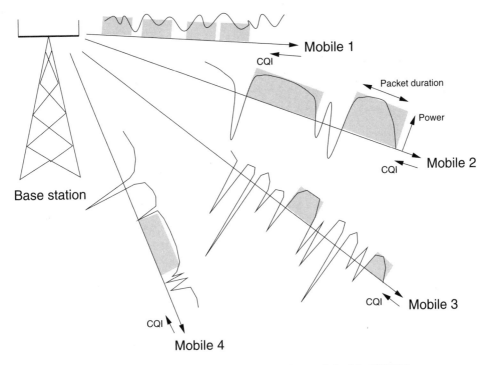

Figure 12.1. Functional principle of multiuser diversity exploited in HSDPA.

been omitted here at this stage for brevity and simplicity. For example, should the serving algorithm guarantee that all users are equally and fairly served with respect to the time from packet request to packet delivery, or should the CQI be the only criterion.

Every *User Equipment* (UE) can support a capability class; refer to Table 12.2 for an example. Each class defines a set of minimum requirements with respect to its capability to support the features offered in R5. The UE in our example has the capability to demodulate QPSK and 16 QAM modulated symbols, to despread up to 15 packet channels (channelization codes) simultaneously at the receiver and is able to handle a number N of different packet processes. The meaning of the terms will be explained later. Higher layers are of course responsible for the configuration of the transport and physical channels at both the Node B and the UE and for the control of the whole process. Higher layers provide each UE with an identification number, UE ID, in order to discriminate between UEs sharing the same packet channel.

We have already mentioned that data packets may only be transmitted when the channel is good, and that the Node B needs to know this information. Thus, we start our HSDPA introduction with the control channel which supplies the Node B with important feedback information. The uplink HSDPA control channel can be configured to carry the CQI and *Acknowledgement* (ACK) information at regular intervals from the UE to the Node B. The CQI is a number which translates to a predefined transmitter modulation and coding scheme (refer to Table 12.1). It tells the Node B which configuration should be employed, i.e. which configuration yields the maximum data throughput given the current state of the propagation channel. How the UE arrives at this number is described later. The state of

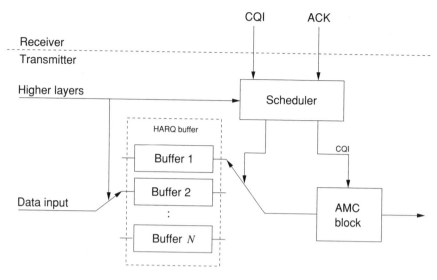

Figure 12.2. Simple illustration of the HSDPA functionality in a Node B responsible for data processing and packet generation.

the ACK information, positive or negative, is the result of the decoded data packet passing the *Cyclic Redundancy Check* (CRC) in the UE.

A data request from an application is passed through the different layers and may finally generate a HARQ process. *Hybrid Automatic Repeat Request* (HARQ) is an HSDPA feature which causes a Node B to retransmit a data packet when the first transmission is not successful. When the decoded packet fails the CRC test, a UE returns a *Negative Acknowledgement* (NACK) to the Node B, together with a CQI. Upon reception of the NACK, the Node B initiates another transmission of the same packet to the UE. A positive acknowledgement (ACK) causes the Node B to send the next packet waiting in the process data queue to the UE. Note that a HARQ process can comprise several packet transmissions and retransmissions.

A UE may require the simultaneous transmission of a large picture, an email with attachments and a web browser update and as a result, several HARQ processes may need to be active at the same time. Each HARQ process is given an identifier, see Figure 12.2. The data which is to be transmitted to a specific UE is stored in the HARQ (process) buffers and every process has its own buffer from which data is drawn for transmission. If the amount of data for a given process does not fit into a single data packet, then the data is transmitted over the air using more than one packet. The UE can follow the sequence of packets belonging to the same HARQ process because there is a downlink control channel which carries a HARQ process identifier and a new data indicator flag. The amount of data carried in a data block (packet) is subject to the state of the UE's propagation channel and can vary. The Node B has the flexibility to choose from a number of different block sizes. The transmission of the Node B can be reconfigured by the UE, via the CQI, by means of adaptive modulation and coding.

In our example, Figure 12.2, the data in Buffer 1 is read and fed into the *Adaptive Modulation and Coding* (AMC) block, where the encoded bits are punctured and modulated

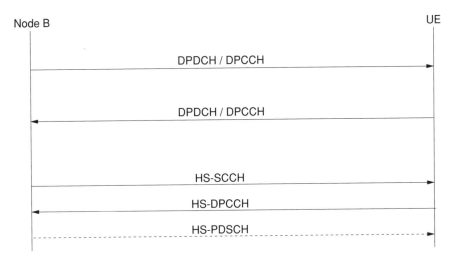

Figure 12.3. Basic channel configuration to receive an HSDPA service. The arrows illustrate the direction of the data transfer.

according to the definition to which the CQI number corresponds. At the same time, another HARQ buffer can be filled. The output of the AMC block is a 2 ms long data block ready for transmission. The scheduler shown in Figure 12.2 is a complex Node B entity and manages the serving process of all UEs in the cell or sector. The scheduler is generally optimized to meet some criteria, like user fairness, for which it is given access to a wide range of information, including CQIs and ACKs.

A prerequisite for a 3GPP HSDPA service is that a *Dedicated Physical Channel* (DPCH) is active. A DPCH connection can be established through a FACH/RACH procedure, see Section 3.9 for details. Therefore, we assume here that a connection between a UE and a Node B exists which supports an HSDPA service. Figure 12.3 shows the physical connections between the Node B and the UE which can support the transmission of the *High Speed Downlink Shared Channel* (HS-DSCH) transport channel. There are the two known dedicated channels, *Dedicated Physical Data Channel* (DPDCH) and *Dedicated Physical Control Channel* (DPCCH), both of which are active in the downlink and uplink. In addition, we have two HSDPA physical channels, the *High Speed Shared Control Channel* (HS-SCCH) and the actual packet channel termed the *High Speed Physical Downlink Shared Channel* (HS-PDSCH). Note that the HS-SCCH can employ up to four channelization codes and hence, strictly speaking, there are up to four HS-SCCHs. The HS-PDSCH connection is drawn with a dashed line because it is a shared data channel and there may not always be data being transmitted. In the uplink, there is the *High Speed Dedicated Physical Control Channel* (HS-DPCCH) which carries the CQI and ACK information.

Figure 12.4 illustrates the state of the physical channels when no packet is sent to the UE in our example, but an HSDPA connection is active and the UE is monitoring the (four) HS-SCCHs. The Node B transmits control information through four channelization codes via the HS-SCCHs over the whole cell or parts thereof to other users. Higher layers configured the UE and the Node B, and as a result, the UE in our example returns a short message every 10 ms through the HS-DPCCH to the Node B which contains a CQI. Clearly,

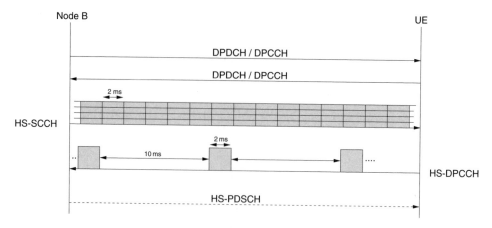

Figure 12.4. Basic data exchange between Node B and UE prior to any packet transmission. The HS-SCCH can employ up to four channelization codes. The third slot of the HS-DPCCH contains zeroes (DTX) if it does not carry an ACK/NACK.

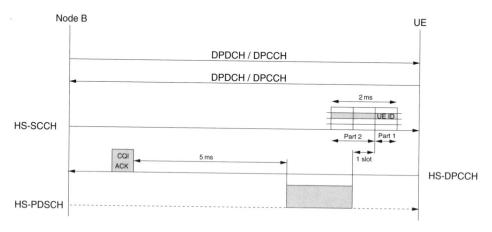

Figure 12.5. The four steps of a packet transmission: (1) send packet description and UE ID on the HS-SCCH; (2) detect valid UE ID and configure receiver; (3) send HS-PDSCH packet and demodulate packet; (4) respond with an acknowledgement message after 5 ms upon reception of the HS-PDSCH packet.

there is no need to have a stationary UE to return the CQI every 10 ms if the channel is static. Therefore higher layers can determine the feedback rate (interval) from the set {2 ms, 10 ms, 20 ms, 40 ms, 80 ms, no feedback} in order to reduce the uplink network traffic and hence interference. Additionally, the HS-DPCCH carries only the CQI in the first two slots if no ACK/NACK is returned, which is achieved by DTXing (zeroing) the third slot. The HS-SCCH is used to indicate to the UE when a packet will follow on the HS-PDSCH, and how the packet can be processed, i.e. demodulated and decoded. Figure 12.5 shows an example of the packet transmission process.

The control information for a given UE may be contained in one of the four different HS-SCCHs. Thus, each UE needs to monitor continuously, i.e. demodulate and decode, four

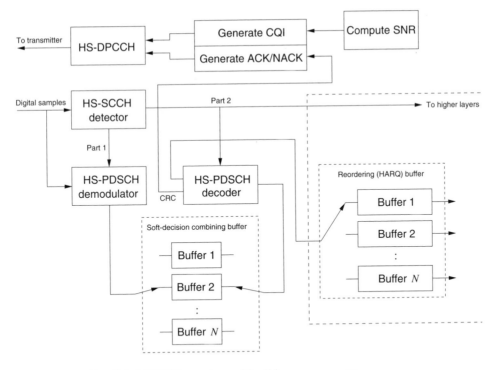

Figure 12.6. Simplified HSDPA receiver with all key components.

HS-SCCHs until it detects its UE identifier in one of the four possible HS-SCCH channels. In our example, Figure 12.5, the third HS-SCCH channel contains the UE identifier (UE ID) which tells the UE that a packet will follow precisely after 2/3 of the 2 ms TTI. Once the UE knows which HS-SCCH channel contains its ID, the UE can restrict its monitoring to the HS-SCCH channel in use. Part 1 of the 2 ms HS-SCCH frame (packet) contains essential information required to demodulate and decode the packet sent on the HS-DSCH. Thus, the UE must reliably detect the first slot, part 1, and configure the HS-PDSCH demodulator. Part 2 contains additional information which is not required for demodulation of the data packet, but later for decoding.

In the receiver, see Figure 12.6, the radio card provides the digitized samples which are fed into a number of receiver blocks. Of interest to us are the HS-SCCH detector, the HS-PDSCH demodulator, the *Signal to Noise Ratio* (SNR) computation block and the buffers. The HS-PDSCH demodulator needs to know the modulation type, QPSK or 16 QAM, and the channelization code number in order to demodulate the data packet successfully. Both pieces of information are supplied by the HS-SCCH detector, from part 1 of the HS-SCCH packet. The HS-PDSCH demodulator despreads the data carried by the HS-PDSCH. This may require processing power for up to 15 HS-PDSCH spreading codes, each with SF = 16, dependent on the UE's capability class. The demodulator block also needs to convert the despread symbols into soft decisions. This process is more complicated when 16 QAM is used. The demodulator output, the soft decisions, are then stored in a soft decision buffer. This buffer is part of the whole HARQ process where soft decision combining is performed. The reordering buffers in Figure 12.6 are used to reorder the data of different

HARQ processes because some applications may need the data in each process in the correct (sequential) order. One process, and hence one series of data blocks, can become available before another process due to the scheduler being influenced by the state of the propagation channel, i.e. CQI. Thus, the HARQ reordering buffers, which are part of the *Medium Access Control* (MAC), take care of this fact and ensure that the data is passed on in sequential order.

Producing the CQI message, i.e. the CQI number, is a complicated process, at least for the modem designer. We assumed in Figure 12.6 that the SNR block generates this message. As shown later in Section 12.2.6, 3GPP specifies tables where each row denotes a set of system parameters, such as coding rate, number of used spreading codes and the modulation alphabet, e.g. Table 12.1. The SNR block will pick the CQI number that guarantees a block error rate lower than 10 %, which has been agreed as the best tradeoff between good error performance and reduced latency due to HARQ retransmissions. In reality, a lookup table could be used which maps an SNR value to a specific CQI number. This lookup table may be computed through simulations.

Besides the error correction coding, HSDPA offers an additional mechanism which helps to compensate for situations when packets are received in error. We have already briefly introduced the HARQ mechanism in conjunction with the Node B and the HS-DSCH, and explain it now in more depth. HARQ is useful because packets cannot always be scheduled, i.e. 'delivered' to the UE, when the channel is good. Also, an error or delay in the CQI feedback loop may cause the selection of an inappropriate modulation and/or coding scheme. Under these circumstances, retransmissions guarantee an acceptable error performance. Note that the envisaged average packet error rate is 10 % or lower but not necessarily zero. This may explain to some extent why the following feature has been incorporated. HARQ includes two different types of soft decision combining scheme of successive retransmissions, with soft decision combining buffers and a HARQ process reordering buffer in the MAC layer. Recall that R99 already has an *Automatic Repeat Request* (ARQ) mechanism in the protocol stack, see Chapter 7.

It has already been stated that the Node B may not transmit the data in the correct order and that it can send data belonging to different HARQ processes, refer to Figure 12.2. The impact of this flexibility complicates the data control and management while the actual receiver blocks are not substantially affected. The HARQ mechanism is active once the UE returns a NACK to the Node B. This is the case when the last HS-DSCH packet could not be decoded correctly. The purpose of the two buffers shown in Figure 12.6 is to:

- allow the MAC to reorder the data from different HARQ processes; and to

- allow the physical layer to perform soft decision combining.

The MAC in the UE needs to be able to discriminate between data of different data flows and pass them on to higher layers in the correct order. Thus the MAC needs to store the data temporarily in a buffer which belongs to a HARQ process, e.g. a large picture message, because the Node B may not have been able to transmit the data within one HS-DSCH packet. This is straightforward to control since each HS-DSCH packet is identified by a HARQ process identifier and a new data indicator. Figure 12.7 shows an example of filling a HARQ process buffer. The receiver already received an HS-DSCH packet but it did not contain all the data because the Node B was not able to choose a packet size which could accommodate all the data bits. Now Figure 12.7 illustrates the case where the receiver has

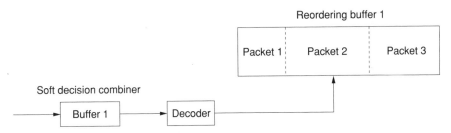

Figure 12.7. The output of the soft decision combiner is fed into a HARQ buffer. The first and the second packet contain a fraction of the total amount of date belonging to the HARQ process. Thus at least another HS-PDSCH packet is required before all data can be forwarded to the higher layers.

just received the second block but again this cannot fill up the HARQ process buffer. What is depicted is that the output of the soft decision combiner is decoded, without error, and that the data bits are stored in the HARQ buffer. If the decoding process is unsuccessful, then the UE returns a NACK to the Node B and the Node B retransmits the packet. However, prior to this retransmission, the Node B may alter the content of the packets in order to exploit one of two soft decision combining methods offered by HSDPA.

The first soft decision combining scheme, called *Chase Combining*, combines the HS-PDSCH soft decisions of different transmissions, where each retransmission contains the same data and parity bits. This yields an SNR improvement and hence improves the system performance. The second scheme, termed *Incremental Redundancy* (IR), is more complex because a retransmitted block contains different parity bits, which were punctured in the previous transmission (packet). This is achieved by applying a different puncturing scheme in the Node B. The performance improvement is obtained by providing the decoder with additional information (increased redundancy) because the actual number of missing (punctured) redundancy bits in the IR buffer is reduced with each retransmission.

Finally, the HS-DPCCH shown in Figure 12.6 is used to indicate to the Node B whether the HS-DSCH packet has been decoded with or without error, and to supply the CQI. In our example, we assume an error free detection and thus the HS-DPCCH carries back a positive acknowledgement (ACK), otherwise it carries a negative acknowledgement (NACK) which triggers a retransmission.

12.2.2 High Speed Downlink Packet Access (HSDPA)

HSDPA relies on a new type of transport channel, the HS-DSCH, which is terminated in the Node B. The *High Speed Downlink Shared Channel* (HS-DSCH) resources are shared among the HSDPA users in both the time and code domains. Key drivers for the HSDPA technology include higher peak data rates, reduced delay and higher throughput. Higher peak data rates of more than 10 Mbps are obtained using higher order modulation and link adaptation by means of *Adaptive Modulation and Coding* (AMC). System throughput and delay performance are improved by fast scheduling (switching between the users every 2 ms) centred at the Node B instead of at the *Radio Network Controller* (RNC). Features supporting fast scheduling are *Hybrid Automatic Repeat Request* (HARQ) and fast rate selection based

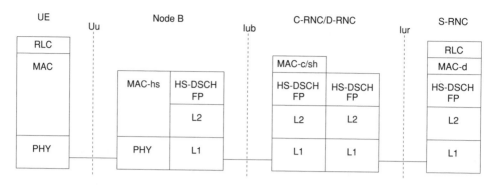

Figure 12.8. Radio interface protocol architecture of HSDPA.

on channel quality feedback. HARQ complements the link adaptation process by masking inaccuracies in the measured channel quality reported by the *User Equipment* (UE). Fast rate selection combined with time domain scheduling on the shared channel takes advantage of the short term variations in the signal power received at the UE, so that each user can be served on a constructive fading, i.e. multiuser diversity. This maximizes the data rate that can be supported by the shared channel.

12.2.3 Protocol Structure

The HSDPA functionality does not affect protocol layers above the *Medium Access Control* (MAC). The HARQ and HSDPA scheduling functions are included in a new MAC entity called MAC-hs, terminated in the Node B, which controls the HS-DSCH transport channel. The MAC-hs will be described in detail in Section 12.2.5, while the operation of HARQ and scheduling will be discussed in Sections 12.2.7 and 12.2.8. The HSDPA radio interface protocol architecture with termination points is shown in Figure 12.8. The MAC-hs in the Node B is located below MAC-c/sh in the *Controlling RNC* (C-RNC). The R5 MAC-c/sh provides functions to HSDPA already included for the *Downlink Shared Channel* (DSCH) in the R99 version. MAC-d is retained in the *Serving RNC* (S-RNC).

The *HS-DSCH Frame Protocol* (HS-DSCH FP) handles the data transport from S-RNC to C-RNC, and between C-RNC and Node B. In an alternative configuration, the C-RNC does not have any user plane function for the HS-DSCH, i.e. in the HS-DSCH user plane the S-RNC is directly connected to the Node B. In this case, MAC-d in the S-RNC is located directly above MAC-hs in the Node B [174].

12.2.4 Physical Layer Structure

The HS-DSCH has the following characteristics:

- the HS-DSCH transport block set consists of one transport block only;
- there is only one HS-DSCH transport channel per UE.

Therefore, the transport format set is completely determined by the *Transport Block Size* (TBS). The HS-DSCH may be transmitted over the entire cell or over only parts of

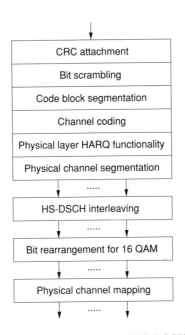

Figure 12.9. Transport channel coding structure for HS-DSCH.

the cell (e.g. using beamforming antennas). At the physical layer, the HS-DSCH transport channel is mapped to the *High Speed Physical Downlink Shared Channel* (HS-PDSCH), and is associated with one downlink DPCH plus one or more *Shared Control Channels* (HS-SCCH).

12.2.4.1 *High Speed Downlink Shared Channel*

The basic time interval for HS-DSCH transmission and its related signalling is the HS-DSCH subframe or *Transmission Time Interval* (TTI), which corresponds to three slots (2 ms). The HS-PDSCH uses one channelization code of fixed spreading factor SF = 16 from the set of channelization codes reserved for HS-DSCH transmission [177]. Multicode transmission is allowed, which corresponds to a single UE being assigned multiple channelization codes in the same TTI, depending on the UE capabilities. It is also possible to perform code domain multiplexing of multiple UEs over the same TTI.

HSDPA implements AMC, where the transport format, including the modulation scheme and code rate, is selected based on feedback information on the downlink channel quality measured at the UE. AMC will be discussed in detail in Section 12.2.6. The HS-DSCH modulation format can be either *Quadrature Phase Shift Keying* (QPSK) or 16 QAM (16-ary Quadrature Amplitude Modulation). HS-DSCH channel coding uses the R99 rate 1/3 turbo code and turbo code internal interleaver. Other code rates, as listed in Table 12.1 for example, are generated from the basic rate 1/3 code through rate matching by means of puncturing or repetition.

The transport channel coding structure for HSDPA is reproduced in Figure 12.9 [178]. Since only one HS-DSCH is allocated per UE, and there is one transport block per HS-DSCH TTI, in HSDPA there is no need for transport block concatenation, and only one interleaving

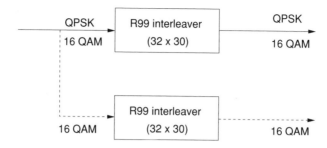

Figure 12.10. Interleaver structure for HSDPA.

step is required. A *Cyclic Redundancy Check* (CRC) sequence with 24 bits is calculated and added per HS-DSCH TTI, reusing the R99 polynomial [178]. After CRC attachment, the bits of the HS-DSCH transport block are bit scrambled. Code block segmentation is performed as in R99. The maximum code block size for turbo coding is 5114 bits. Following channel encoding in Figure 12.9, the physical layer HARQ functionality matches the number of bits at the output of the turbo encoder to the total number of bits in the allocated HS-DSCH physical channels [178]. This operation is performed in two rate matching stages. The first stage provides a number of output bits that match the available UE soft buffering capability (UE buffer rate matching). This stage is transparent if the number of input bits does not exceed the UE soft buffering capability. The second stage matches to the number of physical channel bits in the HS-DSCH TTI (channel rate matching). Both rate matching stages use the R99 rate matching algorithm [178]. The physical layer HARQ functionality requests the retransmission of unsuccessfully decoded transport blocks. As mentioned in Section 12.2.1, the HARQ algorithm may be based on successive retransmissions of the same coded block, termed *Chase Combining* HARQ [179], or on incremental transmission of additional redundancy bits, called *Incremental Redundancy* (IR) HARQ. In both cases, the receiver stores and combines the soft bit values corresponding to the different transmission attempts in a HARQ soft buffer. This soft bit buffering takes place in the UE receiver between the two rate dematching operations, and corresponds to the transmitter virtual IR buffer [178]. Information about the UE soft buffering capability is provided at the transmitter by higher layers.

In the case of multicode HSDPA transmission, physical channel segmentation divides the coded bits among the different physical channels. HS-DSCH interleaving is performed according to the scheme in Figure 12.10, separately for each physical channel. The interleaver is the same as the R99 second interleaver [178], but is fixed in size to 32 rows and 30 columns. For QPSK, a single interleaver is used and the interleaved bits are then mapped to multiple physical channels in the same way as in R99. For 16 QAM, a second identical interleaver is added. Odd and even numbered bits from the output of the physical channel segmentation block are then interleaved by the two interleavers, and the outputs rearranged to produce the mapping into 16 QAM symbols [178, 180].

12.2.4.2 *High Speed Downlink Shared Control Channel*   All downlink signalling related to the HS-DSCH is transmitted in the associated HS-SCCH, which has

a fixed rate of 60 kbps (SF = 128) [177]. The information carried by the HS-SCCH is given by [174, 175, 178]:

Transport Format Resource Indicator (TFRI) which includes:

- channelization codes (7 bits);
- modulation scheme (1 bit);
- transport block size (6 bits).

HARQ information consisting of:

- HARQ process identifier (3 bits);
- HARQ new data indicator (1 bit);
- redundancy and constellation version (3 bits).

The HS-SCCH information is split into two parts: part 1 contains the information on the channelization code set and the modulation scheme, part 2 refers to the transport block size and HARQ-related information.

The HS-SCCH also carries a UE specific CRC attachment that identifies the UE for which it is carrying information. This 16 bit CRC is calculated using the above two fields (TFRI and HARQ information) and the UE ID, and appended to the HS-SCCH information bits. The UE ID is given by the 16 bit *HS-DSCH Radio Network Temporary Identifier* (H-RNTI) defined by the *Radio Resource Controller* (RRC) [181]. Part 1 of the HS-SCCH and part 2 + CRC are separately convolutionally encoded using R99 rate 1/3 code with eight tail bits. After convolutional coding, rate matching is applied separately for the two parts of the coded HS-SCCH, to obtain 120 HS-SCCH channel bits (three slots). The rate matched part 1 is additionally scrambled by the UE ID. The scrambler is based on the UE ID bits, encoded by the R99 rate 1/2 convolutional code with eight tail bits. Eight bits of the resulting 48 bit codeword are then punctured using the same rate matching rule as for the part 1 HS-SCCH bits [178]. Finally, the part 1 sequence is mapped to the first slot of the three slot (2 ms) HS-SCCH subframe, while the part 2 sequence is mapped to the second and third slots.

At the UE, the necessary control information for demodulation and decoding of the HS-PDSCH is obtained from decoding of the HS-SCCH. For each TTI, the indication of which user has been scheduled on the next HS-PDSCH is derived from the HS-SCCH UE ID information. To achieve this, the UE monitors up to four HS-SCCHs. However, if control information intended for a UE has been detected in the previous subframe, it is sufficient for the UE to monitor only the HS-SCCH used in the previous subframe [51]. The timing offset between the start of the HS-SCCH and the start of the corresponding HS-PDSCH TTI is fixed to two slots or 5120 chips. The timing relations between HS-PDSCH and the associated DPCH and HS-SCCH are shown in Figure 12.11.

12.2.4.3 The HSDPA Uplink Channel   The uplink feedback signalling related to the HS-DSCH is carried by the *Dedicated Physical Control Channel* for HS-DSCH (HS-DPCCH) with spreading factor SF = 256, which is code multiplexed with the existing dedicated uplink physical channels. Feedback information consists of a one bit *HARQ*

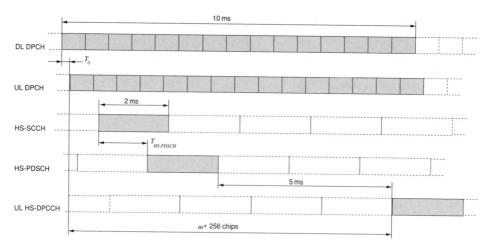

Figure 12.11. Timing of HS-PDSCH control signalling at the UE.

Acknowledgement (HARQ-ACK) and a five bit *Channel Quality Indicator* (CQI). The HARQ-ACK bit is repetition coded to ten bits. The CQI is coded using a (20,5) code [178]. In each subframe of three slots (2 ms), corresponding to 30 bits, HARQ-ACK is transmitted in the first slot, while the remaining two slots are used for CQI [177]. As shown in Figure 12.11, the uplink HS-DPCCH subframe at the UE starts $m \times 256$ chips after the start of the uplink DPCH frame. The HS-DPCCH is symbol aligned but not necessarily slot aligned to the existing uplink DPCH. The calculation of m implies that the HARQ-ACK transmission on the uplink HS-DPCCH has to start within the first 256 chips after 7.5 slots following the end of the received HS-PDSCH. The UE processing time is therefore maintained at 7.5 slots (5 ms) as the offset between downlink DPCH and uplink HS-DPCCH varies. This relaxes the UE implementation complexity requirements, compatibly with the timing constraints imposed by the HARQ mechanism. The transmission rate of the CQI or measurement report from the UE (measurement feedback rate) is configured by the network through higher layer signalling. Possible values of the measurement feedback rate are 2 ms, 10 ms, 20 ms, 40 ms, 80 ms, 160 ms, or no feedback. Both the HARQ-ACK and CQI fields of the HS-DPCCH are gated off when there is no acknowledgement or channel quality information to be sent.

12.2.5 HSDPA MAC Architecture and MAC-hs Functionality

The overall MAC architecture to support HSDPA at the *UMTS Terrestrial Radio Access Network* (UTRAN) side and at the UE side is shown in Figure 12.12 and Figure 12.13 respectively. In both cases, HSDPA specific functions are handled by the MAC-hs. Similarly to MAC-c/sh and MAC-d, the MAC-hs configuration is done via the MAC control *Service Access Point* (SAP) by the RRC.

At the UTRAN side, there is one MAC-d entity for each UE, and one MAC-hs entity for each cell that supports HS-DSCH transmission. The data to be transmitted on the HS-DSCH is transferred from MAC-c/sh to the MAC-hs via the Iub interface, or from the MAC-d via the Iur/Iub in the case of configuration without MAC-c/sh. As illustrated in Figure 12.14, a MAC-hs entity in the Node B has four different logical components [182].

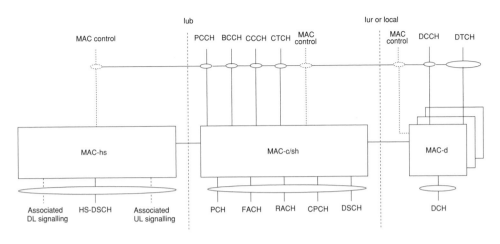

Figure 12.12. UTRAN side MAC architecture with HSDPA.

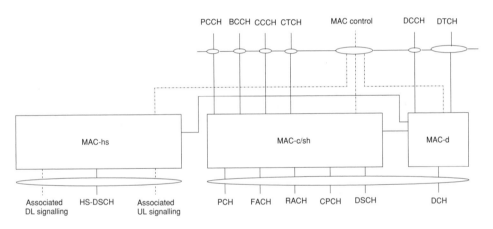

Figure 12.13. UE side MAC architecture with HSDPA.

1. Flow control, which provides a controlled data flow between the MAC-c/sh and the MAC-hs (or between MAC-d and MAC-hs in the case of configuration without MAC-c/sh), taking into account the transmission capabilities of the air interface. This function, provided independently for each priority class, is intended to limit Layer 2 signalling latency and reduce discarded and retransmitted data resulting from HS-DSCH congestion.

2. A scheduling/priority handling entity, which manages HS-DSCH resources between HARQ entities and data flows according to their priority class; based on the status report from the associated uplink signalling, the scheduler determines either a new transmission or a HARQ retransmission. It also sets the priority class identifier, or queue ID, and the *Transmission Sequence Number* (TSN) for each new data block being serviced, to support in-sequence delivery. These parameters are signalled in-band in the MAC-hs header for each queue ID. The TSN is set to zero for the first data block transmitted within an HS-DSCH, and incremented by one for each new

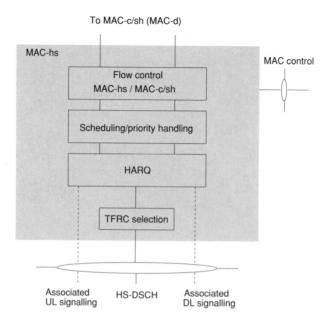

Figure 12.14. MAC-hs details for the UTRAN side MAC architecture.

transmitted block [182]. The scheduler also sets a suitable redundancy version for each block to be transmitted or retransmitted and indicates this parameter to the lower layer.

3. One HARQ entity per user, which is capable of supporting multiple instances of *Stop and Wait* (SaW) HARQ protocols, or HARQ processes, with only one HARQ process active per TTI. The HARQ entity determines the HARQ process for each data block to be serviced, and sets the corresponding HARQ process identifier.

4. The TFRC selection selects the appropriate transport format and resource combination for the data to be transmitted on the HS-DSCH.

One HS-DSCH transport block may consist of several MAC-d *Protocol Data Units* (PDUs), i.e. of several MAC-hs *Service Data Units* (SDUs). The size of the MAC-d PDUs belonging to a transport block is carried in the MAC-hs header. One UE may be associated with one or more MAC-d flows. Each MAC-d flow contains HS-DSCH MAC-d PDUs for one or more priority queues [182].

At the UE side, the MAC-d entity is modified by the addition of a link to the MAC-hs. As shown in Figure 12.15, the MAC-hs in the UE comprises [182]:

1. A HARQ entity which is capable of supporting multiple instances of SaW HARQ protocols, or HARQ processes, with only one HARQ process allowed per TTI.

2. A reordering queue distribution function, which routes the MAC-hs PDUs to the correct reordering buffer based on the queue ID identifying the priority class.

3. One reordering entity for each priority class, which organizes received data blocks according to the received TSN. Data blocks with consecutive TSNs are delivered to

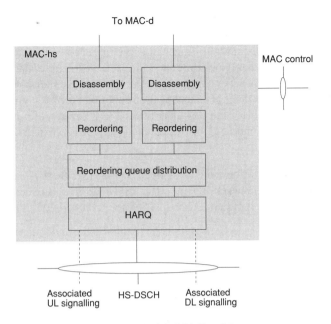

Figure 12.15. MAC-hs details for the UE side MAC architecture.

higher layers after reception, while a timer mechanism determines delivery to higher layers of nonconsecutive data blocks.

4. A disassembly entity, which, for each MAC-hs PDU, removes the MAC-hs header, extracts the MAC-d PDU, removes any padding bits and delivers the MAC-d PDU to higher layers.

The links to MAC-hs and MAC-c/sh cannot be configured simultaneously in one UE. The MAC-c/sh on the UE side is not modified for HSDPA.

12.2.6 Adaptive Modulation and Coding

In cellular systems, the quality of the signal received by a mobile user depends, in general, on the distance from the serving and interfering base stations, pathloss, shadowing and short term multipath fading. In order to improve the system peak data rates and coverage reliability, link adaptation techniques are usually employed, which modify the signal transmitted to and from a particular user to account for variations in the received signal quality. Two link adaptation strategies are fast power control, used in R99, and adaptive modulation and coding. While fast power control in the downlink allocates proportionally more transmitted power to disadvantaged users, with AMC the transmitted power is held constant, and the modulation and coding scheme is changed to match the current channel conditions. In a system implementing AMC, users with favourable channel conditions (e.g. users close to the base station or on a constructive fading) are typically assigned higher order modulation and a higher rate coding scheme, which results in higher transmission data rates. AMC has the advantage of reducing the interference variations due to fast downlink power control, and can exploit the short term variations of signal power received by the UE. Therefore, if supported by fast scheduling and timely channel quality feedback, AMC can provide higher peak data rates and average throughput.

The HS-DSCH uses AMC, and the transport format is selected based on the downlink channel quality. As already mentioned in Section 12.2.5, the selection of the transport format is performed by the MAC-hs located in the Node B. The possible transport formats for the HS-DSCH depend on the number of channelization codes that can be supported by the UE, which is part of the definition of the UE capabilities [183]. Transport format tables for UE categories supporting 5, 10 and 15 HS-DSCH codes [183] are given in [51]. Each transport format is defined by:

1. Turbo encoder information word size, given by the HS-DSCH *Transport Block Size* (TBS), which ranges between 137 and 7168 for 5 code UEs, between 137 and 14 411 for 10 code UEs, and between 137 and 25 558 for 15 code UEs.

2. Number of HS-PDSCH codes, which ranges from 1 to the maximum number of codes supported by the UE.

3. Modulation format, which can be either QPSK or 16 QAM.

The different code rates are obtained through puncturing of the rate 1/3 turbo encoded bits.

The channel quality is mapped to a number from 0 to 30, which provides the transport format that the UE is able to receive, given the current propagation channel conditions. More specifically, the reported CQI identifies the transport format that supports a single transmission *Packet Error Rate* (PER) not greater than 0.1 given the received power level on the HS-PDSCH. For computation of the PER, a packet error is defined as the failure of the 24 bit CRC.

CQI reporting is based on an estimate of the SIR, based on the observation of the CPICH. In particular, for the CQI calculation, the UE assumes a total HS-PDSCH received power (equally distributed among the HS-PDSCH codes):

$$P_{\text{HS–PDSCH}} = P_{\text{CPICH}} + \Gamma + \Delta \quad \text{dB} \quad (12.1)$$

where Γ denotes a measurement power offset signalled to the UE from higher layers, and P_{CPICH} is (an estimate of) the received power on the CPICH. The details of the algorithm for SIR estimation from the CPICH are given in [184].

The possible *Modulation and Coding Schemes* (MCSs) and the corresponding CQIs in the case of a 15 code UE, i.e. UE category 10, are reproduced in Table 12.1. The table also reports as a reference the instantaneous code rate achieved with each transport format. Transmission of CQI = 0 indicates *Out Of Range* (OOR), and is reported by the UE if it cannot support transmission with the lowest rate transport format (CQI = 1). In the case of 5 code and 10 code UEs, the highest rate supported transport format corresponds to CQI = 22 and CQI = 25 respectively. A CQI greater than either 22 or 25 is used to send an HS-PDSCH power reduction factor Δ, which indicates that the UE is able to receive the highest rate transport format, i.e. 22 or 25, at a lower power level, namely reduced by Δ with respect to the current received HS-PDSCH power. The values of the parameter Δ range from -1 dB to -8 dB for 5 code UEs, and from -1 dB to -5 dB for 10 code UEs [51].

For HARQ retransmissions, the transport block set is the same as for the initial transmission, i.e. the number of information bits is the same as for the first transmission. However,

Table 12.1 Transport formats and CQI mapping for a 15 HS-PDSCH code UE, according to [51].

CQI	TBS	Number of codes	Modulation	Code rate
0	N/A	OOR	OOR	OOR
1	137	1	QPSK	0.17
2	173	1	QPSK	0.21
3	233	1	QPSK	0.27
4	317	1	QPSK	0.36
⋮	⋮	⋮	⋮	⋮
11	1483	3	QPSK	0.52
12	1742	3	QPSK	0.61
13	2279	4	QPSK	0.60
14	2583	4	QPSK	0.68
15	3319	5	QPSK	0.70
16	3565	5	16 QAM	0.37
17	4189	5	16 QAM	0.44
⋮	⋮	⋮	⋮	⋮
28	23 370	15	16 QAM	0.81
29	24 222	15	16 QAM	0.84
30	25 558	15	16 QAM	0.89
31	RSVD			

the modulation format may change from the initial transmission to the successive retransmissions. For channel coding, each HARQ retransmission may use a different redundancy version, where each redundancy version is a different subset of the coded bits. Each subset may contain a different number of bits.

As already mentioned, the effectiveness of an AMC scheme depends on the availability of up-to-date channel quality information. In the presence of feedback delay, a possible solution is to employ techniques for the prediction of the downlink channel quality. The downlink channel state can be predicted well at low Doppler frequencies as the envelope correlation is high for a much longer period of time. For example, Figure 12.16 shows the normalized envelope correlation for a Rayleigh flat fading channel. In this case, we note that the first zero occurs for a time lag of about 11 ms when a vehicle at 15 km/h observes a Doppler frequency of about 30 Hz ($\lambda = 0.14$ m). The curve basically shows that the envelope correlation is above 0.5 for time delays of up to 4 ms at a maximum Doppler frequency of 50 Hz. Therefore, we infer that it should be possible to predict the channel well and schedule the served users, especially when they are nomadic. In practice, it is reasonable to assume that Internet browsing takes place while the user is stationary, but we must not forget the browsing child in the back seat of a car.

12.2.7 Hybrid ARQ

Although AMC gives the flexibility to match the MCS to the average channel conditions for each user, it is sensitive to measurement errors and delays in the channel quality feedback from the UE. The operation of AMC is complemented by the additional use of hybrid

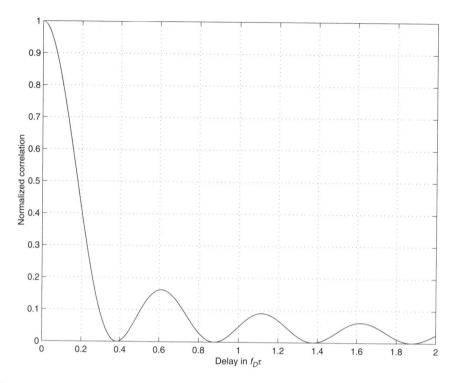

Figure 12.16. The duration of a fade can be guessed according to the correlation of the normalized Rayleigh fading envelope.

automatic repeat request. The term HARQ refers to a combined FEC + ARQ scheme in which unsuccessful attempts are used in the *Forward Error Correction* (FEC) decoder instead of being discarded. HARQ schemes use acknowledgements to make decisions on a retransmission of the data packet. Therefore, HARQ can be viewed as an implicit link adaptation technique, which adapts to the instantaneous channel conditions and is insensitive to measurement errors and delays. On the other hand, HARQ implies multiple transmissions and introduces delays. In this respect, the joint use of HARQ and AMC has the advantage of limiting the occurrence of a retransmission to those cases where the operation of the AMC fails to cope with the actual propagation channel. Different implementations of HARQ schemes have been studied.

Type I HARQ does not use the unsuccessful attempts and can be simply referred to as a combined FEC + ARQ. Better performance is obtained with Chase combining, where the receiver uses retransmission of the same block to perform maximal ratio combining of multiple received versions of the coded bits on the decoder metric (metric combining) [179]. Type II HARQ schemes are incremental redundancy retransmission techniques where additional redundant information is incrementally transmitted in the case of unsuccessful decoding. Type III HARQ refers to IR retransmission schemes where each retransmission is self decodable, which is not the case with Type II HARQ. In Type III HARQ with multiple redundancy versions, different sets of punctured bits are used in each retransmission. HSDPA will support HARQ based on IR (Type II and Type III HARQ). Chase combining is also considered, as a particular case of a Type III HARQ scheme with one redundancy version.

For the actual retransmission procedure, HSDPA specifies the use of an N channel SaW HARQ mechanism. SaW is one of the simplest forms of ARQ. With a SaW strategy, the transmitter operates on the current block until it has been successfully received. Protocol correctness is ensured with a simple one bit sequence number that identifies the current or next block, while indication of successful/unsuccessful decoding is signalled by a single acknowledgement bit. Therefore, SaW HARQ requires very little overhead. Furthermore, since a single block is in transit at any time, UE memory requirements are significantly reduced with respect to the *Selective Repeat* (SR) HARQ mechanism, where the receiver has to store all the data blocks transmitted between two consecutive cumulative acknowledgements. On the other hand, with the basic SaW mechanism the transmitter must wait to receive the acknowledgement of each transmission prior to transmitting the next block. During this time the channel remains idle, which wastes system capacity. A solution to this problem is given by running a separate instantiation of the SaW protocol when the first channel is idle. A generalization of this approach is given by the N channel SaW, where N SaW ARQ processes operate in parallel, each transmitting during a time interval where the others are waiting for the acknowledgement of a previous transmission. The advantage of this strategy with respect to SR HARQ is that a persistent failure in the transmission of one block only affects one channel, allowing data transmission to continue on the other channels. In order to maintain an equivalent stall probability, an SR retransmission should increase the window size, which would result in higher memory requirements at the UE.

The operation of the SaW algorithm, however, imposes a timing constraint on the maximum retransmission delay. In the case of N SaW HARQ processes, to avoid a stall after an HS-PDSCH transmission, the Node B must be ready to perform a possible retransmission with a maximum delay of $(N-1)$ TTIs. For $N = 6$, the HARQ-ACK transmission delay of 7.5 slots leaves approximately 2.8 ms for the Node B to perform scheduling and signal processing functions [176].

Clearly, the above discussion depends on the number of SaW HARQ processes. The maximum number of HARQ processes is configured by higher layers. Increasing the number of HARQ processes relaxes the UE and Node B timing constraints, but increases the required buffering space at the UE. For a given number of HARQ processes, the required soft buffering at the UE also depends on the use of Chase combining or IR HARQ. In R5, the soft buffering capacity of the UE is part of the definition of the UE HS-DSCH capabilities (see Table 12.2[2]), and is known at the Node B. For each transport block to be transmitted, the Node B can therefore dynamically select Chase combining or IR (by setting the appropriate redundancy version parameters), based on the code rate required by AMC and the information on the available UE soft memory.

HARQ downlink signalling (performed on the HS-SCCH) includes a HARQ process identifier (three bits), the new data indicator (one bit), which is incremented for each new data block, and information about the HARQ redundancy version for the support of IR transmission in the HS-DSCH TTI (three bits). For each new data block (MAC-hs PDU) to be transmitted to a given UE, the redundancy version is indicated by the MAC-hs scheduler, while the HARQ process identifier is set by the HARQ entity corresponding to the specific UE, and the new data indicator is set by the relative HARQ process. The new data indicator

---

[2] UEs of categories 11 and 12 of Table 12.2 support QPSK modulation only [51, 183].

Table 12.2   Classification of the HS-DSCH UE capabilities [183].

HS-DSCH category	Maximum number of HS-DSCH codes	Minimum interTTI interval	Maximum number of HS-DSCH transport channel bits within an HS-DSCH TTI	Total number of soft channel bits
Category 1	5	3	7300	19 200
Category 2	5	3	7300	28 800
Category 3	5	2	7300	28 800
Category 4	5	2	7300	38 400
Category 5	5	1	7300	57 600
Category 6	5	1	7300	67 200
Category 7	10	1	14 600	115 200
Category 8	10	1	14 600	134 400
Category 9	15	1	20 432	172 800
Category 10	15	1	28 776	172 800
Category 11	5	2	3650	14400
Category 12	5	1	3650	28800

is set to 0 for the first MAC-hs PDU transmitted by a HARQ process, and incremented by one for each subsequent MAC-hs PDU containing new data. At the UE, the HARQ entity allocates each received MAC-hs PDU to the HARQ process indicated by the HARQ process identifier. If the new data indicator has been incremented, the relevant HARQ process replaces the data in the HARQ soft buffer (for this HARQ process) with the received data. If the new data indicator has not been incremented, the HARQ process combines the received data with the data currently in the soft buffer. The HARQ process then generates a positive/negative acknowledgement, depending on the successful decoding of the data block as detected by the HS-DSCH CRC check. Finally, in the case of successful decoding, it delivers the decoded MAC-hs PDU to the MAC-hs reordering entity, based on the PDU queue ID.

Uplink signalling consists of the one bit HARQ positive/negative acknowledgement (HARQ-ACK/NACK), and the five bit measurement report (CQI). At the Node B, the uplink signalling information or status report is processed by the relative MAC-hs HARQ process and delivered to the MAC-hs scheduler.

12.2.8   HSDPA Scheduling

To achieve improved throughput and delay performance requires fast scheduling of the users on the HS-DSCH. One of the fundamental requirements of high speed packet transmission is the capability of efficiently supporting a mixture of services with different quality of service (QoS), e.g. meeting the data rate and packet delay constraints of the different applications.

The QoS of a data user is defined in a different way depending on the application. For real time users, the delays of most of the data packets need to be kept below a certain threshold. A different sort of QoS is a requirement that the average throughput provided to a given user is not less than a predetermined value. UMTS defines the four different

Table 12.3   UMTS QoS classes.

	Conversational class (Conversational real time)	Streaming class (Streaming real time)	Interactive class (Interactive best effort)	Background class (Background best effort)
Fundamental characteristics	Preserve time relation between information entities of the stream, low delay	Preserve time relation between information entities of the stream	Request response pattern, preserve payload content	Preserve payload content
Example of application	Voice	Streaming video	Web browsing	Background download of emails

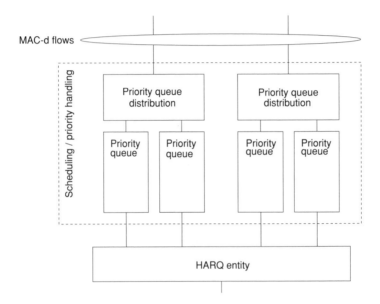

Figure 12.17. MAC-hs scheduling/priority handling function.

QoS classes (or traffic classes) in Table 12.3 [185]. The classification is based on the delay sensitivity of the traffic. Conversational class and streaming class are related to real time traffic flows, where the conversational real time services are usually the most delay sensitive applications. Interactive class and background class are mainly intended for traditional Internet applications. These classes provide better error rates by means of channel coding and retransmission. The main difference between interactive and background classes is that the first guarantees lower delay to interactive applications.

As discussed in Section 12.2.5, the main function of the scheduler is to service priority class queues (Iub logical flows). As shown in Figure 12.17, the scheduler distributes the MAC-d PDUs contained in the incoming MAC-d flows among the different priority queues, and determines the queue to be scheduled for a new transmission or a HARQ retransmission. The need to make efficient use of the wireless channel implies that the scheduling algorithm should meet the above QoS requirements while optimizing the use of the scarce radio resources. In HSDPA, this is achieved by exploiting the time variations of the shared channel, giving some form of priority to users with better channel conditions.

Scheduling algorithms considered in studies of the throughput performance of HSDPA include the maximum carrier to interference ratio (C/I) or *Maximum Rate* (MR) rule, which schedules for transmission the user with the highest instantaneous supportable rate [186, 187]. This approach achieves the maximum system throughput, but is not effective in terms of maximum delay guarantee, especially for low mobility users corresponding to nearly stationary channel conditions. A tradeoff between throughput and fairness can be obtained using the *Proportionally Fair* (PF) scheduling algorithm, see [188, 189] and references therein.

The HSDPA throughput performance with PF scheduling has been studied, for example, in [187]. Denoting by $\mu_i(t)$ and $\overline{\mu}_i$ the instantaneous data rate and average data rate supported by the ith user's channel, respectively, the MR rule schedules for transmission the user:

$$j = \arg\max_i C_i(t), \quad C_i(t) = \mu_i(t) \tag{12.2}$$

while the PF rule schedules the user data flow

$$j = \arg\max_i C_i(t), \quad C_i(t) = \mu_i(t)/\overline{\mu}_i \tag{12.3}$$

PF scheduling attempts to maximize throughput while maintaining some degree of fairness. However, it also does not take into account packet delays, and results in poor delay performance. Better delay performance can be provided by a simple *Round Robin* (RR) policy, which allocates equal time to all users regardless of the respective channel conditions [186]. On the other hand, round robin scheduling is not effective in terms of throughput. A solution to the problem of supporting QoS for real time users while maximizing throughput is given by throughput optimal scheduling schemes such as the *Modified Largest Weighted Delay First* (M-LWDF) algorithm of [188]. Indicating with $\tau_i(t)$ the accumulated head of line packet delay for the ith data flow, M-LWDF schedules the user data flow:

$$j = \arg\max_i C_i(t), \quad C_i(t) = \gamma_i \, \mu_i(t) \, \tau_i(t) \tag{12.4}$$

for an arbitrary set of parameters γ_i. These parameters give control over the packet delay distributions for different users: in fact, increasing the value of γ_i for user i reduces the delays for this flow, at the expense of a delay increase for other flows. It has been shown that a good choice of γ_i is given by $\gamma_i = a_i/\overline{\mu}_i$, where $a_i > 0$ are suitable weights that characterize the desired QoS. The values of a_i determine the tradeoff between reducing the weighted delays and being proportionally fair when delays are small [188]. A similar approach is taken by the so-called exponential rule, in which, given the set of constants $\gamma_i > 0$ and $a_i > 0$, the scheduling algorithm selects the data flow:

$$j = \arg\max_i C_i(t), \quad C_i(t) = \gamma_i \, \mu_i(t) \, \exp\frac{a_i \tau_i(t)}{1 + \sqrt{\overline{a\tau}}} \tag{12.5}$$

where $\overline{a\tau} = (1/Q)\sum_{i=1}^{Q} a_i \tau_i(t)$ [189]. For given values of γ_i and a_i, this policy tries to equalize the weighted delays $a_i \tau_i(t)$ of all the queues when their differences are large. In fact, if one of the queues has a weighted delay larger than the others by more than order $\sqrt{\overline{a\tau}}$, the exponent is close to 1 and the policy approaches the proportionally fair rule.

12.3 LOCATION-BASED SERVICES

UMTS will allow services to be offered based on the location of the mobile user. *Location-based Services* (LCS) have great commercial potential and may be seen as being crucial to the success of 3G. Besides the commercial side, LCS capabilities were additionally pressured by law in the USA which demands that wireless networks are able to support emergency calls [190] with an accuracy of 50 m. In this section, however, we will only focus on the current state of 3GPP specifications.

12.3.1 Introduction

LCS is a service concept that specifies all the necessary network elements and entities for the implementation of a location service functionality in the GSM/UMTS cellular network. UE location information can be utilized by location-based services offered by a service provider. LCS identifies four broad location service categories [191, 192]:

- commercial LCS, which use location information to support value added services;

- internal LCS or *Public Land Mobile Networks* (PLMN) operator LCS, which make use of the UE location information to support access network internal operations (radio resource and mobility management);

- emergency LCS for the support of emergency calls from the subscriber;

- lawful intercept LCS for the support of legally required or sanctioned procedures.

Examples of the variety of services that might be offered based on estimated UE location are given in Table 12.4, together with the associated accuracy requirements [191]. Given the broad range of required accuracies, a number of different UE positioning methods have been standardized, with different implementation requirements.

12.3.2 UTRAN LCS Components

The UTRAN LCS entities are the *Radio Network Controller* (RNC), Node B, *Location Measurement Unit* (LMU), *stand alone Serving Mobile Location Centre* (stand alone SMLC or SAS) and UE [193, 194].

Table 12.4 Accuracy requirements.

Accuracy	Example of offered services
Regional, up to 200 km	Traffic and weather news and warnings
District, up to 20 km	News and traffic reports
Up to 1 km	Congestion avoidance services, vehicle asset management
500 m to 1 km	Information services, rural and suburban emergency calls
100 m to 300 m	Emergency calls with network-based positioning
75 m to 125 m	Information and advertising services, SOS
50 m to 150 m	Emergency calls with UE-based positioning
10 m to 50 m	Navigation, asset management services

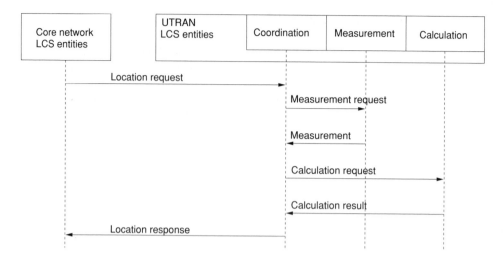

Figure 12.18. General sequence for a UE positioning operation.

The S-RNC may receive a request for UE positioning information from the core network LCS entities across the Iu interface. The RNC UE positioning function then manages the UTRAN resources to estimate the position of the UE, and returns the result to the core network (see Figure12.18). The S-RNC may also make use of the UE positioning function for internal purposes, e.g. position-based handover.

An RNC UE positioning function should:

- request measurements, typically from the UE and one or more Node Bs;

- send the measurement results to the appropriate calculating function within UTRAN;

- perform any necessary coordinate transformation;

- send the result to the core network LCS entities or to application entities within UTRAN.

The LMU makes radio measurements of one of the following two categories [193,194]:

1. Positioning measurements specific to one UE and used to compute its position;

2. Assistance measurements applicable to all UEs in a certain geographic area.

The measurements may be performed in response to a request, e.g. from the D-RNC, or made autonomously and reported regularly or when there are significant changes in the radio conditions. An LMU can be either a stand alone LMU or an associated LMU. A stand alone LMU communicates with an RNC over the UTRAN air interface, and has no connection to any other UTRAN network element (but can communicate with different access networks). An associated LMU communicates with an RNC via the Iub interface, and may make use of the radio equipment of its associated Node B. The SAS, which is not yet specified in 3GPP R99 and R4, communicates with an RNC, acting as a location calculation server and providing assistance data to be delivered to the UE through point to point or broadcast channels [194].

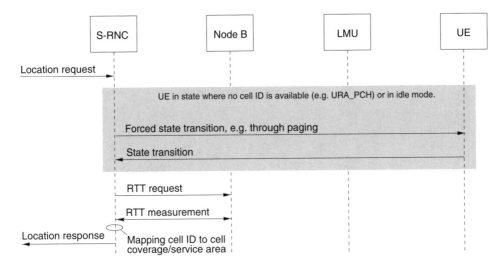

Figure 12.19. The cell ID based location finding principle.

12.3.3 Positioning Techniques and Methods

The standard positioning methods supported within UTRAN are [193, 194]

- cell ID based positioning;

- *Observed Time Difference Of Arrival* (OTDOA) positioning;

- network-assisted *Global Positioning System* (GPS) positioning.

12.3.3.1 Cell ID Based Positioning In the cell ID based positioning method, UE positioning relies on the identification of the cell that provides coverage to the target UE. As shown in Figure 12.19, when the S-RNC receives the LCS request from the core network, it checks the RRC state of the target UE. If the UE is in a state where the cell ID is not available, the UE is paged. Alternatively, the cell ID may be determined as the one that was used during the last active connection to the UE. In the case of soft handover, the cell ID is determined based on the selection of a reference cell [193, 194].

In order to improve the accuracy of the LCS response, the S-RNC may also request *Round Trip Time* (RTT) measurements from the Node B or LMU associated with the cell ID [193, 194]. Depending on the location request, the UTRAN cell ID of the target UE, and/or a *Service Area Identifier* (SAI), is mapped by the S-RNC to geographical coordinates. Antenna beam direction parameters and signal power budget information may be used to improve location accuracy further.

12.3.3.2 OTDOA Positioning The primary OTDOA measurement is the type 2 SFN–SFN (system frame number) observed time difference between two cells as measured at the UE. The measurement is defined as the time difference between the reception of the primary CPICH slot from two different cells. Together with information about the position of the transmitter sites and the *Relative Time Difference* (RTD) of the downlink transmissions,

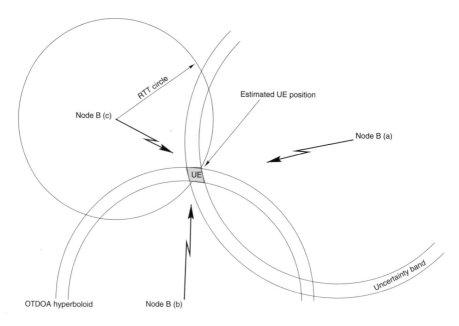

Figure 12.20. The OTDOA positioning principle.

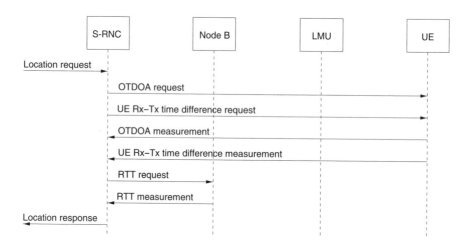

Figure 12.21. OTDOA core network location request process.

the UE position may be calculated. In fact, each OTDOA measurement relative to a pair of downlink transmissions describes a hyperboloid of possible UE locations. The actual UE position can then be determined by the intersection of the hyperboloids of two pairs of Node Bs, see Figure 12.20 [195]. The UE must provide SFN–SFN observed time difference measurements for several cells, including all the cells in the active set and the monitored set.

The OTDOA method may be operated in two modes, see also Figure 12.21.

UE-assisted OTDOA mode where the UE performs the OTDOA measurements and signals the result to the S-RNC. The calculation of the UE position, using the OTDOA measurements, the known position and RTDs of the transmitters, is made in the S-RNC or SAS.

UE-based OTDOA mode where the UE makes the measurements and also carries out the position calculation. In this case the S-RNC should transmit to the UE the information about position and RTD of the measured Node Bs.

The accuracy of the position estimate depends on the precision of the timing measurements made by the UE, the information about transmitter position and RTD, and the relative position of the UE and Node Bs [193, 194].

In order to increase the accuracy of the OTDOA measurements, the Node B may provide *Idle Periods in the Downlink* (IPDL) [41, 196], which can improve the quality of reception of neighbouring Node Bs. In fact, in cases where the UE is close to a Node B, the measurement of CPICH transmissions from distant Node Bs would be masked by the local transmission. This interference is reduced by introducing a series of short gaps in the downlink transmission, during which all the signals transmitted by a single Node B are simultaneously shut down. Unlike compressed mode, the data loss associated with these gaps is not compensated for, and from the perspective of the UE receiver they are viewed as channel erasures to be handled by the channel decoder. In the IPDL method, the idle periods are arranged in a pseudorandom fashion, depending on the value of parameters that define the number of frames between successive idle periods, the length of the idle period measured in CPICH symbols, and a cell specific offset that can be used to synchronize the idle periods from the different cells within a Node B. Since the idle periods have a length of only 5 or 10 CPICH symbols, with IPDL there is not sufficient time for the UE to go through the full three step process normally followed to acquire new Node B pilots. Instead, the UE uses assistance data from the UTRAN to place the searcher within a reasonable uncertainty range and then resorts to the idle period to refine the timing estimate [197]. The IPDL method can be operated in burst mode or continuous mode. In continuous mode the idle periods are inserted pseudorandomly, while in burst mode a given number of idle periods is inserted after a designated starting point, followed by alternate intervals of normal operation and successive bursts of idle periods [41, 196]. Although in principle with IPDL, all the transmissions in a cell must cease during the idle period, in practice, the transmitter will follow a specific on and off time mask, with a power reduction of 20–45 dB in the idle period [198–200]. The support of idle period measurements in the UE is optional.

12.3.3.3 *Network-Assisted GPS Positioning*   The network-assisted GPS positioning methods make use of UEs equipped with GPS radio receivers.

When a UE has access to a GPS receiver, it may perform a stand alone location measurement, without support from the UTRAN [68, 201]. However, if the GPS receiver is designed to interwork with UTRAN, then the network can assist the UE GPS receiver, improving its performance in terms of reduced startup and acquisition times, and increased sensitivity. Assistance data may be generated by means of reference GPS receivers connected to the UTRAN. The network-assisted GPS method may be operated in two modes, namely:

UE-assisted network-assisted GPS mode where the UE performs GPS code phase measurements and signals the result to the S-RNC. The calculation of the UE position is made in the S-RNC or SAS.

UE-based network-assisted GPS mode where the UE performs the GPS measurements and also carries out the position calculation.

In the case of a location request from the core network, the procedure for network-assisted GPS positioning implies the following steps:

1. The S-RNC receives the location request and collects the available network information about the target UE.

2. The S-RNC or SAS computes the GPS assistance data, and sends this information to the UE through the UTRAN air interface using higher layer signalling.

3. The UE makes GPS measurements with the help of the GPS assistance data.

4. In the case of the UE-assisted method, the UE signals the GPS measurements to the S-RNC, and S-RNC or SAS calculates the UE position. In the case of the UE-based method, the UE calculates its position and signals the location information to the S-RNC. The location information includes position, estimated accuracy and time of the estimate.

5. The S-RNC sends the location response back to the core network. In the case of the UE-assisted GPS method, additional non-GPS measurements performed by the UTRAN or UE may be used by the S-RNC to improve the accuracy. These may include RTT measurements, UE Rx–Tx time difference measurements, SFN–SFN observed time difference measurements, or $CPICH\_E_c/N_0$. This information can be made available though RRC signalling for UE measurements or (NBAP) signalling for UTRAN measurements.

12.3.4 LCS Implementation Aspects

The support for OTDOA IPDL, which is optional at the UE, is likely to have the biggest impact on the design of the UE receiver. IPDL affects most of the receiver estimation algorithms, including power control, channel estimation and tracking, automatic frequency correction, automatic gain control, etc. None of these algorithms should be updated with measurements during the idle periods.

The basic measurement period with IPDL is 200 ms. If the assistance data does not allow a sufficiently narrow timing search window for OTDOA IPDL, an integration time of 5–10 *Common Pilot Channel* (CPICH) symbols will not be enough to meet the accuracy requirements of [52, 202]. In this case, IPDL measurements will require integration over several idle periods. In summary:

- The use of the cell ID based positioning method does not require the implementation of additional functionalities in the UE.

- The use of the OTDOA positioning methods requires that the UE supports type 2 SFN–SFN observed time difference measurements, and UE Rx–Tx time difference measurements. In addition, for effective operation of the OTDOA method, the UE

Table 12.5   Implementation requirements for LCS in a FDD UE.

Positioning method	Implementation requirement
Cell ID based	None
	SFN–SFN time difference measurement
OTDOA positioning (UE-assisted)	IPDL support
	UE Rx–Tx time difference measurement
	SFN–SFN time difference measurement
OTDOA positioning (UE-based)	UE Rx–Tx time difference measurement
	IPDL support, UE position computation
	SFN SFN time difference measurement
Network-assisted (UE-assisted)	UE Rx–Tx time difference measurement
	Limited GPS receiver functionality
Network-assisted (UE-based)	Full GPS receiver functionality

should support IPDL. Finally, in the case of UE-based positioning, the UE needs to implement the position calculation function.

- Network-assisted GPS positioning methods require the implementation of partial or full GPS receiver functionality. In addition, in the case of UE-assisted positioning, the UE should support type 2 SFN–SFN observed time difference type 2 measurements, and UE Rx–Tx time difference measurements.

To conclude this section, the UE implementation requirements for LCS (3GPP FDD) are summarized in Table 12.5.

12.4 CPICH INTERFERENCE CANCELLATION AND MITIGATION

The signal power dedicated to the P-CPICH represents a significant fraction of the total *Effective Isotropically Radiated Power* (EIRP) because the pilot channel must cover the cell area well. TS25.101 [9] suggests that 10 % of the I_{or} is dedicated to the pilot channel. This is less than the 20 % we may observe in narrowband CDMA systems.

The effective contribution to the interference level as observed in the receiver is difficult to derive due to, for instance, channel effects, multipath and filter mismatch. The approximate capacity loss is at least 10 % [203], and doubles if a secondary pilot channel is in use broadcasting at the same power level. Assuming further that the neighbouring cells are transmitting at lower power due to lower loading, the observed interference level is larger since the P-CPICH power is defined with respect to the total peak power in order to cover the whole cell.

Based on the aforementioned considerations, the cell capacity can be increased if the interference stemming from the pilot channels could be mitigated or removed. Recent advances in the field of CDMA and multiuser detection for instance, see also Section 12.6.2, and the fact that the pilot code is known at the UE, make it feasible to devise and apply techniques that yield a power saving which translates into increased cell capacity, as more users can be supported simultaneously. CPICH interference mitigation techniques are currently being studied within 3GPP and are reported in [203].

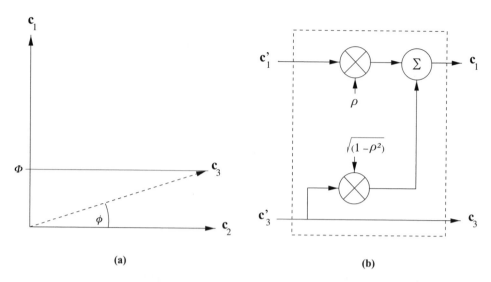

Figure 12.22. (a) The meaning of correlated codes on the signal strength and (b) a model which generates correlated signals.

12.4.1 CPICH Mitigation Principle

Before we proceed and describe the work done to date on CPICH interference cancellation, we present some background on the meaning of correlation and correlation between signals.

Figure 12.22(a) illustrates the effect of correlation on a signal. Three vector signals are depicted, c_1 and c_2 are orthogonal to each other while c_1 and c_3 are nonorthogonal. Orthogonality can be verified by computing the signal's crosscorrelation. Let C be a matrix whose columns are the vectors c_1, c_2 and c_3 respectively. The crosscorrelation is $R = C^H C$, where superscript H denotes the complex conjugate transpose[3]. The off diagonal elements indicate the amount of correlation between the columns, i.e. vector signals, and denote the correlation coefficients. Let ρ be the correlation coefficient between c_1 and c_3 in matrix R. The correlation term ρ is defined in [204, 205] as:

$$\rho = \cos(\Phi) = \frac{E[c_1^H c_3]}{||c_1|| \cdot ||c_3||} \tag{12.6}$$

where $E[c_1^H c_3]$ denotes the expectation of the vector product $c_1^H c_3$. Equation (12.6) tells us that the correlation factor can be visualized as an angle between two vectors.

Figure 12.22(a) shows that the signal c_1 has a contribution from c_3, as large as Φ. This becomes clearer in Figure 12.22(b), where the two uncorrelated signals c_1' and c_3' yield the correlated signal c_1 with a contribution from c_3'.

Figure 12.22(b) also shows a process which could be applied to generate correlation in a Monte Carlo simulation environment. There are several sources which can introduce correlation in a wireless system. The wireless channel destroys the orthogonality between the

---

[3] The covariance is computed according to $C_{var} = C^H C - M^H M$. If the mean term $M^H M$ is zero then the covariance matrix equals the crosscorrelation matrix $C^H C$, e.g. see [204].

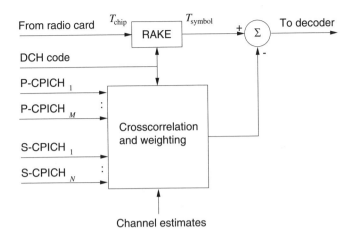

Figure 12.23. Example of a CPICH interference cancellation device.

Orthogonal Variable Spreading Factor (OVSF) codes due to multipath. The case of CDMA spreading codes is not the only situation where correlation is an important parameter. For example, in a space diversity system, see Section 12.5.3, the correlation between the received signals depends on the spatial separation between antennas. The impact of correlation is generally a deterioration of the performance. This can be explained by the fact that correlation is acting like an additional noise source but whose presence does not provide any additional information, in an information theoretic sense. Thus, this leads to a reduction in terms of SIR and hence reduces the performance.

12.4.1.1 *Suggested Interference Canceller*

Figure 12.23 shows the example approach given in [203]. It comprises the typical RAKE receiver block, an interference estimation and weighting block and a cancellation point whose output is fed into the decoder.

The operations carried out in the crosscorrelation and weighting block are described in [206] and revisited below. The output of the ℓth RAKE finger processing the kth user symbol is:

$$y_\ell(k) = \hat{a}_\ell^* \times \sum_{n=1}^{N} r(kN + n)c_{\text{DPDCH}}(kN + n)^* \tag{12.7}$$

where $r(k)$ denotes the correlator input and $c_{\text{DPDCH}}(k)$ is the joint spreading and channelization code, both at chip rate. Term $\hat{a}_\ell^*$ is the associated complex channel estimate, superscript $*$ denotes the complex conjugate, and N denotes the spreading factor SF. The pilot channel code, as any other code, is identified by the subscript label. The *Multiple Access Interference* (MAI) stemming from the pilot channel (CPICH) with respect to the jth multipath is

$$\text{MAI}_{\text{CPICH}} = \hat{a}_\ell^* a_j \times \sum_{n=-\infty}^{\infty} R_{(\delta_{(\ell,j)}+n,k)} \rho(nF - \phi_{(\ell,j)}) \tag{12.8}$$

where $R_{(\alpha,\beta)}$ is the crosscorrelation between the dedicated code and the pilot code, and $\rho(\gamma)$

is the crosscorrelation between the filters (e.g. pulse shaping filter) in the Node B and the UE.

$R_{(\alpha,\beta)}$ is given as:

$$R_{(\alpha,\beta)} = \sum_{n=1}^{N} c_{\text{DPDCH}}(\beta N + n) c_{\text{CPICH}}(\beta N + n - \alpha) \tag{12.9}$$

and $\rho(\gamma)$ is obtained from:

$$\rho(\gamma) = \int_{-\infty}^{\infty} w_{\text{Tx}}\left(\frac{\gamma T_c}{F} - t\right) w_{\text{Rx}}(t)\,dt \tag{12.10}$$

with w_{Tx} and w_{Rx} being the transmit and receiver filter respectively, F denotes the oversampling factor and T_c the chip duration. The delay between the ℓth and the jth multipath is:

$$\tau_\ell - \tau_j = \delta_{(\ell,j)} T_c + \phi_{(\ell,j)} \frac{T_c}{F}$$

and is divided into two parts, namely the integer $\delta_{(\ell,j)}$ and the chip fraction $\phi_{(\ell,j)}$. In reality, it will be difficult to obtain the exact value of $\rho(\gamma)$ from Equation (12.10) because handset manufacturers and network vendors have different implementations with respect to the RRC filter for example. However, this point has not been addressed in [203] and thus one may assume that the impact is negligible. Based on the implementation, a value of $\rho(\gamma)$ can be computed and then stored in the memory.

The next step is to subtract the MAI contribution from the output of the ℓth RAKE finger as illustrated in Figure 12.23. This term is obtained from convolving Equation (12.9) with (12.10), i.e.

$$\text{MAI} = \hat{a}_\ell^* \hat{a}_j \times \sum_{-\infty}^{\infty} R_{(\delta_{(\ell,j)}+n,k)} \rho(nF - \phi_{(\ell,j)}) \tag{12.11}$$

which is equal to Equation (12.8) but based on the estimate of a_j. The integration is carried out over a finite number of steps based on the assumption that 99.5 % of the pilot interference is condensed within ± 3 lobes of the main lobe of $\rho(\gamma)$ and that the Tx and Rx filter are matched RRC filters [206].

12.4.2 Performance Gains

Reference [203] provides a number of simulation results from different sources based on the environment defined in [127]. In an ideal simulation, the weakest cell is cancelled perfectly, while in a realistic simulation, the weakest cell represents additional interference. Comparing the results of both simulations, the average cancellation accuracy (efficiency), defined as:

$$\frac{\text{Gain\_nonideal}}{\text{Gain\_ideal}}$$

lies between 78 % and 97 %. The effects due to timing offset errors, e.g. ± 0.25 chip, have been shown not to affect the gain significantly [203]. Similar results have been obtained for radio card and receiver impairments such as frequency drift, time drift and multipath with rays outside the chip timing. A cancellation accuracy of 90 % has been shown to be still possible.

The capacity gains generally depend on the number of cancelled pilot channels. The biggest gains can be achieved by mitigating the active set, where the capacity gains range from 10 % to 20 % (with an orthogonality factor equal to 0.4) dependent on the cancellation accuracy. Reference [203] also presents results from Sweden, where regulatory requirements dictate that the CPICH power must exceed −85 dBm in order to qualify for cell coverage. This, of course, affects the cell radius and in this case CPICH cancellation is attractive to reduce the number of base stations required. The simulation results show that the cell radius can be increased by 50 m through the use of CPICH cancellation.

In summary, it has been shown that the initially anticipated capacity gains of 10 % (1 dB) are feasible [203]. It was also found that the effects due to imperfect channel estimation, synchronization and timing do not significantly compromise the available capacity gain. As a result of this 3GPP study, the *Dedicated Physical Channel* (DPCH) performance requirements in TS25.101 [9] could be tightened. This means that terminal manufacturers will need to make use of advanced receiver technologies, and an obvious candidate is CPICH interference cancellation.

12.4.3 Terminal Complexity Aspects

A typical CPICH interference canceller is shown in Figure 12.23 and comprises several functional blocks. It was estimated that the total hardware gate count is in the region of 100 k or less, and in a DSP implementation, we may typically expect less than 5 MIPS [203] with a power consumption below 10 mW. The complexity can be further reduced by mitigating the strongest interferers only at little cost with respect to performance degradation. The increase in complexity due to transmit diversity has been estimated to lie between 10–20 %. Multicode operation may add another 50 % to the overall complexity. To summarize, CPICH interference mitigation offers a simple means of improving network capacity at little cost, which makes this technology rather attractive.

12.5 TRANSMIT DIVERSITY FOR MULTIPLE ANTENNAS

3GPP R99 offers *Transmit Diversity* (TxD) schemes for two antennas: two open loop schemes, *Space Time Transmit Diversity* (STTD) and *Time Switched Transmit Diversity* (TSTD), and two closed loop modes based on adaptive antenna arrays.

3GPP is now looking into schemes which provide a further increased downlink capacity. Some work relative to new TxD schemes can be found, for example, in [207], upon which this section is based. There are several issues that need to be addressed before a new scheme can be specified:

- pilot signalling;

- effective aggregate (downlink) capacity;

- feedback data rate and effective cell capacity;

- UE receiver complexity;

- economic aspects.

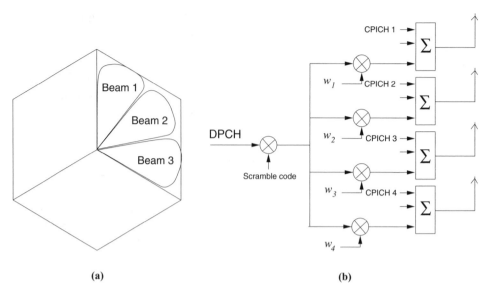

Figure 12.24. (a) Beamforming in a sector and (b) a generic hardware arrangement for eigenbeamforming.

12.5.1 Eigenbeamforming

It has been known for some time that adaptive antennas, smart antennas and space division multiple access yield an additional capacity gain [7, 208]. This may be accomplished by dividing a cell into smaller sectors, i.e. slices, each served by a beam, see Figure 12.24(a). The beams shown in Figure 12.24(a) can be constructed by calculating the weights w_i in Figure 12.24(b) such that the radiation pattern comprises three beams. The ideal solution is obtained if each UE is communicating through its own beam. An economic compromise is to have fixed beams, e.g. three per single 120 degree sector, to reap the benefit of increased capacity by means of reduced interference.

Figure 12.24(b) shows a generic transmitter arrangement where the dedicated channel is scrambled and split into multiple streams, each with its own transmit antenna. Each stream is weighted by a factor w_i and then combined with a pilot signal before transmission.

The eigenbeamforming theory is based on eigenanalysis, computing eigenvectors and eigenvalues [205], and then the beams are defined by the eigenvectors while the transmit power allocation is governed by the eigenvalues. Let $\mathbf{h}_n = [h_{n,1}, h_{n,2}, \dots , h_{n,M}]^\mathrm{T}$ be a channel vector for the nth time sample and M antenna elements, from which we compute the correlation matrix:

$$\mathbf{R} = \sum_{n=1}^{N} \mathbf{h}_n \mathbf{h}_n^\mathrm{H} \tag{12.12}$$

where superscript H denotes the Hermitian (complex conjugate transpose). If Equation (12.12) is obtained from a long term observation window, then we can decorrelate the system applying *Singular Value Decomposition* (SVD) theory, i.e. $\mathbf{R} = \mathbf{USU}^\mathrm{H}$ [205], and arrive at:

$$RU = US \tag{12.13}$$

where matrix U contains the eigenvectors in its columns and S contains the eigenvalues on its main diagonal [207]. We observe in Equation (12.13) that the system is orthogonalized and hence uncorrelated because matrix U is an orthogonal matrix, i.e. $UU^H = I$ where I is the identity matrix. Each eigenvalue in S characterizes the capability of an antenna element to carry power to the UE, or in other words, is an indicator for the channel quality. Hence the transmitter (Node B) should distribute the power amongst the antenna elements which have large eigenvalues, i.e. eigenvalues $\lambda_i \gg 0$.

One drawback of this beamforming scheme, especially in FDD, is that the uplink and the downlink are separated by a wide frequency separation. This means that the uplink and downlink channels are uncorrelated and hence the transmitter (Node B) needs to obtain information on the channel vector from the receiver (UE). Thus it is preferable to compute the weights w_i shown in Figure 12.24(b) in the UE and return the weights via feedback signalling as opposed to sending the channel vector. This reduces the amount of feedback data. To this purpose, one may return either the difference with respect to the last vector or make use of refinements [207]. Further, the total amount of feedback data is subject to the required update rate, i.e. slot by slot or frame by frame, and the overhead due to coding. The use of coding is essential, because the weight vector must be reliable to prevent the beam becoming misaligned in space and the UE not being covered.

12.5.2 Closed Loop Mode for More Than Two Tx Antennas

Another method discussed in [207] is to adapt the existing closed loop transmit diversity scheme of R99 to more than two transmit antennas. This means that the phase or the phase and amplitude is adapted in the Node B subject to the commands sent from the UE. The number of possible phase states may be reduced to cater for the increased number of transmit antennas if the feedback data rate is to be kept constant with respect to R99.

12.5.3 STTD with Four Antennas

The open loop transmit diversity scheme STTD can also be reused and simply adapted to allow for another antenna pair. This means that the Node B has two STTD blocks, one for each antenna pair. This is a rather straightforward approach based on the existing scheme. The interested reader is referred to [209].

12.5.4 Channel Estimation

No matter which of the aforementioned schemes is employed, a significant amount of system complexity will stem from the approach taken to ensure good channel estimates. The design considerations must address many issues [207].

1. The number of pilot channels and the power dedicated to channel estimation.

2. OVSF code exhaustion: every additional pilot channel requires an additional channelization code which could be used for user data.

3. The use of a secondary scrambling code introduces additional interference.

4. Pilot channel symbol length: increasing the spreading factor of each pilot tone in order to preserve code orthogonality means that the channel estimation time at the UE is increased at the expense of reduced response time.

5. UE receiver complexity: a system with four transmit antennas requires the computation of four channel estimates.

6. Backwards compatibility.

A scheme proposed in [207] considers, for example, two OVSF codes, the same pilot pattern as TxD in R99, backwards compatibility with R99, different pilot channel control for two or four transmit antennas, and reduced *Peak to Average Power Ratio* (PAPR). Alternatively, a new pilot scheme for four transmit antennas should only focus on the channels which really benefit from the additional Node B antennas, like DPCH and DSCH. Exploitation of the dedicated pilot bits and slot formats which are favourable to good channel estimation at the terminal side have been suggested. The use of a *Secondary CPICH* (S-CPICH) is only recommended when it does not impact the network performance or R99 terminals in particular.

In summary, there are many aspects which still need to be clarified before the users will be able to benefit from more than one or two transmit antennas at the Node B. The potential advantage of multiple antennas is huge as will be shown in Section 12.6.3.

12.6 IMPROVED BASEBAND ALGORITHMS AND TECHNOLOGY TRENDS

12.6.1 Equalization

Channel equalization is usually employed to counteract the effect of multipath fading on the performance of a digital radio receiver. Optimum equalization strategies include *Maximum Likelihood* (ML) or *Maximum A Posteriori probability* (MAP) trellis processing. However, trellis equalizers often imply an unacceptably high implementation complexity. For this reason, many practical radio receivers employ suboptimum equalization strategies, which include reduced state trellis processing, decision feedback equalization, and linear or transversal equalization [47].

In the WCDMA synchronous downlink transmission, the frequency selective multipath channel destroys the orthogonality of the OVSF codes used to multiplex the different users, causing multiple access interference. Since the downlink user signals transmitted in a given cell propagate through the same multipath channel, channel equalization can restore the code orthogonality, thus suppressing MAI. The use of an equalizer against multipath distortion is particularly important for the case of a UE receiver for HSDPA transmission, which employs a 16 QAM modulation format, as discussed in Section 12.2. In a WCDMA UE receiver, equalization can be performed in principle either at chip level or at symbol level [210]. However, a symbol level implementation implies a higher computational complexity since one equalizer is required for each channelization code. Moreover, operating at symbol level, the equalizer coefficients have to be recomputed every symbol interval, even in the presence of stationary propagation conditions.

The implementation of a chip level equalizer with a simple linear structure has been considered in [211–214]. As shown in Figure 12.25, the principle of chip level equalization is

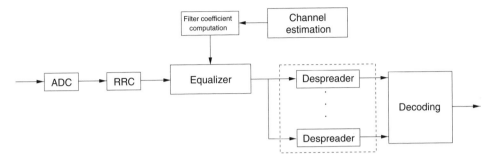

Figure 12.25. WCDMA receiver with chip level equalizer.

equivalent to first estimating the multiuser chips, and then applying conventional despreading and integration over the symbol interval. The chip level equalizer can be designed as a chip spaced or fractionally spaced transversal filter [47]. With fractionally spaced equalization, the input signal sampling interval and the equalizer tap spacing are a fraction of the chip period. The operation of fractionally spaced equalizers is conceptually based on a multichannel representation of the oversampled input channel [215,216]. These equalizers do not suffer from the presence of channel zeroes close to the unit circle, and are insensitive to errors in the sampling instant. Their performance is only sensitive to the presence of common zeroes among the different subchannels. This means that the less complex linear equalizers can be employed without compromising the performance too much.

12.6.1.1 Chip Level Equalization: MMSE Block Linear Equalizer
Consider the downlink transmission, and assume the simple received discrete time signal model

$$r_i = \sum_n d_n h_{i-nM} + n_i \tag{12.14}$$

where $r_i = r(iT_\text{c}/M)$ are the received signal samples taken at rate T_c/M, with T_c denoting the chip interval, $h_\ell = h(\ell T_\text{c}/M)$ are the rate T_c/M samples of the complex equivalent channel impulse response including the transmit root raised cosine pulse shaping filter and the receive root raised cosine matched filters, d_n represent the complex multiuser transmitted chip sequence, and $n_i = n(iT_\text{c}/M)$ are complex additive Gaussian noise samples, which model thermal noise and intercell interference. Also assume that the channel impulse response samples h_ℓ are appreciably different from zero only for $\ell = 0, \ldots, LM - 1$.

The oversampled sequence r_i can be decomposed into M chip rate subsequences relative to M distinct subchannels. In vector notation, we define, for each chip interval,

$$\mathbf{r}_k^{(m)} = [r_{kM+m}\ r_{(k+1)M+m} \cdots r_{(k+N-1)M+m}]^\text{T}, \qquad m = 0, \ldots, M-1 \tag{12.15}$$

where the index k denotes the kth chip interval, and $(\cdot)^\text{T}$ indicates transpose. From Equations (12.14) and (12.15) we can also write:

$$\mathbf{r}_k^{(m)} = \mathbf{H}^{(m)}\mathbf{d}_k + \mathbf{n}_k^{(m)}, \qquad m = 0, \ldots, M-1 \tag{12.16}$$

where $\mathbf{d}_k = [d_{k-L+1} \cdots d_k \cdots d_{k+N-1}]^T$

$$\mathbf{H}^{(m)} = \begin{bmatrix} h_{L-1}^{(m)} & h_{L-2}^{(m)} & \cdots & h_0^{(m)} & 0 & \cdots & 0 \\ 0 & h_{L-1}^{(m)} & \cdots & h_1^{(m)} & h_0^{(m)} & \cdots & 0 \\ \vdots & \vdots & \ddots & \vdots & \vdots & \ddots & \vdots \\ 0 & 0 & \cdots & 0 & 0 & \cdots & h_0^{(m)} \end{bmatrix} \tag{12.17}$$

with $h_\ell^{(m)} = h_{\ell M+m}$, and $\mathbf{n}_k^{(m)} = [n_{kM+m} \ n_{(k+1)M+m} \cdots n_{(k+N-1)M+m}]^T$.

Denoting by $\mathbf{w}_k^{(m)}$ the N dimensional vector of the equalizer coefficients relative to the mth subchannel, $\mathbf{w}_k^{(m)} = [w_{kM+m} \ w_{(k+1)M+m} \cdots w_{(k+N-1)M+m}]^T$, the equalizer output at time k results in:

$$y_k = \mathbf{w}_k^{(0)T} \mathbf{r}_k^{(0)} + \ldots + \mathbf{w}_k^{(M-1)T} \mathbf{r}_k^{(M-1)} = \mathbf{w}_k^T \mathbf{r}_k \tag{12.18}$$

with the $MN \times 1$ vectors

$$\mathbf{w}_k = [\mathbf{w}_k^{(0)T} \cdots \mathbf{w}_k^{(M-1)T}]^T \tag{12.19}$$

$$\mathbf{r}_k = [\mathbf{r}_k^{(0)T} \cdots \mathbf{r}_k^{(M-1)T}]^T = \mathbf{H}\mathbf{d}_k + \mathbf{n}_k \tag{12.20}$$

$$\mathbf{n}_k = [\mathbf{n}_k^{(0)T} \cdots \mathbf{n}_k^{(M-1)T}]^T \tag{12.21}$$

and the $MN \times (N + L - 1)$ matrix

$$\mathbf{H} = [\mathbf{H}^{(0)T} \cdots \mathbf{H}^{(M-1)T}]^T \tag{12.22}$$

With this notation, the output of the linear adaptive equalizer results in:

$$y_k = \mathbf{w}_k^T \mathbf{r}_k \tag{12.23}$$

Note that the above model of a fractionally spaced equalizer with T_c/M spaced coefficients also applies when (all or some of) the M chip rate subchannels correspond to the signal samples obtained from multiple receive antennas.

A common strategy for the design of the MN equalizer coefficients of the vector $\mathbf{w}_k$ is based on the minimization of the *Mean Square Error* (MSE) at the equalizer output. In the case of a block linear equalizer, the optimum vector $\mathbf{w}_{k(opt)}$ according to the *Minimum Mean Square Error* (MMSE) criterion is obtained as:

$$\mathbf{w}_{k(opt)} = \arg \min_{\mathbf{w}_k} E\{|\mathbf{w}_k^T \mathbf{r}_k - d_{k+D}|^2\} \tag{12.24}$$

where $E\{\cdot\}$ denotes statistical expectation, and D is a delay parameter. From Equation (12.24) straightforward calculation gives:

$$\mathbf{w}_{k(opt)} = (E\{\mathbf{r}_k^* \mathbf{r}_k^T\})^{-1} E\{\mathbf{r}_k^* d_{k+D}\} = (\mathbf{H}^* \mathbf{H}^T + \sigma_d^{-2} \mathbf{C}_{nn})^{-1} \mathbf{h}_{k+D}^* \tag{12.25}$$

where $(\cdot)^*$ denotes complex conjugation, $\sigma_d^2 = E\{|d_k|^2\}$, $\mathbf{C}_{nn} = E\{\mathbf{n}_k^* \mathbf{n}_k^T\}$ is the noise covariance matrix, and $\mathbf{h}_{k+D}$ indicates the $MN \times 1$ column of the channel matrix $\mathbf{H}$ corresponding to the multiuser chip d_{k+D}. The calculation of the optimum MMSE equalizer coefficients requires the availability of an estimate of the channel matrix $\mathbf{H}$ and of the noise covariance matrix $\mathbf{C}_{nn}$. The channel estimation can be performed based on either the common or dedicated pilot symbols.

12.6.1.2 Chip Level Equalization: MMSE Adaptive Linear Equalizers

The direct computation of the MMSE equalizer coefficients requires a matrix inversion, which may imply high computational complexity. The matrix inversion can be avoided by an adaptive implementation of the MMSE algorithm, e.g. [47]. However, a problem with the adaptive approach is that no training multiuser sequence is available at the UE receiver. This problem is solved by the adaptive algorithms described in this section.

With the *Griffiths* algorithm [213], the linear equalizer is trained using the estimated channel impulse response. The MMSE fractionally spaced coefficients update is obtained as:

$$\mathbf{w}_{k+1} = \mathbf{w}_k + \gamma E\{\mathbf{r}_k^* d_{k+D} - \mathbf{r}_k^* y_k\} \implies \tag{12.26}$$

$$\implies \mathbf{w}_{k+1} = \mathbf{w}_k + \gamma(\sigma_d^2 \mathbf{h}_{k+D}^* - \mathbf{r}_k^* y_k) \tag{12.27}$$

where $\gamma > 0$ is a step size parameter. With the above adaptation, the need of a training sequence is avoided by using the knowledge of the correlation between the multiuser chip sequence and the received signal and the estimate of the channel response.

A drawback of an adaptation rule based on explicit channel estimation is that, in the presence of fast varying multipath fading, the estimated channel may lag behind the actual propagation channel. Under these conditions, it may be preferable to employ an adaptive equalizer trained directly from the CPICH symbols or dedicated pilot symbols. A straightforward solution is to use a *Normalized Least Mean Square* (NLMS) [214], which gives the equalizer coefficients updating rule:

$$\mathbf{w}_{k+1} = \mathbf{w}_k + \gamma \mathbf{r}_k^* e_k \tag{12.28}$$

where

$$e_k = d_{k+D} - y_k \tag{12.29}$$

$$\gamma = \frac{\alpha}{\beta + \mathbf{r}_k^H \mathbf{r}_k}, \qquad 0 < \alpha < 2, \quad \beta > 0 \tag{12.30}$$

$(\cdot)^H$ denotes conjugate transpose, and d_k is the (spread and scrambled) CPICH sample corresponding to the kth chip interval. Note that with this approach, the equalizer does not suppress the contribution of the user's signals to the error (12.29), due to the pseudorandomness of the spreading codes [214]. This results in a relatively large error signal which requires a small adaptation step size to provide sufficient averaging, and hence slows down the rate of convergence.

12.6.1.3 Practical Equalizer Aspects

Choosing the ideal equalizer can be a cumbersome exercise, trading between economic and engineering aspects. The decisions may be driven by:

- cost and complexity: silicon area, power consumption, MIPS requirements;
- risks: e.g. achieving a numerically stable design;

- time: development time and time to market;

- resources.

On the other hand, there are the engineering considerations which include:

- expected performance gain, e.g. with respect to BER;

- processing rate: chip rate or symbol rate and fractionally spaced or nonfractionally spaced;

- structure: linear or nonlinear equalizer;

- filter weight update algorithm: adaptive or deterministic;

- computational complexity: tap size and filter weight computation.

The choice may also be influenced by work found in the open literature. However, a lot of insight can be gained by investigating the propagation channel [217], since some equalizer structures try to invert the channel. For example, the first step is to derive the distribution of channel zeroes [218], followed by an analysis of the required number of filter (equalizer) taps. Note that it cannot be assumed that the equalizer tap size equals the channel delay spread. Channels with nonminimum phase characteristics cannot be equalized properly by any linear equalizer.

12.6.2 Multiuser Detection

Multiuser Detection (MUD) [219] is a signal processing technique ideally suited to combat MAI. MUD receivers are well suited for the uplink because they need the knowledge of the OVSF code used by each user. Thus, only a suboptimum or pseudo MUD is feasible in a UE unless the protocol is changed to supply the UE with this information, but this would result in the loss of privacy which users may not appreciate. However, the network gains would be greater if MUD techniques were employed in the downlink rather than in the uplink since the downlink is generally considered the capacity bottleneck.

A RAKE receiver is basically a *Matched Filter* (MF) which can exploit temporal diversity (multipath). An MF is optimum when the noise is white and hence treats interference like Gaussian noise. The superposition of codes in a CDMA system can generally be approximated as Gaussian noise except in the case of short spreading codes. Therefore, all high data rate services will suffer from interference (MAI) since the spreading gain is small and the signal power is high.

The error probability P_e for an MF receiver in an *Additive White Gaussian Noise* (AWGN) environment with unity signal power and U users is [220]:

$$P_e^{MF} = Q \left(\sqrt{\frac{G}{\sigma_n^2 + (U - 1)}} \right) \tag{12.31}$$

where G denotes the spreading gain, σ_n^2 the thermal noise variance and $(U - 1)$ denotes MAI, i.e. number of users per cell or sector. Function $Q(.)$ is the Gaussian error function [7].

By inspection of Equation (12.31) we note the following. The probability of error equals the well known *Binary Phase Shift Keying* (BPSK) error probability [47] if one user is active. The thermal noise term σ_n^2 can be neglected because intercell and intracell interference

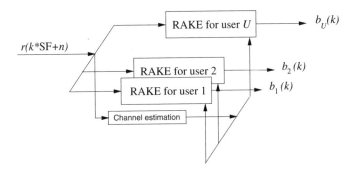

Figure 12.26. A typical first generation Node B receiver structure.

will dominate the interference level. The thermal noise floor lies in the region well be-low $-100\,\text{dBm}$, while the interference power level of a typical medium loaded cell is conjectured to lie $20\,\text{dB}$ or more above the thermal noise floor, e.g. around $-60\,\text{dBm}$ [9].

Assume now that the receiver is able to remove the MAI term $(U - 1)$ through clever signal processing. In this case, Equation (12.31) again represents the single user error probability [47] because the term $(U - 1)$ converges to zero. Consequently, the error ratio P_e of a MUD receiver should remain constant and independent of the cell loading.

The received signal in the uplink is a superposition of many spread signals sent from the UEs to the Node B. However, a Node B, in contrast to a UE, has knowledge of each scrambling code, channelization code and the associated signal timing. Thus, it could use this additional information to cancel the interference from other UEs, in order to improve receiver performance and hence network capacity. The receiver of choice in an R99 FDD UE is the RAKE receiver as discussed in Section 3.4. This is also likely to be the case for first generation Node Bs. A typical Node B receiver is shown in Figure 12.26, and consists of a bank of RAKE receivers, one for each user within a cell or sector. This simple receiver treats each user separately, and so no information is shared between the different RAKE receivers. Introducing a signal processing technique which exploits this additional information in order to improve the receiver performance is the motivation for employing a MUD receiver.

12.6.2.1 On Multiuser Detection

A simplified two user CDMA system is shown in Figure 12.27(a) to demonstrate the working principle of MUD. At the transmitter, the kth information bits for each user, $b_1(k)$ and $b_2(k)$, are spread by the spreading code $\mathbf{c}_1$ and $\mathbf{c}_2$ respectively and combined to form the signal $\mathbf{x}(k)$. The length of each vector is determined by the spreading factor SF.

At the receiver, the signal $\mathbf{y}(k)$ is processed to estimate the originally transmitted bits $b_1(k)$ and $b_2(k)$. Classic multiuser detection may be interpreted as a two stage detection process [221]. First, signal $\mathbf{y}(k)$ is matched, filtered and then processed by the MUD in order to estimate $\hat{b}_1(k)$ and $\hat{b}_2(k)$, the estimated transmitted kth bit for user one and user two respectively. Figure 12.27(a) shows such a simplistic system where we assume perfect conditions, e.g. no time dispersion introduced by the channel and no synchronization effects.

Assume that the kth bits $b_1(k)$ and $b_2(k)$ can be either $+1$ or -1 with equal probability, and that the spreading codes are $\mathbf{c}_1 = [-1 + 1 - 1 - 1]^T$ and $\mathbf{c}_2 = [+1 + 1 + 1 - 1]^T$.

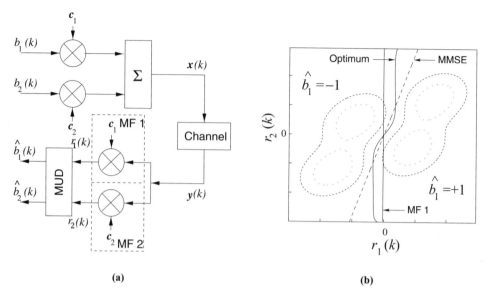

Figure 12.27. The foundation of decision boundaries. (a) An idealized DS-CDMA system; (b) Linear versus nonlinear decision boundary.

Then the combined and transmitted symbol $\mathbf{x}(k)$ is drawn from the set $\mathcal{X}$ which contains all possible chip rate symbol combinations

$$\mathcal{X} = \mathbf{B}\mathbf{C}^{\mathrm{T}} = \begin{bmatrix} +1 & +1 \\ +1 & -1 \\ -1 & +1 \\ -1 & -1 \end{bmatrix} \begin{bmatrix} -1 & +1 & -1 & -1 \\ +1 & +1 & +1 & -1 \end{bmatrix} = \begin{bmatrix} 0 & +2 & 0 & -2 \\ -2 & 0 & -2 & 0 \\ +2 & 0 & +2 & 0 \\ 0 & -2 & 0 & +2 \end{bmatrix}$$

(12.32)

For example, for $b_1(k) = 1$ and $b_2(k) = -1$, the transmitted symbol is $\mathbf{x}(k) = [-2 \ 0 \ -2 \ 0]$.

The received noise free signal is $\mathbf{y}(k) = \mathbf{x}(k)$. More generally, $\mathbf{y}(k) = \mathbf{x}(k) + \mathbf{n}(k)$ where $\mathbf{n}(k)$ denotes the noise term. At the receiver, $\mathbf{y}(k)$ is first matched against each spreading code, $\mathbf{c}_1$ and $\mathbf{c}_2$. A conventional receiver uses the output $r_1(k)$ and $r_2(k)$ to determine $\hat{b}_1(k)$ and $\hat{b}_2(k)$ by means of a slicer, obtaining $\hat{b}_1(k) = 1$ $r_1(k) > 0$ and $\hat{b}_1(k) = -1$ otherwise. This procedure would result in the performance determined by Equation (12.31). However, the idea of MUD is to improve the detection performance of $\hat{b}_1(k)$ through exploiting the information hidden in $r_2(k)$ and vice versa. A positive aspect of the MF preprocessing stage is the reduction of signal dimensionality. The MUD receiver processes $\mathbf{r}(k) = [r_1, r_2, ..., r_U]^{\mathrm{T}}$ with U elements as opposed to vector $\mathbf{y}(k)$ with 256 entries (in the case of SF = 256).

An optimum MUD detector, the maximum likelihood receiver, determines the likelihood between the preprocessed signal $\mathbf{r}(k)$ and all signal hypotheses via the Euclidean metric. Therefore this detector needs to know what signal is to be expected in order to make the best judgement as to whether a +1 or a −1 was transmitted for each user. This signal set is derived from $\mathcal{X}$ by matching them against the employed codes in $\mathbf{C}$. This yields a new set, the hypotheses, denoted $\mathcal{Y}$ according to:

$$\mathcal{Y} = \mathcal{X}\mathbf{C}^{\mathrm{T}} = \begin{bmatrix} 0 & +2 & 0 & -2 \\ -2 & 0 & -2 & 0 \\ +2 & 0 & +2 & 0 \\ 0 & -2 & 0 & +2 \end{bmatrix} \begin{bmatrix} +1 & -1 \\ +1 & +1 \\ +1 & -1 \\ -1 & -1 \end{bmatrix} = \begin{bmatrix} +4 & +4 \\ +4 & -4 \\ -4 & +4 \\ -4 & -4 \end{bmatrix}. \tag{12.33}$$

Each row in $\mathcal{Y}$ represents the possible output at the two matched filters shown in Figure 12.27(a) when no noise is present, and thus each row in the set $\mathcal{Y}$ represents a point in a two dimensional plane. It is easy to see that the points in $\mathcal{Y}$ form a square. A square, cube or hypercube indicates that the used spreading codes are orthogonal, i.e. $\mathbf{CC}^{\mathrm{T}} = \mathbf{I}$ where $\mathbf{I}$ is the identity matrix. If this is the case then the optimum detector is also linear. If correlated spreading codes are used, then the points in $\mathcal{Y}$ form the vertices of a parallelogram, a polyhedron or a polytope, and the decision boundary is generally nonlinear as shown in Figure 12.27(b).

More insight into the performance of a MUD can give us Figure 12.27(b). Each axis represents the output of one of the matched filters, and the loci of the four dotted ellipses represent the points in $\mathcal{Y}$. In a realistic scenario, where noise is present, we will never observe one of the points in $\mathcal{Y}$ but a point somewhere lying in the plane $r_1(k)$ and $r_2(k)$ due to noise. The detection process then becomes a statistical problem. The general assumption made is that the noise is Gaussian, then the dotted lines mark the contour line of equal probability. We further observe in Figure 12.27(b) that the contour lines are elliptical and not circles. This indicates that the used spreading codes are nonorthogonal and the cause of correlation in the signals $r_1(k)$ and $r_2(k)$. The skewness is an indicator of the amount of correlation between the two spreading codes (seven chip Gold codes were used in this example).

The optimum receiver measures the distance between the actual MF outputs $r_1(k)$ and $r_2(k)$, and the four points in $\mathcal{Y}$ and decides in favour of the one which is nearest. The boundary which separates $\hat{b}_1 = +1$ from $\hat{b}_1 = -1$ is termed the decision boundary. Two linear boundaries and a nonlinear boundary have been plotted in Figure 12.27(b). Any receiver or detector generates such a boundary and the shape of the boundary categorizes the type of detector, i.e. linear or nonlinear. The boundary MF 1 in Figure 12.27(b) illustrates the decision boundary for the matched filter receiver for user one. The other two boundaries illustrate the decision boundaries for two types of MUD receiver, MMSE and the optimum Bayes' receiver. Note that the Bayes' receiver uses the Mahalanobis metric and not the Euclidean metric [222, 223]. The optimum boundary yields the minimum bit error ratio.

Obviously, the aim is to approximate the optimum boundary at little cost in terms of detector complexity. For instance, the optimum Bayesian MUD has a complexity which is proportional to 2^U, where U denotes the number of users or codes used. The term 2^U used here to denote the complexity reflects the number of possible signal constellations observed by the receiver, from which it estimates the most likely transmitted bit for each user. Recall that this also represents the number of rows in $\mathcal{Y}$ (12.33). The maximum likelihood MUD, better known as Verdu's MUD [219], approximates the optimum decision boundary well. The boundaries constructed by Verdu's MUD have a Voronoi structure [219] because the detector minimizes the joint bit error probability of all users as opposed to minimizing the individual bit error probability (compare Figure 4.5 with Figure 4.1 in [219]). However, reference [224] shows that the performance difference between Bayes' and Verdu's MUD can be marginal even for short spreading codes.

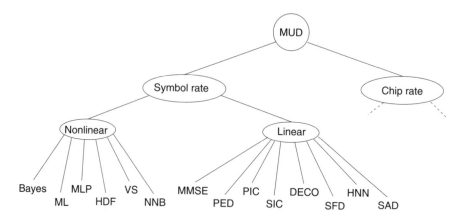

Figure 12.28. A classification of multiuser detector schemes.

12.6.2.2 *MUD Survey* MUD receivers may be categorized into two basic groups, *Chip Rate* (CR) and *Symbol Rate* (SR). The former processes a vector of chip samples and its complexity is linked to the length of the spreading code. The latter deals with preprocessed signals, namely the output of matched filters, and its complexity is subject to the number of active users. Most MUD proposals consider the symbol rate approach because the MF output has sufficient statistics [47, 219]. The advantage of SR MUD receivers is that they generally process smaller matrices, but slightly better performance has been observed with CR receivers. One may further distinguish nonlinear and linear designs which are in turn divided into specific classes. A MUD classification tree is proposed in Figure 12.28. For almost every symbol rate receiver, there exists the equivalent chip rate receiver.

The following key also contains a short list of references for ease of background research.

Bayes Optimum detector [219, 224, 225]

ML Maximum likelihood detector [219]

MLP Multilayer perceptron neural network [226]

HDF Hard decision feedback detector [227]

VS Volterra series detector [228]

NNB Neural network based receivers [229]

MMSE Minimum mean square error detector [219]

PED Polynomial expansion detector [230]

PIC Parallel interference cancellation detector [219, 231, 232]

SIC Serial interference cancellation detector [219, 233, 234]

DECO Decorrelation detector [219]

SDF Soft decision feedback detector [227]

HNN Hopfield neural network receiver [235, 236]

RNN Recurrent neural network receiver [237, 238]

SAD Simulated annealing detector [239]

Note that the classification is not as strict as our figure may suggest. For example, the interference cancellation receivers may be considered to be nonlinear but it is known that the PED converges to the MMSE solution if an infinite number of stages is considered. It has also been shown that some interference cancellation schemes follow the same trend [240] and converge to the MMSE performance bound. It seems that some detectors converge to the MMSE bound while others converge to the zero forcing performance bound [241].

In conclusion, a good MUD receiver design, with respect to the bit error probability, must be able to approximate closely the optimum decision boundary at little cost in terms of complexity. Thus it is conjectured, referring to Figure 12.27(b), that both requirements can only be fulfilled through a step wise linear approximation of the optimum decision boundary with linear segments.

12.6.3 STC and MIMO Technologies

It is generally acknowledged that the bottleneck of a mobile network is the downlink, since it usually requires higher data rates for Internet based applications. It should also be noted that in the downlink the user is retrieving data with a device (UE) which has limited processing resources. The downlink capacity for a constant Baud rate system can be increased either by sending more information bits per symbol or transmitting more than one symbol at a time. The former method implies the use of higher order modulation and/or a reduction in coding protection (for example by means of increased puncturing), while the latter strategy may be implemented by employing more than one transmit antenna.

The use of multiple transmit and receive antennas can increase the system capacity [242, 243]. If the transmitter has more than one antenna, then the transmission data rate may be increased by transmitting a substream of $\log_2(M_A)$ bit symbols on each of the M_{Tx} transmit antennas, where M_A denotes the size of the modulation alphabet, e.g. $M_A = 16$ for 16 QAM. With this operation, the number of transmitted bits increases linearly with the number of transmit antennas.

At the receiver, the antenna observes a signal constellation with $(M_A)^{M_{Tx}}$ points. We observe that

- the signals from the other transmit antennas complicate our efforts in applying maximum likelihood detection because the receiver complexity increases exponentially;

- the other signals can be seen as interference.

Thus, exploiting the capacity gains from multiple transmit antennas needs to be carefully weighted against the added receiver complexity. The current state of technology may prevent us from implementing optimum receivers in a commercial UE in the near future.

With the above approach, the increased channel capacity obtained by multiple transmit and receive antennas ($M_{Tx} > 1$, $M_{Rx} > 1$) is exploited by the use of transmit diversity, while receive diversity is used to increase the receive SNR and attain the required error performance [244]. Figure 12.29(a) illustrates that with two transmit antennas and QPSK, any receive antenna observes 16 constellation points which is a superposition of two QPSK symbols, both phase shifted and attenuated by the channels h_1 and h_2 respectively. The

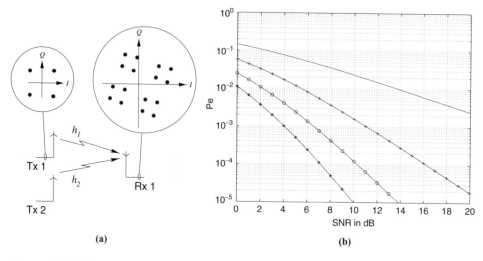

(a)   (b)

Figure 12.29. The effect of transmit and receive diversity. (a) TxD: increased constellation
pattern density as seen at a receive antena; (b) RxD: improved BER due to
SNR gain from addition antennas, 1(−), 2(+), 3(o), 4(∗).

theoretical BER improvement due to the SNR gain from additional receive antennas is
reported in Figure 12.29(b), from Section 14-4 in [47]. A comparison between transmit and
receive diversity can be found in Wittneben [245], which introduces a downlink diversity
model and studies the performance of a Viterbi based detector.

Space Time Coding (STC) may be seen as an approach to combine the benefits of
transmit and receive diversity. STC is a technique which, in essence, divides the channel
into many subchannels over which data is transmitted in parallel. This leads to an increase
in channel capacity over the common 1-to-1 antenna channel [242, 243, 246, 247], which
clever algorithms can exploit. However, to achieve these huge potential gains, the number
of antenna elements at the transmitter and receiver must be increased. Thus, the channel
becomes a *Multiple Input Multiple Output* (MIMO) channel.

The channel capacity of the 1-to-1 antenna channel can be calculated according to
Shannon [248]. The channel capacity for MIMO channels has been derived in [243].
Figure 12.30 shows the theoretical capacities for a variety of $M_{Tx}:M_{Rx}$ antenna combinations. The MIMO channel states are collected in a matrix **H**, where the channel connecting
transmit antenna i to receive antenna j is denoted by $h_{i,j}$. The propagation channel is here
assumed to be Rayleigh fading with independent subchannels $h_{i,j}$ [242].

To capitalize on the aforementioned theoretically available channel capacity, several
methods have been proposed which may be broadly categorized into three families:

Space time block coding was introduced by Alamouti [209] and analysed by Tarokh [249,
250].

Space time trellis coding was introduced by Tarokh [251, 252].

Layered space time coding was introduced by Foschini [253].

12.6.3.1 Space Time Block Coding   Space time block coding has been incorporated
in the 3GPP specifications and is known there as STTD. Thus STTD will not be described

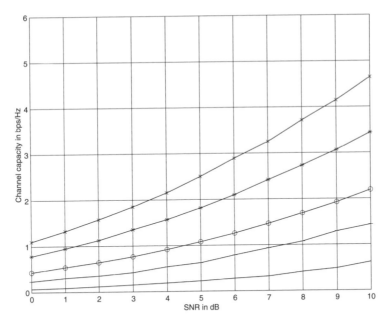

Figure 12.30. Theoretical channel capacity (with 5 % outage) for: 1:1 $(-)$, 2:1 $(+)$, 1:2 (o), 2:2 (*) and 4:2 ($\times$) transmit and receive antennas respectively.

any further. Tarokh *et al.* [249, 250] took the work of Alamouti [209] further and found additional codes for more than two transmit antennas. However, none of the block codes found for more than two Tx antennas can achieve the same efficiency as STTD.

Below follows a short discussion of other published work related to STTD. The STTD schemes proposed by Tarokh [249] were derived for coherent detection and require accurate channel estimates. This adds to the computational complexity at the receiver if there are many MIMO subchannels. A remedy is to employ a differentially encoded STC scheme, as proposed by Tarokh [254] and Hughes [255]. Tarokh's proposed transmission scheme is 3 dB worse than the equivalent STTD scheme from Alamouti [209]. Lo [256] analysed the radiation patterns of the STTD block codes, and found that if the transmit antennas are spaced by $\lambda/2$, then the beam pattern consists of one major lobe. The number of lobes increases with increasing antenna spacing. Block codes were also assessed from the capacity point of view by Sandhu and Paulraj [257] for example. The authors found that space time block coding is suboptimal for MIMO channels. In particular: (a) a full rate STTD code used over any channel with one receive antenna yields optimum capacity; (b) an STTD code of any rate used over the i.i.d.[4] Rayleigh channel with multiple receive antennas always incurs a loss in capacity. Simulation results show a significant capacity loss, even with a moderate number of receive antennas. Finally, a comparison between different diversity techniques, such as *Delay Diversity* (DD) and *Orthogonal Transmit Diversity* (OTD), of concatenated coding schemes can be found in [258]. The simulation results show that Alamouti's simple diversity technique outperforms DD and OTD.

---

[4] independent identically distributed

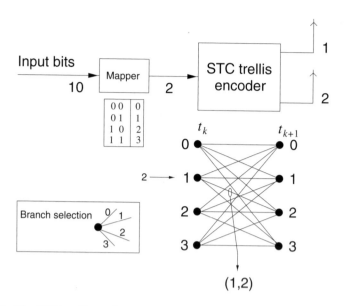

Figure 12.31. The STC trellis coding scheme as proposed by Tarokh *et al.*

12.6.3.2 Space Time Trellis Coding

A simple but suboptimum multiantenna transmission scheme has been proposed in [259], where the output of a standard convolutional encoder, e.g. Proakis [47], is demultiplexed onto M_{Tx} parallel data streams, one for each transmit antenna. At the receiver, the authors propose a Viterbi processor performing maximum likelihood decoding. This encoding scheme yields a code trellis which can be easily implemented in a Viterbi decoder. However, the resulting new code is not optimum because the code is not optimized for a multiantenna (space time) transmission scheme.

There are two optimization criteria, diversity gain and coding gain.

Tarokh published the first diversity optimized space time codes in [251]. Figure 12.31 illustrates his proposed technique for an $M_{Tx} = 2$ antenna system and with a trellis structure suited for a QPSK modulator. In the QPSK example of Figure 12.31, the mapper block converts a bit pair onto a QPSK symbol according to the table shown beneath the block. The resulting symbol from the QPSK alphabet $\in \{0, 1, 2, 3\}$ is then fed into the trellis encoder. The encoder moves from a current state at time index t_k to another state at time index t_{k+1} along a branch. Each state has four branches because the alphabet has four elements, and the branch to take is governed by a branch selection rule: if an i enters the current state, then the $i + 1$th branch leads to the next state. In our example, the encoder is in state 1 at time t_k, and a symbol 2 is fed into the encoder. Then the encoder moves one step in time to t_{k+1} to state 2 and outputs a symbol pair, e.g. (1,2), where the first number denotes the QPSK symbol to be transmitted from antenna one and the second symbol from antenna two. Figure 12.31 does not reveal which symbol pair belongs to which branch. This information is given, for example, in references [250–252,260] which define the codes explicitly.

An optimum receiver would employ ML detection to estimate the transmitted bits, however, this turns out to be very complex. For example, a two transmit antenna system with QPSK modulation and a four state space time code requires an ML detector which can match the received signal against 4^2 signals (hypotheses), while a 16 QAM antenna system

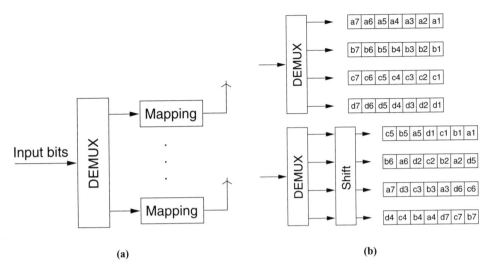

Figure 12.32. Two examples of Foschini's proposed layered STC architectures. (a) Basic 'layered' transmitter. (b) Two types of layered scheme: vertical and diagonal.

with eight transmit antennas demands an ML detector with 16^8 hypotheses. So, despite the promising capacity gains, space time trellis coding is today handicapped by the complexity of the ML receiver. Thus alternative space time schemes or receiver designs, such as iterative detectors or linear receivers, are desirable.

12.6.3.3 *Layered Schemes*
The principle behind layered STC is firstly to generate parallel streams of data, one for each transmit antenna, and then to process them at the receiver at minimum cost in terms of complexity. Foschini suggests several methods, which are known as BLAST (*Bell laboratories Layered Space Time*) [253]. Figure 12.32(a) shows the basic layered approach, where input bits are demultiplexed onto several parallel streams and then mapped onto the M-ary modulation points. A single encoder and interleaver can be introduced before the demultiplexer, or alternatively, each stream may have its own encoder and interleaver.

Figure 12.32(b) depicts the two best known layered schemes, *Vertical BLAST* (VBLAST) and *Diagonal BLAST* (DBLAST). VBLAST is simple because the streams are kept independent. DBLAST offers additional redundancy because the symbols of each stream are distributed in the same diagonal manner amongst the different streams.

We shall now discuss in more detail the VBLAST detector with $M_{Tx} = 8$ and $M_{Rx} = 12$ antennas as described in [261]. The transmitter has no encoding stage and demultiplexes the input bits onto eight streams, each stream maps the bits onto a QAM point as in Figure 12.32(a).

At any time instant at the receiver, an M_{Rx} element complex vector is received, which contains eight symbols to be estimated. The receiver employed is based on *Zero Forcing* (ZF). The transmitted symbol vector is denoted $\mathbf{a} = [a_1, a_2, \ldots, a_{M_{Tx}}]^T$ and the received signal vector is $\mathbf{r} = [r_1, r_2, \ldots, r_{M_{Rx}}]^T$. Matrix $\mathbf{H}$ contains the channel estimate for all $M_{Tx} \times M_{Rx}$ channels between the transmitter and the receiver.

The whole detection process may be described as searching first for the set of channel states, a column in $\mathbf{H}$, which allows a symbol from one of the transmit antennas to be carried to all receive antennas with the highest cumulative *Signal to Noise Ratio* (SNR). The probability that this symbol can be estimated correctly from all transmitted symbols is of course maximized. Then we iteratively determine all the other symbols, according to their likelihood of being correctly detected based on the channel strength. Then contributions from the stronger symbols onto the weaker ones are removed, i.e. cancelled, by subtracting the already detected symbols from the received signal. This process is iteratively repeated until all transmitted symbols are estimated. When the last symbol has been determined, only thermal noise is left because we have removed the contributions from the other symbols, i.e. interference. This of course assumes perfect symbol detection of all the previously detected symbols and cancellation.

The detailed receiver operation consists of the following 7 steps:

1. Initialize: compute the Moore–Penrose[5] pseudoinverse [262] of $\mathbf{H}$: $\mathbf{G}_i = \mathbf{H}^+, i = 1$;

2. Compute the index k_i of the row in $\mathbf{G}_i$ with minimum norm:
   $k_i = \text{argmin}_{j \notin \{k_1 \ldots k_{i-1}\}} ||(\mathbf{G}_i)_j||^2$;

3. Extract the k_ith row of $\mathbf{G}_i$:
   $\mathbf{w}_{k_i} = (\mathbf{G}_i)_{k_i}$;

4. Suppress interference with nulling vector $\mathbf{w}_{k_i}$: $y_{k_i} = \mathbf{w}_{k_i}^{\text{T}} \mathbf{r}_i$;

5. Estimate the most likely transmitted symbol from the symbol alphabet Q: $\hat{a}_{k_i} = Q(y_{k_i})$;

6. Cancel interference: $\mathbf{r}_{i+1} = \mathbf{r}_i - \hat{a}_{k_i} (\mathbf{H})_{k_i}$;

7. Zero channel column and update channel inverse: $\mathbf{G}_{i+1} = \mathbf{H}_{\overline{k_i}}^+$

8. Increment i and repeat steps 2 to 8 until $i = M_{\text{Tx}}$.

In step 2, we add up the sum of inverse channel powers in each row of $\mathbf{G}$ and then determine which row has the lowest accumulated sum. The row number is then stored in k_i. This number is also stored in a set S. In step 4, we actually carry out the zero forcing process using a row from the channel matrix inverse $\mathbf{G}$. The zero forcing detector forces all terms to zero except one, and this holds here since $\mathbf{HG} = \mathbf{I}$. In step 5, we find the most likely transmitted QAM symbol applying a maximum likelihood search. In step 7 we zero the column k_i in $\mathbf{H}$ which corresponds to the weakest row in $\mathbf{G}$.

The reported effective capacity gain for the discussed layered scheme is 20.7 bits/s/Hz [261]. In a real system, a significant amount of data must be dedicated to channel estimation. For example, in the aforementioned system, a raw spectral efficiency of 25.9 bits/s/Hz was reduced by the fact that 20 % of the symbols per burst are used as pilot symbols. In principle, a better receiver is the maximum likelihood detector [263] but its complexity makes it only feasible for small antenna arrangements. Alternative suboptimum receivers have been proposed in [264–266].

12.6.3.4 Summary   MIMO and STC schemes are very attractive for meeting future capacity demands. However, a significant amount of data dedicated to channel estimation reduces the potentially available capacity gains [252, 267–272]. From this point of view,

---

[5] Denoted A^+ as opposed to A^{-1}

it may be advantageous to consider alternative solutions like noncoherent types of STC scheme, e.g. [254], or more advanced receiver structures which exploit advanced signal processing techniques such as joint and iterative decoding schemes [273].

12.6.3.5 Beamforming

As already mentioned in Section 12.6.2, CDMA is an interference limited system in terms of capacity. A method of mitigating interference is to limit its creation. This can be achieved by exploiting the space dimension and cutting a cell or sector into even smaller slices or beams. An advantage of beamforming is also that an interferer can be suppressed by placing a zero (null) in the antenna pattern at the receiver.

A viable method is known as *Space CDMA* (SCDMA) [7]. In SCDMA, the base station can produce beams which serve hot spots. Ideally, we want to be able to create narrow beams and have the capability to track and steer them. This functionality may be realized with smart antenna systems and beamformers [208], which are used in radar technology. The number of beams that can be formed and accurately placed in space is dependent on the number of available antenna elements. Unfortunately, large antenna arrays require more of the expensive RF circuitry. A cheaper alternative is to have a number of fixed beams, say three per sector or cell.

12.6.4 Technology Trends

Orthogonal Frequency Division Multiplexing (OFDM) has been recognized during the last few years as an efficient wireless interface for high speed application [274]. Its main advantage is given by the possibility of implementing fast signal processing algorithms at the receiver, with significant computational savings with respect to *Time Division Multiple Access* (TDMA) or CDMA receivers.

The drive towards higher data rates and energy efficient terminal equipment has recently determined the adoption of an OFDM physical layer for the *Wireless Local Area Network* (WLAN) IEEE 802.11a standard, which operates at 5 GHz providing data rates of up to 54 Mbps [275]. IEEE 802.11a has replaced the older 802.11b based on *Direct Sequence Spread Spectrum* (DSSS) transmission, operating in the 2.4 GHz band with a maximum rate of 11 Mbps. For compatibility with 802.11b, IEEE is additionally producing a new OFDM based standard at 2.4 GHz, called 802.11g.

In OFDM, a high rate data sequence is split into lower rate sequences, which are transmitted over separate orthogonal subcarriers generated using an *Inverse Fast Fourier Transform* (IFFT). Since the lower rate parallel subcarriers have an increased symbol duration, the relative amount of time dispersion caused by multipath is decreased, apart from the symbol guardband which mitigates the effects of multipath. Therefore, OFDM can be seen as an alternative approach to channel equalization to counteract multipath fading. At the receiver side, the subcarriers are demodulated using a simple *Fast Fourier Transform* (FFT). As a result, an OFDM 802.11a receiver has a much lower complexity than a typical 802.11b receiver, while operating at a rate up to five times higher.

A feasibility study is currently ongoing within 3GPP for the introduction of the OFDM physical layer in UTRAN [276, 277]. As a starting point, OFDM will be considered in the downlink only, in a 5 MHz spectrum allocation, i.e., coupled with WCDMA in the uplink for a two times 5 MHz deployment scenario. The objective is to achieve data rates above 10

Mbps in urban and indoor environments [276]. The study item will also identify solutions for mobility between OFDM carriers and between OFDM and WCDMA carriers [276].

A separate 3GPP working item is investigating the integration of lower layers of OFDM based WLAN systems into UTRAN.

12.6.5 Interworking Between 3GPP and WLAN Systems

A problem with IEEE 802.11 WLANs concerns security and support for quality of service. A feasibility study on interworking between UTRAN and WLAN systems has been conducted within 3GPP, with the purpose of extending 3GPP services and functionality to the different WLAN access standards [278]. The ultimate goal is to allow the user to access the same set of services, irrespective of the radio access technology used. The study considers both single mode WLAN UEs and multimode WLAN-3GPP UEs. In general, 3GPP system functionalities provided to a WLAN user may reside behind the WLAN, or in parallel to the WLAN. WLANs may be deployed by 3GPP or non3GPP operators [279], and may partially or completely overlap [278]. In a user state with overlapping 3GPP and WLAN coverage, the radio access technology may be automatically selected based on predetermined choices set by the UE, user, network, location or context. Possible interworking scenarios include:

Common billing where the connection between the WLAN and the 3GPP system is that there is a single customer relationship, for example in the case of a 3GPP subscriber who wants to access the WLAN service provided by his home operator.

3GPP access control and charging where authentication, authorization, and accounting are provided by the 3GPP system, for example for a subscriber using dual mode equipment.

Service continuity where the user can continue to access the same set of services switching between 3GPP and WLAN coverage areas without having to establish a new session, but with possible brief interruptions of data flow.

Seamless services where a multimedia session is maintained in a handover between 3GPP and WLAN without noticeable interruption.

Network protocol functionalities that allow subscribers to move between (WLAN) 802.11, WCDMA, cdma2000, or GPRS networks without losing their connection are already commercially available [280].

12.6.6 Ultra Wideband Radio

Ultra WideBand (UWB) radio technology [281, 282] has properties that make it a suitable option for short range communications in dense multipath environments. In a UWB system, information is modulated by the position of a string of sub nanosecond pulses with low duty cycle (pulse position modulation). The signal energy is correspondingly spread over a bandwidth that goes from near DC to several GHz, with extremely low power spectral densities. This ensures that UWB transmission does not interfere with narrowband radio systems operating in dedicated bands. Multiple access is achieved by time hopping according to a user specific pseudorandom pattern [281]. In essence, UWB can be seen as an overlay system which lives on the noise margin of the other air interface standards operating within the UWB frequency band, which reminds us of the early days of CDMA overlay network concepts.

Major advantages of UWB systems are:

- very low transmission power;

- resistance to interference due to receiver processing gains in the order of 40 dB;

- robustness with respect to multipath fading due to the capability to resolve multipath components with differential delays of less than 1 ns, which can be constructively combined in a rake receiver;

- improved time of arrival resolution [281, 282].

UWB technology [283] is currently receiving a positive evaluation from the US *Federal Communications Commission* (FCC). A European Union research project has also been recently established to further study UWB system aspects related to implementation, coexistence with different radio systems, ad hoc networking and UE positioning [284].

Pulse position modulation in an infinite bandwidth channel theoretically has the capability to achieve maximum power efficiency, and (single user) UWB radios have already been implemented and demonstrated [281]. The practical application and deployment is likely to be realized only at a later stage with respect to the technologies discussed in the previous sections.

Appendix A
ML detection for uncoded QPSK

Chapter 4 introduced maximum likelihood detection for encoded *Quadrature Phase Shift Keying* (QPSK). This appendix presents a derivation for detecting the most likely transmitted bit in an uncoded QPSK symbol, and is an example showing that the solution is the same as the process of decomposing QPSK into two BPSK symbols given in most textbooks.

Figure A.1. shows a QPSK example for a received complex signal $r(t)$. Each QPSK symbol comprises two bits, the *Least Significant Bit* (LSB) and the *Most Significant Bit* (MSB), which represent the real and the imaginary signal components. The ML detector makes a bit estimate for each bit based on the ML criterion.

It is assumed that the LSB can only be a 1 or a 0, hence the posterior probabilities sum up to unity, i.e. $P(\text{Re}(r)) = P(\text{LSB} = 0|\text{Re}(r)) + P(\text{LSB} = 1|\text{Re}(r)) = 1$. The LSB can have a value 0 or 1, or in other words, we have two classes of posterior probability. The conditional probability for the first class is

$$P(\text{LSB} = 0|\text{Re}(r)) = \frac{P(\text{Re}(r)|\text{LSB} = 0)P(\text{LSB} = 0)}{P(\text{Re}(r))} \tag{A.1}$$

and similarly for $P(\text{LSB} = 1|\text{Re}(r))$. The value of the LSB is estimated by comparing the two probabilities $P(\text{LSB} = 0|\text{Re}(r))$ and $P(\text{LSB} = 1|\text{Re}(r))$ and choosing the larger one

WCDMA – Requirements and Practical Design. Edited by R. Tanner and J. Woodard.
© 2004 John Wiley & Sons, Ltd. ISBN: 0-470-86177-0.

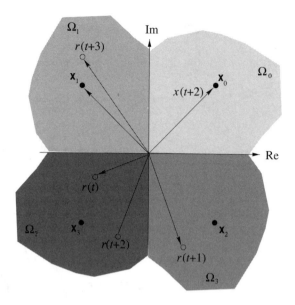

Figure A.1. The QPSK constellations $\{x_0, x_1, x_2, x_3\}$ and the decision boundaries.

because it is most likely correct. In *Binary Phase Shift Keying* (BPSK), this selection process is accomplished with a 'slicer' defined by:

$$\text{sgn}(z) = \begin{cases} 1 \text{ if } z \geq 0 \\ 0 \text{ if } z < 0 \end{cases} \tag{A.2}$$

Returning to the example, estimating the most likely transmitted LSB in the transmitted symbol x starts with the relationship:

$$\hat{x}_{\text{LSB}} = \text{sgn}\left(\frac{P(\text{LSB} = 1|\text{Re}(r))}{P(\text{LSB} = 0|\text{Re}(r))}\right) \tag{A.3}$$

The ratio in the brackets of Equation (A.3) is calculated according to:

$$\frac{P(\text{LSB} = 1|\text{Re}(r))}{P(\text{LSB} = 0|\text{Re}(r))} = \frac{\frac{P(\text{Re}(r)|\text{LSB} = 1)P(\text{LSB} = 1)}{P(\text{Re}(r))}}{\frac{P(\text{Re}(r)|\text{LSB} = 0)P(\text{LSB} = 0)}{P(\text{Re}(r))}} \tag{A.4}$$

$$= \frac{P(\text{Re}(r)|\text{LSB} = 1)}{P(\text{Re}(r)|\text{LSB} = 0)} \tag{A.5}$$

where Equation (A.5) has been obtained assuming $P(\text{LSB} = 1) = P(\text{LSB} = 0) = 0.5$. The term $P(\text{Re}(r)|\text{LSB} = 0)$ is the sum of two probabilities because there are two QPSK alphabet points where the LSB is zero, refer to Figure A.1, namely x_0 and x_2, hence we can write $P(\text{Re}(r)|\text{LSB} = 0) = P(\text{Re}(r)|x_0) + P(\text{Re}(r)|x_2)$. The remaining open question is the computation of $P(\text{Re}(r)|x_0)$.

The received signal

$$r = x + n \tag{A.6}$$

must correspond to one of the four transmitted QPSK constellation points, which are well defined. Further, it is assumed that the received signal is corrupted by additive Gaussian noise. The latter assumption is commonly made in communication systems. If the noise distribution is not Gaussian, then the subsequent equations need modifying accordingly. The probability density function for a zero mean Gaussian signal is:

$$P(z) = \frac{1}{\sqrt{2\pi\sigma^2}} \exp\left(-\frac{|z|^2}{2\sigma^2}\right) \tag{A.7}$$

The expression (A.6) denotes the fact that the received signal is a superposition of the transmitted signal corrupted by the noise term n, and is therefore itself Gaussian. The location of the centre of the Gaussian distribution function of r is shifted away from zero (recall the assumption of zero mean Gaussian noise), since signal x introduces an offset. For instance, if $x_0 = 1 + j$, then the Gaussian density function will be located around $1 + j$, provided we transmit x_0. Therefore, the probability $P(r|x_0)$ can be written as:

$$P(r|x_0) = \frac{1}{\sqrt{2\pi\sigma^2}} \exp\left(-\frac{|r - x_0|^2}{2\sigma^2}\right) \tag{A.8}$$

which now takes account of the mean. In the same manner, we can further expand the term in brackets in Equation (A.3) as:

$$\frac{P(\text{LSB} = 1|\text{Re}(r))}{P(\text{LSB} = 0|\text{Re}(r))} = \frac{\exp(\frac{-|\text{Re}(r)-\text{Re}(x_3)|^2}{2\sigma^2}) + \exp(\frac{-|\text{Re}(r)-\text{Re}(x_1)|^2}{2\sigma^2})}{\exp(\frac{-|\text{Re}(r)-\text{Re}(x_0)|^2}{2\sigma^2}) + \exp(\frac{-|\text{Re}(r)-\text{Re}(x_2)|^2}{2\sigma^2})} \tag{A.9}$$

Through back substitution of $\text{Re}(x_0) = +1$, $\text{Re}(x_1) = -1$, $\text{Re}(x_2) = +1$, $\text{Re}(x_3) = -1$ (see Figure A.1), the terms:

$$\exp\left(\frac{-\text{Re}(r)^2}{2\sigma^2}\right) \text{ and } \exp\left(\frac{-1}{2\sigma^2}\right)$$

cancel and we arrive at:

$$\frac{P(\text{LSB} = 1|\text{Re}(r))}{P(\text{LSB} = 0|\text{Re}(r))} = \exp\left(\frac{-2\text{Re}(r)}{2\sigma^2} - \frac{2\text{Re}(r)}{2\sigma^2}\right) \tag{A.10}$$

$$= \exp\left(\frac{-2\text{Re}(r)}{\sigma^2}\right) \tag{A.11}$$

This term contains an exponential function which one wants to avoid by taking the logarithm on both sides, hence

$$\ln\left(\frac{P(\text{LSB} = 1|\text{Re}(r))}{P(\text{LSB} = 0|\text{Re}(r))}\right) = \frac{-2\text{Re}(r)}{\sigma^2} \tag{A.12}$$

The log function is monotonically increasing and thus we can back substitute this term into Equation (A.3) as seen in textbooks, e.g. [47] and arrive finally at:

$$\hat{x}_{\text{LSB}} = \text{sgn}\left(\ln\left(\frac{P(\text{LSB} = 1|\text{Re}(r))}{P(\text{LSB} = 0|\text{Re}(r))}\right)\right) \tag{A.13}$$

$$= \text{sgn}\left(\frac{-2\text{Re}(r)}{\sigma^2}\right) \tag{A.14}$$

The final result Equation (A.14) confirms that the LSB is only dependent on the sign of $\text{Re}(r)$, or in other words, on the location of the received signal r, i.e. left or right of the ordinate in Figure A.1. The ordinate is said to be the decision boundary for the LSB, while the abscissa is the decision boundary of the MSB.

Appendix B
SIR computation

In this section we derive the *Signal to Interference Ratio* (SIR). The result could be used to implement an SIR measurement device. The analysis does not cover the special case when a transmit diversity scheme is present.

At the transmitter, we denote the composite signal, which comprises all active physical channels, as:

$$\mathbf{c} = \mathbf{c}_0 b_0 + \alpha \mathbf{c}_1 b_1 + \ldots, \tag{B.1}$$

where vector $\mathbf{c}_0 = (c_{0,1}, c_{0,2}, \ldots, c_{0,256})^{\mathrm{T}}$ represents the pilot channel (e.g. P-CPICH) with 256 chips while $\mathbf{c}_1, \mathbf{c}_2, \ldots$ denote other intracell channels, of which code $\mathbf{c}_1$ is the dedicated traffic channel of interest. Terms b_k in Equation (B.1) denote either information or pilot bits in complex notation, i.e. $(b_k = b_{k,\mathrm{I}} + j b_{k,\mathrm{Q}})$. The initial assumption is that $\mathbf{c}_k^{\mathrm{H}} \mathbf{c}_k = \mathrm{SF}_k$, where superscript H denotes the Hermitian form (complex conjugate transpose) and SF_k is the spreading factor for the kth code. In the following notation, the time or a symbol number index has been omitted for ease of notation. We normalize the dedicated channel power with respect to the P-CPICH.

The 3GPP specifications do suggest that the SIR measurement is carried out on the DPCCH (DPCH). However, the pilot symbol sequence is rather short and thus one may want to exploit the P-CPICH instead. Obviously, computing the SIR, for the power control loop for instance, needs some thought with regard to the expected accuracy. Signal power

WCDMA – Requirements and Practical Design. Edited by R. Tanner and J. Woodard.
© 2004 John Wiley & Sons, Ltd. ISBN: 0-470-86177-0.

and noise power estimations obtained from small data sample sets exhibit a large spread (variance) while averaging over a long period of time outdates the measurements with time and also leads to an increased accuracy spread. A good compromise must be found between averaging length and accuracy spread. A disadvantage is that the power ratio between the DPCCH and the P-CPICH is unknown. For example, the SIR estimation error for medium and high SIR figures is around 4 dB if the SIR is derived from four CPICH samples. Alternatively, the noise may be estimated on the P-CPICH and the signal power on the *Time Multiplexed Pilot* (TMP) bits.

The propagation channel will vary the signal envelope due to fading and multipath will introduce interchip interference. The true channel state is denoted a and is assumed here to be constant for the duration of one P-CPICH symbol, i.e. 256 chips or 66.7μs. Again, we omit the time index for simplicity because we look only at one symbol. The multipath channel states are denoted by a_ℓ, where the index refers to the ℓth ray, i.e. channel path. The RAKE receiver itself can assign up to L fingers, one finger for every ray, which is deemed worthwhile combining to increase the signal SNR at the RAKE output.

At the receiver, we can compute the SIR from the DPCCH signal power while the interference and noise power may be measured on the P-CPICH. In the sequel, every signal vector is assumed to have 256 entries, equivalent to one P-CPICH symbol, in order to ease the notation.

The received chip rate vector signal for the ℓth ray is:

$$\mathbf{y}_\ell = a_\ell \mathbf{c} + \mathbf{n}_\ell \tag{B.2}$$

where $\mathbf{n}$ denotes the combined noise and interference vector. The common assumption is that the noise is uncorrelated and Gaussian, due to the central limit theorem [204], hence $E[\mathbf{nn}^H] = \sigma_n^2 \mathbf{I}_{SF_0 \times SF_0}$, where σ_n^2 is the noise variance (power) per chip and $\mathbf{I}_{SF_0 \times SF_0}$ is the identity matrix of size SF_0.

As discussed in Chapter 3, $\mathbf{y}_\ell$ will be derotated in the RAKE receiver by the ℓth finger using the associated channel estimate $\hat{a}_\ell$. Then the MRC combined RAKE receiver output r_1 for the channelization code $\mathbf{c}_1$ is:

$$r_1 = \sum_{\ell=1}^{L} \mathbf{c}_1^H \hat{a}_\ell^* \mathbf{y}_\ell \tag{B.3}$$

where superscript $*$ denotes the complex conjugate. Through back substitution, we arrive at:

$$r_1 = \sum_{\ell=1}^{L} \left[\mathbf{c}_1^H \hat{a}_\ell^* a_\ell \mathbf{c}_1 b_1 + \mathbf{c}_1^H \hat{a}_\ell^* a_\ell \mathbf{s}' + \mathbf{c}_1^H \hat{a}_\ell^* \mathbf{n}_\ell \right] \tag{B.4}$$

where $\mathbf{s}' = \mathbf{c}_0 b_0 + \mathbf{c}_2 b_2 + \ldots$ comprises all remaining signals coming from the same transmitter, the self interference or intracell interference.

The SIR can be derived from Equation (B.4) for the traffic channel of interest, with subscript 1, as:

$$\mathrm{SIR}(r_1) = \frac{E\left[\left| \sum_{\ell=1}^{L} \left[\mathbf{c}_1^H \hat{a}_\ell^* a_\ell \mathbf{c}_1 b_1 \right] \right|^2 \right]}{E\left[\left| \sum_{\ell=1}^{L} \left[\mathbf{c}_1^H \hat{a}_\ell^* a_\ell \mathbf{s}' + \mathbf{c}_1^H \hat{a}_\ell^* \mathbf{n}_\ell \right] \right|^2 \right]} \tag{B.5}$$

where $E[.]$ denotes the expectation. The numerator yields:

$$E\left[\left|\sum_{\ell=1}^{L}[\alpha\hat{a}_\ell^* a_\ell]\right|^2\right] = \alpha^2 E\left[\left|\sum_{\ell=1}^{L}\hat{a}_\ell^* a_\ell\right|^2\right] \tag{B.6}$$

where $\alpha = \sqrt{E[b_1^2]SF_1}$.

The denominator in Equation (B.5) requires a more elaborate analysis. First we expand the denominator into three terms:

$$E\left[\left|\sum_{\ell=1}^{L}[\mathbf{c}_1^H\hat{a}_\ell^* a_\ell \mathbf{s}' + \mathbf{c}_1^H\hat{a}_\ell^*\mathbf{n}_\ell]\right|^2\right] = E\left[\left|\mathbf{c}_1^H\mathbf{s}'\right|^2\left|\sum_{\ell=1}^{L}\hat{a}_\ell^* a_\ell\right|^2 + Z + \left|\sum_{\ell=1}^{L}\mathbf{c}_1^H\hat{a}_\ell^*\mathbf{n}_\ell\right|^2\right]$$

$$\tag{B.7}$$

Of interest is first the middle term Z, which becomes zero because the noise term $\mathbf{n}_\ell$ can be assumed to be uncorrelated with the remaining terms, hence:

$$Z = E\left[\left(\mathbf{c}_1^H\mathbf{s}'\right)\left(\sum\hat{a}_\ell^* a_\ell\right)\left(\sum\mathbf{n}_\ell^H\hat{a}_\ell\mathbf{c}_1\right) + \left(\sum\mathbf{c}_1^H\hat{a}_\ell^*\mathbf{n}_\ell\right)(\mathbf{s}'^H\mathbf{c}_1)\left(\sum\hat{a}_\ell a_\ell^*\right)\right] = 0.$$

The third term in Equation (B.7) yields:

$$E\left[\left|\sum_{\ell=1}^{L}\mathbf{c}_1^H\hat{a}_\ell^*\mathbf{n}_\ell\right|^2\right] = \mathbf{c}_1^H\left(\sum_{\ell=1}^{L}|\hat{a}_\ell|^2\right)E\left(\mathbf{n}_\ell\mathbf{n}_\ell^H\right)\mathbf{c}_1$$

where the channels are treated as a deterministic quantity. The product $E[\mathbf{n}_\ell\mathbf{n}_\ell^H]$ denotes the combined noise and interference covariance matrix. Thus we can write $E[\mathbf{n}_\ell\mathbf{n}_\ell^H] = \frac{\sigma_n^2}{SF_1}\mathbf{I}$, where σ_n^2 represents the variance of the noise plus interference.

By substituting the intermediate results back into Equation (B.5), we arrive at:

$$SIR(r_1) = \frac{\alpha^2\left|\sum_{\ell=1}^{L}\hat{a}_\ell^* a_\ell\right|^2}{\left|\mathbf{c}_1^H\mathbf{s}'\right|^2\left|\sum_{\ell=1}^{L}\hat{a}_\ell^* a_\ell\right|^2 + \sigma_n^2\sum_{\ell=1}^{L}|\hat{a}_\ell|^2} \tag{B.8}$$

This expression can be written as:

$$SIR(r_1) = \frac{\alpha^2}{\left|\mathbf{c}_1^H\mathbf{s}'\right|^2 + \sigma_n^2\frac{\sum_{\ell=1}^{L}|\hat{a}_\ell|^2}{\left|\sum_{\ell=1}^{L}\hat{a}_\ell^* a_\ell\right|^2}} \tag{B.9}$$

and

$$SIR(r_1) = \frac{\alpha^2}{\left|\mathbf{c}_1^H\mathbf{s}'\right|^2 + \sigma_n^2\frac{1}{\sum_{\ell=1}^{L}|\hat{a}_\ell|^2}} \tag{B.10}$$

where we assume that the estimated channel state information equals the true channel state, $\hat{a}_\ell \simeq a_\ell$. Obviously, Equation (B.10) still contains an 'annoying' term, $|\mathbf{c}_1^H\mathbf{s}'|^2$, which represents the crosscorrelation between the spreading code of the traffic channel of interest and all other spreading codes. This term is difficult to resolve analytically because the crosscorrelation depends on the chosen spreading code itself and the multipath channel profile. It can be assumed that long spreading codes are employed, e.g. with 256 chips, and that the channel is benign, i.e. does not significantly destroy the orthogonality between the spreading codes, then the crosscorrelation term $(\mathbf{c}_1^H \mathbf{s}' \mathbf{s}'^H \mathbf{c}_1)$ is negligible because it is zero, and Equation (B.10) becomes Equation (B.11). This result is very similar to the expression found in [285] but uses a different notation and approach.

$$\mathrm{SIR}(r_1) = \frac{b_1^2 \cdot SF_1}{\sigma_n^2 \frac{1}{\sum_{\ell=1}^{L} |\hat{a}_\ell|^2}} \tag{B.11}$$

References

1. P. H. Young, *Electronic Communication Techniques*. Upper Saddle River, USA: Prentice Hall, 3rd edition, 1994.

2. M. E. Frerking, *Digital Signal Processing in Communication Systems*. Boston: Kluwer Academic Publishers KAP, 1994.

3. H. Holma and A. Toskala, *WCDMA for UMTS*. Chichester, UK: John Wiley & Sons, Ltd., 1st edition, 2000.

4. "*http://www.3gpp.org,*".

5. "*http://www.3gpp2.org,*".

6. H. Holma and A. Toskala, *WCDMA for UMTS*. Chichester, UK: John Wiley & Sons, Ltd., 2nd edition, 2002.

7. T. S. Rappaport, *Wireless Communications*. Upper Saddle River, USA: Prentice Hall, 1st edition, 1996.

8. A. Springer and R. Weigel, *UMTS The Universal Mobile Telecommunications Systems*. Berlin: Springer Verlag, 1st edition, 2002.

9. 3GPP, *TS25.101, V3.11.0, UE Radio Transmission and Reception (FDD)*. Sophia Antipolis, F: ETSI, 2002.

10. 3GPP, *TS34.121, V3.11.0, Terminal Conformance Specification; Radio Transmission and Reception (FDD)*. Sophia Antipolis, F: ETSI, 2002.

WCDMA – Requirements and Practical Design. Edited by R. Tanner and J. Woodard.
© 2004 John Wiley & Sons, Ltd. ISBN: 0-470-86177-0.

11. O. K. Jensen, T. E. Kolding, C. R. Iversen, S. Laursen, R. V. Reynisson, J. H. Mikkelsen, E. Pedersen, M. B. Jenner, and T. Larsen, 'RF receiver requirements for 3G W-CDMA mobile terminals,' *Microwave Journal*, **43**, 22–46, 2000.

12. P. Madsen, T. Amtoft, R. V. Reynisson, J. H. Mikkelsen, S. Laursen, C. R. Iversen, T. E. Kolding, T. Larsen, and M. B. Jenner, 'RF requirements for UTRA/FDD transceivers,' in *International Symposium on Wireless Personal Multimedia Communications (WPMC)*, **1**, 197–202, 2001.

13. ITU Geneva, *http://www.itu.int/ITU-R/conferences/wrc/wrc-00/index.asp*, 2000.

14. UMTS forum, *Candidate Extension Bands for UMTS / IMT-2000 Terrestrial Component*, 1999.

15. K. Holladay and D. Burman, 'Design loop filters for PLL frequency synthesizers,' *Microwaves & RF*, **38**, 1999.

16. L. Litwin, 'Matched filtering and timing recovery in digital receivers,' *RF Design*, pp. 32–48, 2001.

17. S. K. Reynolds, B. A. Floyd, T. J. Beukema, T. Zwick, U. R. Pfeiffer and H. A. Ainspan, 'A direct-conversion receiver integrated circuit for WCDMA mobile systems,' *IBM Journal Research & Development*, **47**, 337–352, 2003.

18. J. Wilson and M. Ismail, 'CMOS multi-standard transceivers for wireless communications,' in *Proceedings of Wireless Design Conference*, London, **1**, 83–87, 2002.

19. A. Pärssinen, J. Jussila, J. Ryynänen, L. Sumanene and K. A. Halonen, 'A 2-GHz wide-band direct conversion receiver for WCDMA applications,' *IEEE Journal of Solid-State Circuits*, **34**, 1893–1903, 1999.

20. A. A. Abidi, 'Direct-conversion radio transceivers for digital communications,' *IEEE Journal of Solid-State Circuits*, **30**, 1399–1410, 1995.

21. B. J. Minnis and P. A. Moore, 'Estimating the IP-2 requirement for a zero-IF UMTS receiver,' in *Proceedings of Wireless Design Conference*, London, **1**, 25–29, 2002.

22. C. J. Grebenkemper, *Local Oscillator Phase Noise and its Effect on Receiver Performance*. Watkins–Johnson Inc., 2001.

23. M. Valkama, M. Renfors and V. Koivunen, 'Compensation of frequency-selective I/Q imbalance in wideband receivers: models and algorithms,' in *Workshop on Advances in Signal Processing SPAWC'01*, **1**, 42–45, 2001.

24. D. Boyd, 'Calculate the uncertainty of NF measurements,' *Microwaves & RF*, **38**, 93–102, 1999.

25. S. Bible, 'Selecting crystals for stable oscillators,' *Microwaves & RF*, **41**, 53–61, 2002.

26. S. Alechno, 'Oscillators: a new look at an old model, part 1,' *Microwaves & RF*, **41**, 51–67, 2002.

27. S. Alechno, 'Oscillators: a new look at an old model, part 2,' *Microwaves & RF*, **42**, 85–90, 2003.

28. T. C. Hofner, 'Measuring and evaluating dynamic ADC parameters,' *Microwaves & RF*, **39**, 78–94, 2000.

29. 3GPP, *TS25.213, V3.8.0, Spreading and Modulation (FDD)*. Sophia Antipolis, F: ETSI, 2002.

30. P. Stadnik, 'Baseband clipping can lead to improved WCDMA signal quality,' *Wireless Systems Design*, pp. 40–44, September 2000.

31. Rohde & Schwarz, *User Manual: Software WinIQSIM*™ *for Calculating I/Q Signals for Modulation Generator AMIQ*, 2000.

32. T. Ojanperä and R. Prasad, *WCDMA: Towards IP Mobility and Mobile Internet*. Boston: Artech House, 1st edition, 2001.

33. M. Kolber, 'Predict phase-noise effects in digital communications systems,' *Microwaves & RF*, **38**, 1999.

34. B. Goldberg, 'The effects of clock jitter on data conversion devices,' *RF Design*, pp. 26–32, August 2002.

35. J. B. Groe and L. E. Larson, *CDMA Mobile Radio Design*. Norwood, USA: Artech House, 1st edition, 2000.

36. K. I. Pedersen and M. J. Flanagan, 'Quantisation Loss in a Pilot-aided Rake Receiver,' *Proceedings of the ACTS Mobile Communication Summit*, Aalborg, Denmark, pp. 617–622, October 1997.

37. B. N. Vejlgaard, P. Mogensen and J. B. Knudsen, 'Performance Analysis for UMTS Downlink Receiver with Practical Aspects,' *Proceedings of the International Conference on Vehicular Technology*, Amsterdam, NL, **2**, 998–1002, September 1999.

38. Agilent, *Application Note 1335: HPSK Spreading for 3G*.

39. A. Netsell, 'Interpret and apply EVM to RF system design,' *Microwaves & RF*, **40**, 83–94, 2001.

40. M. Banerjee and R. Desquiotz, 'Generate wide-dynamic-range WCDMA/3GPP signals,' *Microwaves & RF*, **42**, 101–108, 2003.

41. 3GPP, *TS25.214, V3.10.0, Physical Layer Procedures (FDD)*. Sophia Antipolis, F: ETSI, 2002.

42. P. B. Kenington, *High-Linearity RF Amplifier Design*. Boston: Artech House, 1st edition, 2000.

43. F. Zavosh, M. Thomas, C. Thron, T. Hill, D. Artusi, D. Anderson, D. Ngo, and D. Runton, 'Digital predistortion techniques for RF power amplifiers with CDMA applications,' *Microwave Journal*, October 1999.

44. B. Berglund, T. Nygren and K. Sahlman, 'RF multicarrier amplifier for third-generation systems,' *Ericsson Review*, pp. 184–189, October 2001.

45. Altera Corp., *Implementing Digital IF & Digital Predistortion Linearizer Functions with Programmable Logic*, 2003.

46. D. Liu, B. P. Gaucher, E. B. Flint, T. W. Studwell, H. Usui and T. J. Beukema, 'Developing integrated antenna subsystems for laptop computers,' *IBM Journal Research & Development*, **47**, 1–12, 2003.

47. J. G. Proakis, *Digital Communications*. New York, USA: McGraw Hill, 3rd edition, 1993.

48. A. J. Viterbi, *CDMA*. Reading, Massachusetts, USA: Addison Wesley Publishing Company, 1st edition, 1995.

49. 3GPP, *TS25.211, V3.11.0, Physical Channels and Mapping of Transport Channels onto Physical Channels (FDD)*. Sophia Antipolis, F: ETSI, 2002.

50. 3GPP, *TS25.212, V3.10.0, Multiplexing and Channel Coding (FDD)*. Sophia Antipolis, F: ETSI, 2002.

51. 3GPP, *TS25.214, V5.2.0, Physical Layer Procedures (FDD)*. Sophia Antipolis, F: ETSI, 2002.

52. 3GPP, *TS25.133 V3.10.0, Requirements for Support of Radio Resource Management (FDD)*. Sophia Antipolis, F: ETSI, 2002.

53. 3GPP, *TS25.215, V3.10.0, Physical Layer Measurements (FDD)*. Sophia Antipolis, F: ETSI, 2002.

54. J. S. Lee and L. E. Miller, *CDMA System Engineering Handbook*. Boston: Artech House, 1st edition, 1998.

55. N. Benvenuto and G. Cherubini, *Algorithms for Communications Systems and Their Applications*. Chichester, UK: John Wiley & Sons, Ltd., 1st edition, 2002.

56. R. Price and P. E. Green, 'A communication technique for multipath channels,' *Proceedings of the IRE*, **46**, 555–570, 1958.

57. R. C. Dixon, *Spread Spectrum Systems with Commercial Applications*. New York, USA: John Wiley & Sons Inc., 3rd edition, 1994.

58. S. Fukumoto, M. Sawahashi and F. Adachi, 'Matched filter-based RAKE combiner for wideband DS-CDMA mobile radio,' *IEICE Transactions on Communications*, **E81-B**, 1384–1390, 1998.

59. A. Aziz, *Application Note AN2253/D: Channel Estimation for a WCDMA Rake Receiver*. Motorola.

60. M. Sakamoto, J. Huoponen and I. Niva, 'Adaptive channel estimation with velocity estimator for W-CDMA receiver,' in *Proceedings International Conference on Vehicular Technology*, **4**, 2626–2629, IEEE, 2001.

61. D. G. Brennan, 'Linear diversity combining technique,' *Proceedings of the IRE*, **43**, 1530–1555, 1955.

62. K.-C. Gan, *Application Note AN2251/D: Maximum Ratio Combining for a WCDMA Rake Receiver*. Motorola.

63. K.-C. Gan, *Application Note AN2252/D: Path Searcher for a WCDMA Rake Receiver*. Motorola.

64. S. B. Wicker, *Error Control Systems for Digital Communication and Storage*. Upper Saddle River, USA: Prentice Hall, Inc., 1st edition, 1995.

65. W. H. Press, S. A. Teukolsky, W. T. Vetterling and B. P. Flannery, *Numerical Recipes in C – The Art of Scientific Computing*. Cambridge, UK: Cambridge University Press, 2nd edition, 1992.

66. A. Viterbi, 'Error bounds for convolutional codes and an asympotically optimum decoding algorithm,' *IEEE Transactions on Information Theory*, pp. 260–269, April 1967.

67. C. Berrou, A. Glavieux and P. Thitimajshima, 'Near Shannon limit error-correcting coding and decoding: turbo codes,' *Proceedings of the International Conference on Communications*, pp. 1064–1070, May 1993.

68. 3GPP, *TS25.306, V3.6.0, UE Radio Access Capabilities*. Sophia Antipolis, F: ETSI, 2002.

69. R. Knopp and P. A. Humblet, 'On coding for block fading channels,' *IEEE Transactions on Information Theory*, **46**, 189–205, 2000.

70. S. Haykin, *Digital Communications*. New York, USA: John Wiley & Sons Inc., 1st edition, 1988.

71. S. G. Wilson, *Digital Modulation and Coding*. Upper Saddle River, USA: Prentice Hall, 1st edition, 1999.

72. L. H. C. Lee, *Convolutional Coding*. New York, USA: Artech House, 1st edition, 1997.

73. R. Johannesson and K. Zigangirov, *Fundamentals of Convolutional Coding*. New York, USA: IEEE, 1st edition, 1999.

74. I. A. Glover and P. M. Grant, *Digital Communications*. London, UK: Prentice Hall, 1st edition, 1998.

75. G. D. Forney, 'The Viterbi Algorithm,' *Proceedings of the IEEE*, **61**, 268–278, 1973.

76. J. F. Hayes, 'The Viterbi algorithm applied to digital data transmission,' *IEEE Communications Magazine*, pp. 26–32, May 2002.

77. B. Sklar, 'How I learned to love the trellis,' *IEEE Signal Processing Magazine*, pp. 87–102, May 2003.

78. A. P. Hekstra, 'An alternative to metric rescaling in Viterbi decoders,' *IEEE Transactions on Communications*, pp. 1220–1222, November 1989.

79. L. R. Bahl, J. Cocke, F. Jelinek and J. Raviv, 'Optimal decoding of linear codes for minimising symbol error rate,' *IEEE Transactions on Information Theory*, pp. 284–287, March 1974.

80. J. P. Woodard and L. Hanzo, 'Comparative study of turbo decoding techniques: an overview,' *IEEE Transactions on Vehicular Technology*, **49**, 2208–2233, November 2000.

81. C. E. Shannon, 'A mathematical theory of communication,' *Bell Systems Technical Journal*, pp. 379–423, 623–656, October 1948.

82. P. Robertson, E. Villebrun and P. Hoeher, 'A comparison of optimal and sub-optimal map decoding algorithms operating in the log domain,' *Proceedings of the International Conference on Communications*, pp. 1009–1013, June 1995.

83. C. Berrou, P. Adde, E. Angui and S. Faudeil, 'A low complexity soft-output Viterbi decoder architecture,' *Proceedings of the International Conference on Communications*, pp. 737–740, May 1993.

84. S. Benedetto and G. Montorsi, 'Design of parallel concatenated convolutional codes,' *IEEE Transactions on Communications*, **44**, 591–600, 1996.

85. L. C. Perez, J. Seghers and D. J. Costello, 'A distance spectrum interpretation of turbo codes,' *IEEE Transactions on Information Theory*, **42**, 1698–1709, 1996.

86. S. Benedetto and G. Montorsi, 'Unveiling turbo codes: some results on parallel concatenated coding schemes,' *IEEE Transactions on Information Theory*, **42**, 409–428, 1996.

87. J. Hagenauer, E. Offer and L. Papke, 'Iterative decoding of binary block and convolutional codes,' *IEEE Transactions on Information Theory*, **42**, 429–445, 1996.

88. R. Pyndiah, 'Iterative decoding of product codes: block turbo codes,' *Proceedings of the International Symposium on Turbo Codes & Related Topics*, pp. 71–79, September 1997.

89. C. Douillard, M. Jézéquel, C. Berrou, N. Brengarth, J. Tousch and N. Pham, 'The turbo code standard for DVB-RCS,' *Proceedings of the 2nd International Symposium on Turbo Codes & Related Topics*, pp. 535–538, September 2000.

90. 3GPP2, *C.S0002-0-2, Physical Layer Standard for cdma2000 Spread Spectrum System, Release 0*. 2001.

91. A. Burr, 'Turbo-codes: the ultimate error control codes?,' *IEE Electronics and Communication Engineering Journal*, **13**, 155–165, 2001.

92. B. Sklar, 'A primer on turbo code concepts,' *IEEE Communications Magazine*, pp. 94–102, December 1997.

93. C. Berrou and A. Glavieux, 'Near optimum error correction coding and decoding: turbo codes,' *IEEE Transactions on Communications*, **44**, 1261–1271, 1996.

94. W. Koch and A. Baier, 'Optimum and sub-optimum detection of coded data disturbed by time-varying inter-symbol interference,' *IEEE Globecom*, pp. 1679–1684, December 1990.

95. J. A. Erfanian, S. Pasupathy and G. Gulak, 'Reduced complexity symbol detectors with parallel structures for ISI channels,' *IEEE Transactions on Communications*, **42**, 1661–1671, 1994.

96. J. Hagenauer and P. Hoeher, 'A Viterbi algorithm with soft-decision outputs and its applications,' *IEEE Globecom*, pp. 1680–1686, 1989.

97. J. Hagenauer, 'Source-controlled channel decoding,' *IEEE Transactions on Communications*, **43**, 2449–2457, 1995.

98. A. J. Viterbi, 'An intuitive justification and a simplified implementation of the MAP decoder for convolutional codes,' *IEEE Journal on Selected Areas in Communications*, pp. 260–264, February 1998.

99. P. Robertson, 'Illuminating the structure of code and decoder of parallel concatenated recursive systematic (turbo) codes,' *IEEE Globecom*, pp. 1298–1303, 1994.

100. H. R. Sadjadpour, N. J. A. Sloane, M. Salehi and G. Nebe, 'Interleaver design for turbo codes,' *IEEE Journal on Selected Areas in Communications*, pp. 831–836, May 2001.

101. G. Montorsi and S. Benedetto, 'Design of fixed-point iterative decoders for concatenated codes with interleavers,' *IEEE Journal on Selected Areas in Communications*, **19**, 871–882, 2001.

102. T. K. Blankenship and B. Classon, 'Fixed-point performance of low-complexity turbo decoding algorithms,' *Proceedings of IEEE Conference on Vehicular Technology*, pp. 1483–1487, May 2001.

103. H. Michel and N. Wehn, 'Turbo-decoder quantization for UMTS,' *IEEE Communications Letters*, **5**, 55–57, 2001.

104. G. Masera, G. Piccinini, M. R. Roch and M. Zamboni, 'VLSI architectures for turbo codes,' *IEEE Transactions on VLSI Systems*, pp. 369–379, September 1999.

105. A. Shibutani, H. Suda and F. Adachi, 'Complexity reduction of turbo decoding,' *Proceedings of IEEE Conference on Vehicular Technology*, pp. 1570–1574, 1999.

106. J. Nikolic-Popovic, 'Using TMS320C6416 coprocessors: turbo coprocessor (TCP),' *Texas Instruments Application Report SPRA749*, June 2001.

107. Samsung, 'Harmonization impact on TFCI and new optimal coding for extended TFCI with almost no complexity increase,' in *TSGR1-99970*, 3GPP, July 1999.

108. M. Mouly and M.-B. Pautet, *The GSM System for Mobile Communications*. Telecom Publishing, 1st edition, 1992.

109. T. R. Bednar, P. H. Buffet, R. J. Darden, S. W. Gould and P. S. Zuchowski, 'Issues and strategies for the physical design of system-on-a-chip ASICs,' *IBM Journal of Research & Development*, **46**, 661–674, 2002.

110. J. A. Darringer, R. A. Bergamaschi, S. Bhattacharya, D. Brand, A. Herkersdorf, J. K. Morrell, I. I. Nair, P. Sagmeister and Y. Shin, 'Early analysis tool for system-on-a-chip design,' *IBM Journal of Research & Development*, **46**, 691–706, 2002.

111. 'Special report on system on a chip design at the nanometer scale,' *IEE Electronics Systems and Software*, **1**, February 2003.

112. M. C. Jeruchim, P. Balaban and K. S. Shanmugan, *Simulation of Communication Systems*. New York, USA: Plenum Press, 1st edition, 1992.

113. B. Parhami, *Computer Arithmetic*. New York, USA: Oxford, 1st edition, 2000.

114. M. Rupp, A. Burg and E. Beck, 'Rapid prototyping for wireless designs: the five-ones approach,' *EURASIP Signal Processing*, **83**, 1427–1444, 2003.

115. P. Almers, A. Birkedal, S. Kim, A. Lundqvist and A. Milen, 'Experiences of the live WCDMA network in Stockholm, Sweden,' *Ericsson Review*, pp. 204–215, 2000.

116. A. Birkedal, E. Corbett, K. Jamal and K. Woodfield, 'Experiences of operating a pre-commercial WCDMA network,' *Ericsson Review*, pp. 50–61, 2002.

117. G. Alsenmyr, J. Bergström, M. Hagberg, A. Milen, W. Müller, H. Palm, H. van der Velde, P. Wallentin and F. Wallgren, 'Handover between WCDMA and GSM,' *Ericsson Review*, pp. 6–41, 2003.

118. 3GPP, *TS34.108, V3.12.0, Common Test Environments for User Equipment (UE) Conformance Testing*. Sophia Antipolis, F: ETSI, 2003.

119. 3GPP, *TS34.109, V3.9.0, Terminal Logical Test Interface; Special Conformance Testing Functions*. Sophia Antipolis, F: ETSI, 2003.

120. 3GPP, *TS34.123-1, V3.5.0, User Equipment (UE) Conformace Specification; Part 1: Protocol Conformace Specification*. Sophia Antipolis, F: ETSI, 2001.

121. 3GPP, *TS34.123-2, V5.4.0, User Equipment (UE) Conformance Specification; Part 2: Implementation Conformance Statement (ICS) Specification*. Sophia Antipolis, F: ETSI, 2001.

122. 3GPP, *TS34.123-3, V3.1.0, User Equipment (UE) Conformance Specification; Part 3: Abstract Test Suite (ATS)*. Sophia Antipolis, F: ETSI, 2003.

123. 3GPP, *TS34.124, V3.4.0, Electromagnetic Compatibility (EMC) Requirements for Mobile Terminals and Ancillary Equipment*. Sophia Antipolis, F: ETSI, 2003.

124. 3GPP, *TR34.901, V3.0.0, Test Time Optimisation Based on Statistical Approaches; Statistical Theory Applied and Evaluation of Statistical Significance*. Sophia Antipolis, F: ETSI, 2003.

125. 'Ad hoc meeting results,' in *TSGR4-AD01(99)758*, 3GPP, October 1999.

126. 'Summary of UE DL performance results,' in *TSGR4-8(99)759*, 3GPP, October 1999.

127. 3GPP, *TS25.942, V3.3.0, RF System Scenarios*. Sophia Antipolis, F: ETSI, 2001.

128. R. Tanner and G. Williams, '3GPP functional and performance testing of user equipment,' in *IEE Proceedings Int. Conference on 3G*, pp. 5–9, IEE, May 2002.

129. T. Reichel, *Voltage and Power Measurements*. Rohde und Schwarz, Munich.

130. R. Desquiotz, *WCDMA Signal Generator Solutions by Rohde&Schwarz*. Rohde und Schwarz, Munich, September 2000.

131. Agilent, *Application Note 1311: Understanding CDMA Measurements for Base Stations and Their Components*.

132. Mini-Circuits, *Understanding Power Splitters*.

133. Agilent, *Application Note 150: Spectrum Analysis Basics*.

134. 3GPP, *TS25.943, V4.1.0, Deployment Aspects*. Sophia Antipolis, F: ETSI, 2001.

135. J. D. Parsons, *The Mobile Radio Propagation Channel*. Chichester, UK: John Wiley & Sons, Ltd., 2nd edition, 2000.

136. Elektrobit Ltd., *PROPSim C2 Wideband Radio Channel Simulator Operational Manual*, February 2003.

137. ISO, *Guide to the Expression of Uncertainty in Measurements*. Geneva, CH: International Standardisation Organisation, 1st edition, 1995.

138. B. Melis and G. Romano, 'UMTS W-CDMA: evaluation of radio performance by means of link level simulations,' *IEEE Personal Communications Magazine*, **7**, 42–49, 2000.

139. 3GPP, *TS25.321, V3.12.0, MAC Protocol Specification*. Sophia Antipolis, F: ETSI, 2002.

140. 3GPP, *TS25.322, V3.11.0, Radio Link Control (RLC) Protocol Specification*. Sophia Antipolis, F: ETSI, 2002.

141. 3GPP, *TS25.331, V3.11.0, Radio Resource Control: Protocol Specification*. Sophia Antipolis, F: ETSI, 2002.

142. 3GPP, *TS25.301, V3.10.0, Radio Interface Protocol Architecture*. Sophia Antipolis, F: ETSI, 2002.

143. 3GPP, *TS33.102, V3.12.0, 3G Security; Security Architecture*. Sophia Antipolis, F: ETSI, 2002.

144. 3GPP, *TS25.302, V3.13.0, Services Provided by the Physical Layer*. Sophia Antipolis, F: ETSI, 2002.

145. 3GPP, *TS25.323, V3.11.0, Packet Data Convergence Protocol (PDCP) Specification*. Sophia Antipolis, F: ETSI, 2002.

146. 3GPP, *TS25.303, V3.12.0, Interlayer Procedures in Connected Mode*. Sophia Antipolis, F: ETSI, 2002.

147. *IETF RFC 2507, IP Header Compression*.

148. *IETF RFC 3095, IP Header Compression*.

149. 3GPP, *TS25.324, V3.5.0, Broadcast/Multicast Control BMC*. Sophia Antipolis, F: ETSI, 2002.

150. 3GPP, *TS25.925, V3.5.0, Radio Interface for Broadcast/Multicast Services*. Sophia Antipolis, F: ETSI, 2002.

151. 3GPP, *TS25.304, V3.11.0, UE Procedures in Idle Mode and Procedures for Cell Reselection in Connected Mode*. Sophia Antipolis, F: ETSI, 2002.

152. 3GPP, *TS25.402, V3.10.0, Synchronisation in UTRAN Stage 2*. Sophia Antipolis, F: ETSI, 2002.

153. 3GPP, *TS35.201, V3.2.0, 3G Security; Specification of the 3GPP Confidentiality and Integrity Algorithms; Document 1: f8 and f9 Specification*. Sophia Antipolis, F: ETSI, 2001.

154. 3GPP, *TS35.202, V3.1.2, 3G Security; Specification of the 3GPP Confidentiality and Integrity Algorithms; Document 2: KASUMI Specification.* Sophia Antipolis, F: ETSI, 2001.

155. H. T. How, T. H. Liew, E. L. Kuan and L. Hanzo, 'An AMR coding, RRNS channel coding and adaptive JD-CDMA system for speech communications,' submitted for publication *in IEEE Journal on Selected Areas in Communications,* 2000.

156. N. Jayant and P. Noll, *Digital Coding of Waveforms, Principles and Applications to Speech and Video.* Englewood Cliffs, NJ: Prentice Hall, 1984.

157. L. Hanzo, F. Somerville and J. Woodard, *Voice Compression and Communications: Principles and Applications for Fixed and Wireless Channels.* IEEE Press and John Wiley & Sons, Ltd., 2001. (For detailed contents and sample chapters please refer to http://www-mobile.ecs.soton.ac.uk.).

158. A. M. Kondoz, *Digital Speech: Coding for Low Bit Rate Communications Systems.* John Wiley & Sons, Ltd., 1994.

159. 3GPP, *TS26.071, V3.0.1, AMR Speech Codec; General Description.* Sophia Antipolis, F: ETSI, 1999.

160. R. Steele and L. Hanzo, (Ed), *Mobile Radio Communications.* IEEE Press and John Wiley & Sons, Ltd., 2nd edition, 1999.

161. W. B. Kleijn and K. K. Paliwal, *Speech Coding and Synthesis.* Elsevier, 1995.

162. 3GPP, *TS26.171, V5.0.0, AMR Wideband Speech Codec; General Description.* Sophia Antipolis, F: ETSI, 2001.

163. M. R. Schroeder and B. S. Atal, 'Code excited linear prediction (CELP) : high quality speech at very low bit rates,' in *Proceedings of ICASSP* (Tampa, Florida), pp. 937–940, March 1985.

164. R. A. Salami, C. Laflamme, J. P. Adoul and D. Massaloux, 'A toll quality 8 kbit/s speech codec for the personal communications system (PCS),' *IEEE Transactions on Vehicular Technology,* **43**, 808–816, 1994.

165. Digital Cellular Telecommunications System (Phase 2+), *GSM 06.90, V7.0.0, Adaptive Multi-Rate (AMR) Speech Transcoding.* ETSI, 1998.

166. S. Bruhn, E. Ekudden and K. Hellwig, 'Adaptive multi-rate: a new speech service for GSM and beyond,' in *Proceedings of 3rd ITG Conference on Source and Channel Coding* (Technical University Munich, Germany), pp. 319–324, January 2000.

167. S. Bruhn, P. Blocher, K. Hellwig and J. Sjoberg, 'Concepts and solutions for link adaptation and inband signalling for the GSM AMR speech coding standard,' in *Proceedings of VTC,* Houston, Texas, USA, **3**, 2451–2455, 1999.

168. R. A. Salami and L. Hanzo, 'Speech coding,' in *Mobile Radio Communications* (R. Steele and L. Hanzo, Eds.), pp. 187–335, IEEE Press and John Wiley & Sons, Ltd., 1999.

169. F. Itakura, 'Line spectrum representation of linear predictor coefficients of speech signals,' *Journal of Acoustical Society of America*, **57**, no. S35(A), 1975.

170. K. K. Paliwal and B. S. Atal, 'Efficient vector quantization of LPC parameters at 24 bits/frame,' *IEEE Transactions on Speech and Audio Processing*, **1**, 3–14, 1993.

171. L. Hanzo and J. P. Woodard, 'An intelligent multimode voice communications system for indoors communications,' *IEEE Transactions on Vehicular Technology*, **44**, 735–749, 1995.

172. ITU, 'Recommendation G.729, Coding of Speech at 8 kbps using Conjugate-Structure Algebraic Code-Excited Linear Prediction (CS-ACELP),' Geneva, 1995.

173. J. H. Chen and A. Gersho, 'Adaptive postfiltering for quality enhancement of coded speech,' *IEEE Transactions on Speech and Audio Processing*, **3**, 59–71, 1995.

174. 3GPP, *TS25.308, V5.2.0, High Speed Downlink Packet Access (HSDPA); Overall Description*. Sophia Antipolis, F: ETSI, 2002.

175. 3GPP, *TR25.855, V5.0.0, High Speed Downlink Packet Access; Overall UTRAN Description*. Sophia Antipolis, F: ETSI, 2001.

176. 3GPP, *TR25.858 V5.0.0, High Speed Downlink Packet Access; Physical Layer Aspects*. Sophia Antipolis, F: ETSI, 2002.

177. 3GPP, *TS25.211, V5.2.0, Physical Channel and Mapping of Transport Channels onto Physical Channels (FDD)*. Sophia Antipolis, F: ETSI, 2002.

178. 3GPP, *TS25.212,V5.2.0, Multiplexing and Channel Coding (FDD)*. Sophia Antipolis, F: ETSI, 2002.

179. D. Chase, 'Code combining: a maximum-likelihood decoding approach for combining an arbitrary number of noisy packets,' *IEEE Transactions on Communications*, **33**, 593–607, 1985.

180. 3GPP, *TS25.213, V5.2.0, Spreading and Modulation (FDD)*. Sophia Antipolis, F: ETSI, 2002.

181. 3GPP, *TS25.331, V5.2.0, Radio Resource Control (RRC) Protocol Specification*. Sophia Antipolis, F: ETSI, 2002.

182. 3GPP, *TS25.321, V5.2.0, MAC Protocol Specification*. Sophia Antipolis, F: ETSI, 2002.

183. 3GPP, *TS25.306, V5.2.0, UE Radio Access Capabilities*. Sophia Antipolis, F: ETSI, 2002.

184. 3GPP, 'HSDPA UE measurement report,' in *TSGR4-21(02)0005*, February 2002.

185. 3GPP, *TS23.107, V5.2.0, QoS Concept and Architecture*. Sophia Antipolis, F: ETSI, 2002.

186. 3GPP, 'Evaluation methods for high speed downlink packet access (HSDPA),' in *TSGR1-14(00)909*, July 2000.

187. 3GPP, 'Throughput simulations for MIMO and transmit diversity enhancements to HSDPA,' in *TSGR1-17(00)1388*, November 2000.

188. M. Andrews, K. Kumaran, K. Ramanan, A. Stolyar, P. Whiting and R. Vijayakumar, 'Providing quality of service over a shared wireless link,' *IEEE Communications Magazine*, **39**, 150–154, 2001.

189. S. Shakkottai and A. Stolyar, 'Scheduling algorithms for a mixture of real-time and non-real-time data in HDR,' *Bell Labs Technical Memorandum*, pp. 150–154, August 2000.

190. M. A. Birchler, 'E911 phase II location technologies,' *Vehicular Technology Society News*, **49**, 4–9, 2002.

191. 3GPP, *TR22.071, V5.1.1, Location Services (LCS)*. Sophia Antipolis, F: ETSI, 2002.

192. 3GPP, *TS23.271, V5.3.0, Functional Stage 2 Description of LCS*. Sophia Antipolis, F: ETSI, 2002.

193. 3GPP, *TS25.305, V3.8.0, Stage 2 Functional Specification of User Equipment (UE); Positioning in UTRAN*. Sophia Antipolis, F: ETSI, 2002.

194. 3GPP, *TS25.305, V5.4.0, Stage 2 Functional Specification of User Equipment (UE); Positioning in UTRAN*. Sophia Antipolis, F: ETSI, 2002.

195. J. J. Caffery and G. L. Stüber, 'Overview of radiolocation in CDMA cellular systems,' *IEEE Personal Communications Magazine*, **36**, 38–45, 1998.

196. 3GPP, *TS25.214, V5.1.0, Physical Layer Procedures (FDD)*. Sophia Antipolis, F: ETSI, 2002.

197. 3GPP, 'UE positioning with OTDOA-IPDL,' in *TSGR4-19(01)1306*, September 2001.

198. 3GPP, 'Time mask proposal for IPDL,' in *TSGR4-20(01)1476*, November 2001.

199. 3GPP, 'UE positioning measurements and performance,' in *TSGR4-21(02)0331*, January 2002.

200. 3GPP, 'IPDL requirements and impacts,' in *TSGR4-23(02)0849*, May 2002.

201. 3GPP, *TS25.306, V5.1.0, UE Radio Access Capabilities*. Sophia Antipolis, F: ETSI, 2002.

202. 3GPP, *TS25.133, V5.3.0, Requirements for Support of Radio Resource Management (FDD)*. Sophia Antipolis, F: ETSI, 2002.

203. 3GPP, *TS25.991, V5.1.0, Feasibility Study on the Mitigation of the Effect of the Common Pilot Channel (CPICH) Interference at the User Equipment*. Sophia Antipolis, F: ETSI, 2002.

204. A. Papoulis, *Probability, Random Variables and Stochastic Processes*. Singapore: McGraw Hill, 3rd edition, 1991.

205. C. D. Meyer, *Matrix Analysis and Applied Linear Algebra*. Philadelphia, USA: SIAM, 1st edition, 2000.

206. 'CPICH interference cancellation as a means for increasing DL capacity,' in *TSGR1-00-1371*, 3GPP, 2000.

207. 3GPP, *TR25.869, V1.1.0, Tx Diversity Solutions for Multiple Antennas*. Sophia Antipolis, F: ETSI, 2002.

208. J. Litva and T. Lo, *Digital Beamforming in Wireless Communications*. Boston, USA: Artech House, 1st edition, 1996.

209. S. M. Alamouti, 'A simple transmit diversity technique for wireless communications,' *IEEE Transactions on Selected Areas in Communications*, **16**, 1451–1458, 1998.

210. T. P. Krauss, W. J. Hillery and M. D. Zoltowski, 'MMSE equalization for forward link in 3G CDMA: symbol-level versus chip-level,' in *Proceedings Workshop on Statistical Signal and Array Processing*, Pocono Manor, PA, USA, pp. 18–22, IEEE, August 2000.

211. A. Klein, 'Data detection algorithms specially designed for the downlink of CDMA mobile radio systems,' in *Proceedings International Conference on Vehicular Technology*, Phoenix, AZ, USA, **1**, 203–207, IEEE, 1997.

212. T. P. Krauss, M. D. Zoltowski and G. Leus, 'Simple MMSE equalizer for CDMA downlink to restore chip sequence: comparison to zero-forcing and rake,' in *Proceedings International Conference on Acoustics, Speech and Signal Processing*, Istanbul, Turkey, **5**, 2865–2868, IEEE, 2000.

213. M. Heikkila, P. Komulainen and J. Lilleberg, 'Interference suppression in CDMA downlink through adaptive channel equalization,' in *Proceedings International Conference on Vehicular Technology*, Amsterdam, NL, **2**, 978–982, IEEE, 1999.

214. K. Hooli, M. Latva-aho and M. Juntti, 'Performance evaluation of adaptive chip-level channel equalizers in WCDMA downlink,' in *International Conference on Communications*, Helsinki, **6**, 1974–1979, IEEE, 2001.

215. J. R. Treichler, I. Fijalkow and C. R. Johnson, 'Fractionally spaced equalizers,' *IEEE Signal Processing Magazine*, **13**, 65–81, 1996.

216. T. P. Krauss and M. D. Zoltowski, 'Oversampling diversity versus dual antenna diversity for chip-level equalization on CDMA downlink,' in *Proceedings Workshop on Sensor Array and Multichannel Signal Processing*, Cambridge, MA, USA, pp. 47–51, IEEE, March 2000.

217. R. Schober and W. H. Gerstacker, 'The zeros of random polynomials: further results and applications,' *IEEE Transactions on Communications*, **50**, 892–896, 2002.

218. R. Tanner, C. Luschi and G. Howard, 'Design considerations for high-data-rate UMTS FDD user equipment,' in *IEE Proceedings International Conference on 3G*, pp. 343–347, IEE, June 2003.

219. S. Verdu, *Multiuser Detection*. Cambridge, UK.: Cambridge University Press, 1st edition, 1998.

220. K. S. Gilhousen, I. M. Jacobs, R. Padovani, A. J. Viterbi, L. A. Weaver and C. E. Wheatley, 'On the capacity of a cellular CDMA system,' *IEEE Transactions on Vehicular Technology*, **40**, 303–312, 1991.

221. S. Verdu, 'Minimum probability of error for asynchronous Gaussian multiple-access channels,' *IEEE Transactions on Information Theory*, **IT-32**, 85–96, 1986.

222. R. O. Duda and P. E. Hart, *Pattern Classification and Scene Analysis*. New York, USA: John Wiley & Sons, Inc., 1st edition, 1973.

223. C. M. Bishop, *Neural Networks for Pattern Recognition*. Oxford, UK: Oxford University Press, 1st edition, 1995.

224. R. Tanner and D. Cruickshank, 'Radial basis function receiver for DS-CDMA,' *European Transactions on Telecommunications (ETT)*, **13**, 211–219, 2002.

225. U. Mitra and H. V. Poor, 'Neural network techniques for adaptive multiuser demodulation,' *IEEE Transactions on Selected Areas in Communications*, **12**, 1460–1470, 1994.

226. B. Aazhang, B. Paris and G. Orsak, 'Neural networks for multiuser detection in code-division multiple-access communications,' *IEEE Transactions on Communications*, **40**, 1212–1222, 1992.

227. S. Moshavi, 'Multi-user detection for DS-CDMA communications,' *IEEE Communications Magazine*, **34**, 124–136, 1996.

228. R. Tanner and D. G. M. Cruickshank, 'Volterra based receivers for DS-CDMA,' in *Proceedings, International Symposium on Personal, Indoor and Mobile Radio Communications*, Helsinki, Finland, **3**, 1166–1170, IEEE, 1997.

229. S. Chen, A. K. Samingan and L. Hanzo, 'Support vector machine multiuser receiver for DS-CDMA signals in multipath channels,' *IEEE Transactions on Neural Networks*, **12**, 604–611, 2001.

230. S. Moshavi, E. G. Kanterakis and D. L. Schilling, 'Multistage linear receivers for DS-CDMA systems,' *International Journal of Wireless Information Networks*, **3**, 1–17, 1996.

231. D. Divsalar and M. Simon, 'Improved CDMA performance using parallel interference cancellation,' *JPL publication 95-21*, 1995.

232. A. Nahler, R. Irmer and G. Fettweis, 'Reduced and differential parallel interference cancellation for CDMA systems,' *IEEE Transactions on Selected Areas in Communications*, **20**, 237–247, 2002.

233. P. H. Tan and L. K. Rasmussen, 'Linear interference cancellation in CDMA based on iterative techniques for linear systems,' *IEEE Transactions on Communications*, **48**, 2099–2108, 2000.

234. A. Lampe and J. B. Huber, 'Iterative interference cancellation for DS-CDMA systems with high system loads using reliablity-dependent feedback,' *IEEE Transactions on Vehicular Technology*, **51**, 445–452, 2002.

235. T. Miyajima, T. Hasegawa and M. Haneishi, 'On the multiuser detection using a neural network in code-division multiple-access communications,' *IEICE Transactions: Communication*, **E76-B**, 961–968, 1993.

236. G. I. Kechriotis and E. S. Manolakos, 'Hopfield neural network implementation of the optimal CDMA multiuser detector,' *IEEE Transactions on Neural Networks*, **7**, 131–141, 1996.

237. W. G. Teich, M. Seidl and M. Nold, 'Multiuser detection for DS-CDMA communication systems based on recurrent neural networks,' in *Proceedings, International Symposium on Spread Spectrum Techniques and Applications*, Sun City, South Africa, **3**, 863–867, IEEE, 1998.

238. G. I. Kechriotis, E. Zervas and E. S. Manolakos, 'Using recurrent neural networks for adaptive communication channel equalization,' *IEEE Transactions on Neural Networks*, **5**, 96–104, 1994.

239. S. H. Yoon and S. S. Rao, *Multiuser detection in code division multiple access communications using annealed neural networks*,' Report. University of Villanova, Dept. of Electrical Engineering, Villanova, PA, USA, 1997.

240. D. Guo, L. K. Rasmussen, S. Sun, T. J. Lim and C. Cheah, 'MMSE-based linear parallel interference cancellation in CDMA,' in *Proceedings, International Symposium on Spread Spectrum Techniques and Applications*, Sun City, South Africa, **3**, 917–921, IEEE, 1998.

241. L. K. Rasmussen and I. J. Oppermann, 'Ping-pong effects in linear parallel interference cancellation for CDMA,' *IEEE Transactions on Communications*, **2**, 357–363, 2003.

242. I. E. Telatar, 'Capacity of multi-antenna Gaussian channels,' *European Transactions on Telecommunications*, **10**, 585–595, 1995.

243. G. J. Foschini and M. J. Gans, 'On limits of wireless communications in a fading environment when using multiple antennas,' *Wireless Personal Communications Journal*, **6**, 311–335, 1998.

244. A. Lozano and A. M. Tulino, 'Capacity of multiple-transmit multiple-receive antenna architectures,' *IEEE Transactions on Information Theory*, **48**, 3117–3128, 2002.

245. A. Wittneben, 'A new bandwidth efficient transmit antenna modulation diversity scheme for linear digital modulation,' *International Conference on Communications*, Geneva, **3**, 1630–1634, May 1993.

246. T. L. Marzetta and B. M. Hochwald, 'Capacity of a mobile multiple-antenna communication link in Rayleigh flat fading,' *IEEE Transactions on Information Theory*, **45**, 139–157, 1999.

247. D. Shiu, G. J. Foschini, M. J. Gans and J. M. Kahn, 'Fading correlation and its effect on the capacity of multielement antenna systems,' *IEEE Transactions on Communications*, **48**, 502–512, 2000.

248. T. M. Cover and J. A. Thomas, *Elements of Information Theory*. New York, USA: John Wiley & Sons, Inc., 1st edition, 1991.

249. V. Tarokh, H. Jafarkhani and A. R. Calderbank, 'Space-time block codes from orthogonal designs,' *IEEE Transactions on Information Theory*, **45**, 1456–1467, 1999.

250. V. Tarokh, H. Jafarkhani, and A. R. Calderbank, 'Space-time block coding for wireless communications: performance results,' *IEEE Transactions on Selected Areas in Communications*, **17**, 451–460, 1999.

251. V. Tarokh, N. Seshadri and A. R. Calderbank, 'Space-time codes for high data rate wireless communication: performance criterion and code construction,' *IEEE Transactions on Information Theory*, **44**, 744–765, 1998.

252. V. Tarokh, A. Naguib, N. Seshadri and A. R. Calderbank, 'Space-time codes for high data rate wireless communication: performance criteria in the presence of channel estimation errors, mobility and multiple paths,' *IEEE Transactions on Communications*, **47**, 199–206, 1999.

253. G. L. Foschini, 'Layered space-time architecture for wireless communication in a fading environment when using multi-element antennas,' *Bell Systems Technical Journal*, pp. 41–59, Autumn 1996.

254. V. Tarokh and H. Jafarkhani, 'A differential detection scheme for transmit diversity,' *IEEE Transactions on Selected Areas in Communications*, **18**, 1169–1174, 2000.

255. B. L. Hughes, 'Differential space-time modulation,' *IEEE Transactions on Information Theory*, **46**, 2567–2578, 2000.

256. T. Lo and V. Tarokh, 'Space-time block coding – from a physical perspective,' in *Proceedings, Wireless Communications and Networking Conference*, New Orleans, USA, **1**, 150–153, IEEE, September 1999.

257. S. Sandhu and A. Paulraj, 'Space-time block codes: a capacity perspective,' *IEEE Communication Letters*, **4**, 384–386, 2000.

258. J. Guey, 'Concatenated coding for transmit diversity systems,' in *Proceedings, International Conference on Vehicular Technology*, Amsterdam, NL, **5**, 2500–2504, IEEE, 1999.

259. K. Ban, M. Katayama, W. E. Stark, T. Yamazato and A. Ogawa, 'Multi-antenna transmission scheme for convolutionally coded DS/CDMA system,' *IEICE Transactions: Fundamentals*, **E80-A**, 2437–2443, 1997.

260. V. Tarokh, A. Naguib, N. Seshadri and A. R. Calderbank, 'Combined array processing and space-time coding,' *IEEE Transactions on Information Theory*, **45**, 1121–1128, 1999.

261. G. D. Golden, G. J. Foschini, R. A. Valenzuela and P. W. Wolniansky, 'Detection algorithm and initial laboratory results using V-BLAST space-time communication architecture,' *IEE Electronics Letters*, **35**, 14–16, 1999.

262. G. H. Golub and C. F. V. Loan, *Matrix Computations*. Baltimore, MD, USA: Johns Hopkins University Press, 3rd edition, 1996.

263. R. van Nee, A. van Zelst and G. Awater, 'Maximum likelihood decoding in a space division multiplexing system,' in *Proceedings, International Conference on Vehicular Technology*, Boston, **1**, 6–10, IEEE, September 2000.

264. R. L. Cupo, G. D. Golden, C. C. Martin, K. L. Sherman, N. R. Sollenberger, J. H. Winters and P. W. Wolniasky, 'A four-element adaptive antenna array for IS-136 PCS base stations,' in *Proceedings, International Conference on Vehicular Technology*, Phoenix, AZ, USA, **3**, 1577–1581, May 1997.

265. C. Z. Hassell, J. S. Thompson, B. Mulgrew and P. M. Grant, 'A comparison of detection algorithms including BLAST for wireless communication using multiple antennas,' in *Proceedings, International Symposium on Personal, Indoor and Mobile Radio Communications*, London, UK, **1**, 698–703, IEEE, September 2000.

266. G. Awater, A. van Zelst and R. van Nee, 'Reduced complexity space division multiplexing receivers,' in *Proceedings, International Conference on Vehicular Technology*, Boston, **1**, 11–15, IEEE, September 2000.

267. B. Hassibi and B. M. Hochwald, 'How much training is needed in multiple-antenna wireless links?,' *IEEE Transactions on Information Theory*, **49**, 951–963, 2003.

268. T. L. Marzetta, 'BLAST training: estimating channel characteristics for high capacity space-time wireless,' in *Annual Allerton Conference on Communications, Control, and Computing*, Monticello, Ill., pp. 958–966, IEEE, September 1999.

269. B. Hassibi and B. Hochwald, 'Optimal training in space-time systems,' in *Proceedings, International Conference*, Asilomar, **1**, 743–747, IEEE, October 2000.

270. H. Vikalo, B. Hassibi, B. Hochwald and T. Kailath, 'Optimal training for frequency-selective fading channels,' in *Proceedings, International Conference on Acoustics, Speech and Signal Processing*, Salt Lake City, USA, **4**, 2105–2108, IEEE, November 2001.

271. A. C. Koutalos and J. S. Thompson, 'Pilot signal effects on adaptive antenna arrays in FDD wideband CDMA,' in *Proceedings, International Symposium on Spread Spectrum Techniques and applications*, **2**, 531–535, IEEE, 2002.

272. C. Fragouli, N. Al-Dahir and W. Turin, 'Reduced-complexity training schemes for multiple-antenna broadband transmissions,' in *Proceedings, Wireless Communications and Networking Conference*, **1**, 78–83, IEEE, March 2002.

273. A. Grant, 'Joint decoding and channel estimation for linear MIMO channels,' in *Proceedings, Wireless Communications and Networking Conference*, Chicago, USA, **3**, 1009–1012, IEEE, September 2000.

274. R. van Nee and R. Prasad, *OFDM for Wireless Multimedia Communications*. London: Artech House, 1st edition, 2000.

275. R. van Nee, G. Awater, M. Morikura, H. Takanashi, M. Webster and K. Halford, 'New high rate wireless LAN standards,' *IEEE Communications Magazine*, **37**, 82–88, 1999.

276. 'Assumptions and objectives for analysis of OFDM in UTRAN enhancement,' in *TSGR1-27(02)0932*, 3GPP, July 2002.

277. 3GPP, 'Revised study item description on analysis of OFDM for UTRAN enhancement,' in *TSGR1-28(02)1023*, August 2002.

278. 3GPP, *TR22.934, V6.0.0, Technical Specification Group Services and System Aspects; Feasibility Study on 3GPP System to Wireless Local Area Network (WLAN) Interworking*. Sophia Antipolis, F: ETSI, 2002.

279. P. R. Chevillat and W. Schott, 'Broadband radio LANs and the evolution of wireless beyond 3G,' *IBM Journal Research & Development*, **47**, 327–336, 2003.

280. "*http://www.ipwireless.com/press_101602.html*," tech. rep.

281. M. Z. Win and R. A. Scholtz, 'Ultra-wide bandwidth time-hopping spread-spectrum impulse radio for wireless multiple-access communications,' *IEEE Transactions on Communications*, **48**, 679–689, 2000.

282. "*http://www.time-domain.com/products/ourtech/whitepapers.html*," tech. rep.

283. D. Porcino and W. Hirt, 'Ultra-wideband radio technology: potential and challenges ahead,' *IEEE Communications Magazine*, **41**, 66–74, 2003.

284. F. Sestini, J. da Silva and J. Fernandes, 'Expanding the wireless universe: EU research on the move,' *IEEE Communications Magazine*, **40**, 132–140, 2002.

285. P. Komulainen and V. Haikola, 'Adaptive filtering for fading channel estimation in WCDMA downlink,' in *Proceedings, International Symposium on Personal, Indoor and Mobile Radio Communications*, London, UK, **1**, 549–553, IEEE, September 2000.

Index